T0348618

Advances in

# THE STUDY OF BEHAVIOR

VOLUME 44

# Advances in THE STUDY OF BEHAVIOR

*Edited by*

H. Jane Brockmann
*Department of Biology, University of Florida, Gainesville, Florida*

Timothy J. Roper
*School of Life Sciences, University of Sussex, Falmer, Brighton Sussex, United Kingdom*

Marc Naguib
*Behavioural Ecology Group, Department of Animal Sciences, Wageningen University, The Netherlands*

John C. Mitani
*Department of Anthropology, University of Michigan, Ann Arbor, Michigan*

Leigh W. Simmons
*Center for Evolutionary Biology, The University of Western Australia, Crawley, WA, Australia*

*VOLUME 44*

ELSEVIER

AMSTERDAM • BOSTON • HEIDELBERG • LONDON
NEW YORK • OXFORD • PARIS • SAN DIEGO
SAN FRANCISCO • SINGAPORE • SYDNEY • TOKYO
Academic Press is an imprint of Elsevier

Academic Press is an imprint of Elsevier
32 Jamestown Road, London NW1 7BY, UK
225 Wyman Street, Waltham, MA 02451, USA
525 B Street, Suite 1900, San Diego, CA 92101-4495, USA
Radarweg 29, PO Box 211, 1000 AE Amsterdam, The Netherlands

First edition 2012


ISBN: 978-0-12-394288-3
ISSN: 0065-3454

For information on all Academic Press publications
visit our website at www.elsevierdirect.com

# Contents

Causes and Consequences of Differential Growth in Birds: A Behavioral Perspective

MARK C. MAINWARING AND IAN R. HARTLEY



Increasing Awareness of Ecosystem Services Provided by Bats

SIMON J. GHANEM AND CHRISTIAN C. VOIGT

# Contributors

Numbers in parentheses indicate the pages on which the authors' contributions begin.

MATHIEU AMY (183) *Laboratoire d'Ethologie et de Cognition Comparées, Université Paris Ouest Nanterre La Défense, Nanterre Cedex, France*

LETICIA AVILÉS (99) *Department of Zoology and Biodiversity Research Centre, University of British Columbia, Vancouver, British Columbia, Canada*

NATHALIE BÉGUIN (183) *Laboratoire d'Ethologie et de Cognition Comparées, Université Paris Ouest Nanterre La Défense, Nanterre Cedex, France*

DALILA BOVET (183) *Laboratoire d'Ethologie et de Cognition Comparées, Université Paris Ouest Nanterre La Défense, Nanterre Cedex, France*

TUDOR ION DRAGANOIU (183) *Laboratoire d'Ethologie et de Cognition Comparées, Université Paris Ouest Nanterre La Défense, Nanterre Cedex, France*

SIMON J. GHANEM (279) *Leibniz Institute for Zoo and Wildlife Research, and Department of Behavioral Biology, Freie Universität Berlin, Berlin, Germany*

FRÉDÉRIQUE HALLÉ (183) *Laboratoire d'Ethologie et de Cognition Comparées, Université Paris Ouest Nanterre La Défense, Nanterre Cedex, France*

IAN R. HARTLEY (225) *Lancaster Environment Centre, Lancaster University, Lancaster, United Kingdom*

MICHEL KREUTZER (183) *Laboratoire d'Ethologie et de Cognition Comparées, Université Paris Ouest Nanterre La Défense, Nanterre Cedex, France*

GÉRARD LEBOUCHER (183) *Laboratoire d'Ethologie et de Cognition Comparées, Université Paris Ouest Nanterre La Défense, Nanterre Cedex, France*

SARA LEWIS (53) *Department of Biology, Tufts University, Medford, Massachusetts, USA*

MARK C. MAINWARING (225) *Lancaster Environment Centre, Lancaster University, Lancaster, United Kingdom*

LAURENT NAGLE (183) *Laboratoire d'Ethologie et de Cognition Comparées, Université Paris Ouest Nanterre La Défense, Nanterre Cedex, France*

SUSAN PERRY (135) *Department of Anthropology, Behavior, Evolution and Culture Program, University of California, Los Angeles, California, USA*

JESSICA PURCELL (99) *Department of Zoology and Biodiversity Research Centre, University of British Columbia, Vancouver, British Columbia, Canada, and Department of Ecology and Evolution, Batiment Biophore, University of Lausanne, Lausanne, Switzerland*

RICHARD SHINE (1) *Biological Sciences A08, University of Sydney, Sydney, New South Wales 2006, Australia*

ADAM SOUTH (53) *Department of Biology, Tufts University, Medford, Massachusetts, USA*

ERIC VALLET (183) *Laboratoire d'Ethologie et de Cognition Comparées, Université Paris Ouest Nanterre La Défense, Nanterre Cedex, France*

CHRISTIAN C. VOIGT (279) *Leibniz Institute for Zoo and Wildlife Research, and Department of Behavioral Biology, Freie Universität Berlin, Berlin, Germany*

# Sex at the Snake Den: Lust, Deception, and Conflict in the Mating System of Red-Sided Gartersnakes

Richard Shine

BIOLOGICAL SCIENCES A08, UNIVERSITY OF SYDNEY,
SYDNEY, NEW SOUTH WALES 2006, AUSTRALIA

## I. Introduction

We know a lot more about some types of organisms than others; even for well-studied groups, our knowledge is far more comprehensive for some topics than for others. Logistical challenges are the primary reason for these imbalances; for example, it is easier to study common species than rare species and easy to study animals in times and places that facilitate observation. Thus, for example, most scientific studies on sea-turtles focus on the <1% of their lives that is spent out of water, as eggs (Zug et al., 2009). Clearly, the resultant uneven coverage of topics and taxa impedes any kind of broad overview of the diversity and evolution of traits. Behavioral studies pose particular problems, because observing the behavior of a free-ranging animal is likely to be more difficult logistically than obtaining a genetic sample from that animal, or collecting it so that its physiology can be studied in the laboratory. Also, the complexity and context dependence of many behavioral traits, and of their links to reproductive success, often make laboratory studies a poor substitute for field-based observations and experiments. Inevitably, then, detailed knowledge of behavioral biology is woefully lacking in the case of difficult-to-study organisms, including (in some cases) some types of animals that have a high profile with the general public.

For many questions in behavioral ecology, the worst possible study species would be ones that are rare, highly mobile, alert, and likely to flee from your approach and that spend much of their time hidden away and/or inactive. Many species of snakes fulfill these criteria, and some of them have another disadvantage: they are capable of killing an unlucky or incompetent investigator. In contrast, the sister group to snakes, the lizards, has long



0065-3454/12 $35.00
DOI: 10.1016/B978-0-12-394288-3.00001-0

been classical "model organisms" in behavioral ecology (Huey et al., 1983; Olsson and Madsen, 1995). Although the development of miniature radiotransmitters has made snake ecology a more feasible scientific pursuit (Shine, 2003; Shine and Bonnet, 2000), obtaining quantitative data on the behavior of free-ranging snakes has remained challenging in the extreme. In 1929, J. R. Kinghorn wrote:

> The study of the habits of snakes in their natural surroundings is not only difficult, but in many instances practically impossible, on account of the extreme caution they display. It would need weeks of undivided attention . . . to observe the habits of any of our snakes, and . . . our knowledge of the habits of all but the commoner snakes remains practically where it was many years ago (Kinghorn, 1929, p. 3).

Almost a century later, not much has changed: behavioral data for most snake species are limited to anecdotal reports of specific activities such as male–male combat, courtship, or feeding.

Fortunately, the diversity of snakes (>2500 species: Zug et al., 2009), and their abundance in many systems, means that at least a few taxa occur in situations where detailed behavioral research is feasible. These include a range of taxa and locations, from arboreal Chinese pit-vipers (Shine and Sun, 2002; Sun et al., 2001) to terrestrial European adders (Madsen and Shine, 1993; Madsen et al., 1992) and aquatic seasnakes on and around coral islands in the Pacific (Shetty and Shine, 2002; Shine and Shetty, 2001; Shine et al., 2002, 2004e). For ease of studying reproductive behavior in snakes, however, one system stands out for sheer numbers of animals, and for ease of observation and manipulation. In the Manitoba prairies of central Canada, tens of thousands of red-sided gartersnakes (*Thamnophis sirtalis parietalis*) overwinter in communal hibernacula, and court and mate around the entrances to those dens before dispersing to their summer ranges to feed and give birth (Gregory, 1974, 1977, 2009; Shine and Mason, 2001a). The snakes virtually ignore human observers, creating unparalleled opportunities for research on mating tactics. In this review, I will summarize the results of my collaborative studies on this remarkable system.

## II. The Study System

Natricine colubrids are nonvenomous snakes, widely distributed through Europe, Asia, Africa, and the Americas (Zug et al., 2009). In North America, the most common natricines are gartersnakes (*Thamnophis*) and watersnakes (*Nerodia*) (Rossman et al., 1996). Delineating species can be challenging in these groups, reflecting their complex and dynamic recent biogeographic history. Northern parts of the Northern Hemisphere experienced major glaciations <2 million years ago, in the Pleistocene (Gradstein et al., 2004; Hewitt, 2004; Joger et al., 2007), eliminating most vertebrates from wide areas, as well as modifying the landscape (e.g., by scraping loose surface soil away as the glaciers progressed). As the glaciers retreated, plants and animals expanded their ranges northward to colonize the newly exposed lands (Cwynar and MacDonald, 1987). Although intuition suggests that ectothermic ("cold-blooded") animals would be poorly suited to the severely cold climates at the margins of those retreating glaciers, a small suite of cold-adapted ectotherms thrives under such conditions. The red-sided gartersnake is one such species (Janzen et al., 2002; Placyk et al., 2007). At our study areas, adult female gartersnakes typically average around 60 cm snout-vent length and 70 g (maximum to 80 cm, 180 g: Shine et al., 2003e). Males are smaller (average 45 cm, 30 g; maximum around 60 cm, 70 g: Shine and Mason, 2005b). Dark in color with lighter longitudinal stripes, the "red sides" that give the snake its common name are largely hidden beneath overlapping scales and exposed only when the snake expands its body in a threat display (Fig. 1A).

The use of communal overwintering sites is common even in snake species that live in areas with relatively mild climates (Gregory, 1984). In colder climates, where sites deep enough to escape freezing are limited (Viitanen, 1967), snakes from extensive summer range (feeding) areas are forced to use the same winter hibernaculum (Brown and Parker, 1976; Gregory, 1984, 2009; Klauber, 1956; Viitanen, 1967). In Manitoba, several factors combine to create truly enormous aggregations of gartersnakes. First, the continental climate (hot summers, cold winters) creates favorable thermal conditions during the (brief) summer but means that snakes must penetrate deep below the soil surface to avoid lethal freezing in winter. Second, food (e.g., earthworms, frogs) for these generalized predators is abundant, allowing the snakes to attain high population densities. Third, the removal of topsoil during periods of glacial advance means that bedrock is close to the soil surface, and the only deep fissures are the result of underground streams that have dissolved cavities in the limestone (Shine and Mason, 2001a). Mark–recapture studies suggest that most but not all adult gartersnakes return to the same dens in successive winters (Shine et al., 2001a,c).

At least two other features are critical to the use of these snakes as a model for mating-system research. One is that (like many natricines) mating occurs immediately after the snakes emerge from winter inactivity, rather than (as in many snakes) being delayed for a few weeks or months until males undergo spermiogenesis (Seigel, 1996). In gartersnakes, sperm are produced the preceding summer and stored until use in spring ("dissociated" cycle: Crews and Garstka, 1982). Thus, mating occurs while the

FIG. 1. (A) Female red-sided gartersnake *Thamnophis sirtalis parietalis* in threat display, revealing red lateral coloration (photo by Jon Webb); (B) map showing locations of the dens at which behavioral studies were conducted; (C) landscape around one of the smaller dens (photo by Rick Shine). (For color version of this figure, the reader is referred to the online version of this chapter.)

snakes are still concentrated around the den entrance (and hence, readily accessible for study), rather than being delayed until after they have dispersed to their summer ranges. Second, the summer activity season is so short (4 months above ground) that snakes simply cannot afford to be distracted during the brief mating period. Although handling stress induces a corticosterone spike in red-sided gartersnakes, the animal's behavior is unaffected (Moore et al., 2000) and thus, they virtually ignore human observers and continue to court and mate even if translocated to small experimental enclosures (e.g., Shine et al., 1999, 2000b). That tolerance facilitates manipulative studies.

Although most research on the Manitoba gartersnakes has been conducted at a few large dens (for logistical reasons; Fig. 1B), the surrounding countryside contains hundreds (probably thousands) of smaller dens also.

The general landscape is flat (Fig. 1C), with numerous low-lying patches (muskeg) that frequently contain water; trembling aspen (*Populus tremuloides*) dominates the tree flora. Typically, a snake den lies within some kind of depression, with access to crevices in the underlying soil to which snakes retreat in autumn and from which they emerge in spring. After emergence in spring, snakes continue to enter these crevices at night and in inclement weather. The below-ground topography of dens remains unknown and would be an exciting project with ground-penetrating radar. Some dens fill with melting snow water, which suddenly disappears as soil temperatures increase in spring—presumably reflecting melting of an underground ice plug. As soon as the water drains out, the snakes emerge. The seasonal timing of emergence of snakes is regular—late April through May—but annual variation in weather conditions means that in some years the snakes emerge into a bleak landscape with snowdrifts, whereas in other years, the aspen has already begun budding out and the landscape is green, shaded, and warm (Shine et al., 2006b). Thermal differentials are extreme during the spring mating period, both day to day (when heavy snowfall can abolish snake activity for days) and at a small spatial scale (e.g., operative temperatures of snake models can be $>10$ °C higher inside the den area than on the wind-exposed prairie a few meters away; R. Shine, unpublished data).

The dynamics of snake emergence and dispersal have been studied by sampling and mark–recapture, but such studies are logistically demanding because even a relatively small den may contain $>30{,}000$ snakes (Gregory, 1974, 2011; Shine et al., 2006b). The most detailed data come from May 2003, when Tracy Langkilde, Mike Wall, and I set up a 60-m-long drift fence 50 m from a den near Inwood (Fig. 1B) and measured all of the 6653 snakes that were captured in funnel traps along that fence for the next 19 days (Shine et al., 2004c, 2006b). We also captured and measured 909 snakes at the den, on the days that they emerged. This large data set revealed strongly nonrandom trends in emergence patterns and dispersal away from the den. Males emerged before females and stayed longer at the den. Most snakes remained at the den for $<5$ days (Shine et al., 2006b); most then dispersed, but some of the "dispersing" males actually remained at the den periphery for days or weeks (because they focused courtship on dispersing rather than newly emerging females: Shine et al., 2001a, 2005b; Fig. 2).

In red-sided gartersnakes, courtship consists of a male (or commonly, several males) aligning his body with that of the female, with his tailbase positioned close to hers. The marked sexual size dimorphism means that the head and neck of a large female are unencumbered even when the rear part of her body is festooned with males (Fig. 2B). Males rub their chins along the female's dorsal and lateral surfaces, tongue flicking to detect skin lipids.

FIG. 2. Male red-sided gartersnakes in Manitoba court and mate either (A) in large groups close to the communal den, where females are emerging from their overwinter activity, or (B) in smaller groups in aspen clearings as females are dispersing from the den. Male tactics differ in these two contexts. Photographs by Jon Webb. (For color version of this figure, the reader is referred to the online version of this chapter.)

Methyl ketones serve as sex-identifying pheromones in this species and attract vigorous courtship from male snakes (Mason, 1992, 1993). By pressing his chin firmly downward as he moves anteriorly along the female's body, the male's darting tongue tip can access lipids hidden in the pockets beneath each overlapping scale (Shine et al., 2000f). Once aligned with the female's body, males attempt to keep their own cloaca close to that of the female, which necessitates considerable agility if she moves and/or if many other males simultaneously attempt the same feat. Competing males tail-wrestle, and a male's ability to maintain his tailbase close to the female's cloaca is a strong predictor of his eventual mating success (Shine et al., 2004a). Typically, each male's body is looped to embrace the female's body tightly. Courting males exhibit vigorous posterior-to-anterior waves of muscular contraction along their bodies. These caudocephalic waves impede female attempts to breathe, forcing anoxic air out of the rear (saccular) part of her lung (like most snakes, gartersnakes have a single large functional lung: Shine et al., 2003f). Stressed females attempt to escape but are often unable to do so because of the weight of courting males (Shine et al., 2004b). Highly stressed females eventually gape the cloaca open to expel the nauseous secretions from glands in this area, a widespread antipredator mechanism in these and related snake species (Ford, 1996; Greene, 1988; Zug et al., 2009). Snakes of other species (and gartersnakes, at other times of year) defecate as part of this response, but female gartersnakes at the dens in springtime have empty digestive tracts after 8 months of inactivity. Because the digestive and reproductive tracts in snakes share a common orifice, cloacal gaping by the female presents a copulatory opportunity for males (Shine et al., 2003f, 2004b, 2005e,f).

Females often writhe at this point, perhaps in response to the male's attempts at copulation (the hemipenis bears large spines: Rossman et al., 1996). Snakes often rotate their bodies rapidly in this way if the animal is seized by a human hand, so it is presumably a generic escape response (Shine et al., 2003f). Usually, a single male succeeds in inserting a hemipenis; occasionally, two males achieve this feat simultaneously (Shine et al., 2000d). As soon as copulatory fluids begin to flow from the male to the female (often visible as fluids on the female's body near the cloaca), other males lose interest and move away to court other females (Shine et al., 2000d). The pheromone responsible for discouraging courtship in this way is airborne and is part of the copulatory fluids (Shine and Mason, 2012a). Copulation lasts about 15 min (Shine et al., 2000c) and ends with the transfer of milky fluids that harden into a gelatinous mating plug (Devine, 1975) which occludes the female's cloaca and prevents remating for 2 or 3 days (Shine et al., 2000d).

In suitable (warm) weather in spring, an observer at the Manitoba dens sees a seething mass of snakes that consists primarily of mate-searching males in constant activity, tongue flicking every other snake they see, with occasional flurries of activity as males locate an emerging female and court her *en masse* (Fig. 2A). If the observer walks a few tens of meters away to clearings in the surrounding aspen forest, the scene is very different. Here, the snakes occur either solitarily (basking females or males either trail following or with heads raised well above the substrate, surveying the area for potential mates) or in small groups (typically, a female being courted by less than a dozen males; Fig. 2B). After leaving the dens, snakes move rapidly toward their summer feeding ranges, which can be >18 km away (Gregory and Stewart, 1975; Larsen, 1987; Lawson, 1989). Dispersing males are still willing to court, and dispersing females retain their attractivity to males (Shine et al., 2003c); these patterns suggest that mating also occurs away from the den. The summer ecology and behavior of red-sided gartersnakes are poorly known (Larsen, 1987; Lawson, 1989). Females are viviparous; they bear their young (about 5–25 per litter, mean = 16.4) in late summer, before the females (but not their offspring) migrate back to the den in autumn (Ford, 1996; Gregory, 1977; Gregory and Stewart, 1975).

## III. Methods

The opportunities for ecological research at the Manitoba dens were first recognized by Patrick T. Gregory, whose PhD studies on the ecology of these snakes set the scene for future work (Gregory, 1974, 1977; Gregory and Stewart, 1975). Physiological studies were also conducted by local investigators during this time (e.g., Aleksiuk, 1971; Aleksiuk and Gregory, 1974; Aleksiuk and Stewart, 1971; Hawley and Aleksiuk, 1975). Several investigators from David Crews' research group exploited those opportunities in ensuing years (e.g., Camazine et al., 1981; Crews, 1983, 1984, 1985; Crews and Garstka, 1982; Mendonca and Crews, 1989, 1990; Whittier et al., 1985, 1987a,b). Although the work included some behavioral research (e.g., Joy and Crews, 1985, 1988), the major focus was on physiology and chemical ecology. By far, the most important contributor has been Robert T. Mason. His work identified a range of sophisticated chemical communication systems and explored behavioral and endocrine parameters (e.g., Mason, 1992, 1993, 1994; Mason et al., 1987, 1989, 1990). In more recent years, Mason has continued these investigations with successive cohorts of students and collaborators (e.g., Huang et al., 2006; LeMaster and Mason, 2001a,b, 2002, 2003; LeMaster et al., 2001, 2007; Lutterschmidt and Mason, 2005, 2008, 2009; Lutterschmidt et al., 2004, 2006; Moore and Mason, 2001; Moore et al., 2000, 2001; Parker and Mason, 2009). As a result, we know more about the chemical communication systems of red-sided gartersnakes than is the case for any other reptile species (e.g., Mason and Parker, 2010). The resultant understanding of communication systems provided an unparalleled opportunity for more detailed work on behavior and ecology and was a primary motivation for me to conduct research at the dens.

The vast number of snakes at a Manitoba den, and the ease with which they can be collected and observed, makes many approaches possible. Paradoxically, the consequent freedom from logistical constraints creates novel challenges for snake researchers. For example, the primary obstacle for research on snakes generally is to find a way to utilize limited numbers of animals, and limited opportunities for observation. Because those constraints do not apply at the Manitoba dens, investigators require a very different approach. For example, one can abandon repeated-measures approaches in favor of using each individual only once; the rapidity with which robust tests can be conducted means that one can frame a novel idea one evening, test it the following day, and devise a follow-up experiment over dinner the next evening. That timeframe, plus frequent and unpredictable weather shifts that preclude outdoor trials, made the Manitoba fieldwork both exhilarating and frustrating. I traveled to Manitoba in May (the snake emergence season) for 3-week trips (sandwiched between my wedding anniversary and my eldest son's birthday) over 7 years, 1997 through 2004 (except 2000). The research output (resulting in > 40 papers from that 21 weeks' work) underlines the value of a study system where numbers of animals are not the limiting factor.

Most of our methods were simple, and many were modified from those developed by Bob Mason over preceding years. As examples, some of the most important methods involved:

1. *Collecting*. This simply involved picking up snakes, often in groups. Snakes at the den rarely flee from observers and bite even more rarely; these behaviors change dramatically when the animals begin to disperse from the den (Shine et al., 2003c).
2. *Marking*. By writing individual numbers on their bodies or heads using nontoxic pens, or (for cohort marking) simple color stripes along the lighter lateral lines on the snake's body.
3. *Measuring*. By holding snakes beside a meter rule. For experiments where we needed to control size distributions, we used cool weather (when it was futile to run trials) to collect and size-sort animals into outdoor pens—such as small (< 42 cm SVL), medium (43–45 cm), large (46–48 cm), and extra-large (> 49 cm) males, and then give each male a color-code corresponding to his size class (Shine et al., 2000b). High densities of snakes in enclosures were not stressful to the animals, because they occurred at equally high densities outside the enclosures. We could thus exploit the short windows of suitable thermal conditions to run trials constantly, without having to pause to collect, measure, and mark additional snakes. Similarly, detailed measurements and weighing were done in the evening or early morning, or on cold days.

4. *Enclosure trials.* Open-topped nylon boxes, 1.0 × 1.2 m and with walls 0.9 m high, were held in place by metal rods pounded into the soil in an area protected from the wind but often within a few meters of the den. Snakes readily courted and mated in these enclosures, allowing us to manipulate snake densities, sex ratios, body sizes, temperatures, skin lipids (pheromones), and the like. For video-taping courting groups to quantify male tactics and determinants of mating success, we used smaller circular pop-up enclosures (designed for clothes storage) 0.56 m high and 0.48 m in diameter (Shine et al., 2004a).

5. *Responses of free-ranging snakes.* The males' single-minded focus on courtship allowed us to measure response intensities of free-ranging animals by sitting in the den while holding a "target" snake by the tail and scoring the intensity of response elicited by that animal (typically, from the first 5 or 10 males that encountered it). Similarly, we assayed responses to chemical cues (such as those collected on paper towels soaked in hexane and wiped across the dorsal surface of a specific type of "target" snake) by laying the towel on the ground (with a wire frame to hold it in place) and scoring responses of males traveling over it (e.g., Shine et al., 2003e). In trials to examine airborne cues, we wafted air containing specific scent cues (e.g., from a female snake or a mating pair) across the heads of courting males, by holding a female's tail and then exposing her suitors to the specific stimuli (Fig. 3).

6. *Effects of courtship on female behavior.* Our methods evolved over the course of the work. For example, we used a range of techniques to clarify how females are affected by being courted (and especially, to measure costs due to male harassment). The first involved physically attaching a male to a female and measuring the decrement in her speed, but later work was more sophisticated (Shine et al., 2004b, 2005f). This included manipulating courtship intensity to a female by smearing her with courtship-discouraging pheromones to assess effects on her dispersal rate (Shine et al., 2004b, 2005f).

7. *Trail following.* In several studies, we allowed a female to lay down a pheromonal trail and then explored the ability of males to follow that trail. By setting up courting groups inside a plastic bin, and allowing the female to exit through a small hole, we could control factors such as female attributes, male attributes, the number of males previously following a trail, and substrate type to determine how they affected

FIG. 3. The massive numbers of snakes, and their tolerance of human observers, enable researchers to use methods that would be impossible in other study systems. For example, the effects of airborne cues on courting male snakes can be quantified simply by holding a female by the tail, then exposing her suitors to specific treatments. In this photograph (by Tracy Langkilde), Mike Wall is holding a female snake (the largest snake in the photograph) by the tail and is using a portable aquarium pump to run air through a chamber (a margarine tub) containing treatment stimuli (such as a mating pair of snakes), then out a flexible plastic hose. The air current can thus be positioned so that is encountered by targeted male snakes, and their reactions to those scents can be measured quickly and reliably. (For color version of this figure, the reader is referred to the online version of this chapter.)

trail-following behavior. Also, by picking up a female after she had dispersed a few meters, we could explore the tactics of a trail-following male who encountered the end of a trail (Shine et al., 2005g).

## IV. Behavioral Tactics of Male Gartersnakes

At the risk of anthropomorphizing, a convenient framework in which to describe our results is to look at the sequential "decisions" made by reproducing snakes. I will treat the sexes separately, because of profound divergences in their reproductive tactics.

### A. Where to Overwinter?

Because mating occurs immediately after springtime emergence from the den, males will benefit from overwintering in the same places as females. The large body size of adult females restricts their choice of overwintering sites (i.e., they need crevices large enough for them to penetrate). The potential danger of shallow hibernation was demonstrated by the complete annihilation of major den populations when atypically shallow snow cover during one winter allowed deeper-than-usual freezing of the soil (Shine and Mason, 2004). Thus, although the smaller body sizes of males allow them to overwinter in a wider range of sites than can females, the reproductive advantages of overwintering with females force males to use the large dens. Juvenile males (and juvenile females) derive no such reproductive benefit (and indeed, may pay a cost in terms of sexual harassment), and thus overwinter elsewhere—presumably widely dispersed across the landscape, in smaller crevices (Shine et al., 2006b). On a proximate level, both males and females follow scent trails of conspecifics to locate dens in autumn (LeMaster and Mason, 2001a; LeMaster et al., 2001), a trait that has been exploited by managers to recolonize dens after winterkill events (Macmillan, 1995).

### B. When to Emerge in Springtime?

Because the frenetic courting of males causes a rapid loss in energy reserves (Shine and Mason, 2005b), most males disperse from the den $<2$ weeks after emerging (but sometimes after up to 5 weeks: Gregory, 2011). Males with higher initial reserves, and lower rates of energy loss, emerged earlier and remained longer before dispersing (Gregory, 2011; Shine and Mason, 2005b; Shine et al., 2005b). Mark–recapture studies show that males tend to emerge before females and that larger animals tend to emerge before smaller ones (Shine et al., 2005b).

### C. How to Accelerate Recovery from Overwintering?

At their initial emergence after 8 months of inactivity, snakes of both sexes are weak, slow, and lethargic (Shine et al., 2000g,h, 2005f). They regain full locomotor ability after a few days' basking, but until they do so, the snakes are vulnerable to predation by crows (*Corvus brachyrhynchos*), especially during cold weather (Shine et al., 2001b). In this period shortly after emergence, males will attempt to court if they encounter a solitary female (an unlikely event at the dens) but are unlikely to obtain a mating; they cease courting if other males arrive (Shine et al., 2000g,h). The postemergence period, thus, is one of considerable dangers but offers little reproductive benefit to male gartersnakes.

Male red-sided gartersnakes accelerate their postemergence recovery (and thus, reduce the duration of this costly period) by female mimicry. At emergence, males exhibit skin lipids (methyl ketones) that usually act as female identifiers in this system (Mason and Parker, 2010; Mason et al.,

1990; Shine et al., 2000g,h). By producing female-like skin lipids, males suppress their own (unproductive) inclination to court females; experimental application of female pheromones stops a male from courting (Shine et al., 2000g). Female mimicry has another benefit also: it enables males to exploit the reproductive activities of earlier-emerging males. The newly emerged "she-males" attract courtship from other males, resulting in a free massage, transfer of heat from their suitors, and protection against crow predation (a solitary snake likely is more vulnerable than one covered by suitors: Shine et al., 2001d). Courtship also comes at a cost; it may impede the courted snake's breathing, even causing suffocation (Shine et al., 2001b, 2003f) and occasionally, may result in a male being copulated (Pfrender et al., 2001). These risks are low, however, because the "she-male's" pheromonal mix is less attractive than that of females, so he is likely to be courted by small rather than large groups, and by small rather than large he-males (Shine et al., 2000h). Also, copulation is unlikely because "she-males" are very reluctant to gape their cloacas when courted or otherwise stressed (Shine et al., 2003f).

Attracting courtship ceases to be beneficial to the "she-male" when he attains a high body temperature (because he is then capable of escaping from a crow without assistance); manipulative experiments show that the pheromonal composition of his sex-attractant lipids (high proportions of saturated to unsaturated methyl ketones) results in thermally dependent volatility, such that he attracts courtship only when he is cold (Shine et al., 2012). Within a day or two after emergence, males cease to produce female-like pheromones, and thereafter begin to court females and cease to attract courtship from other males (Shine et al., 2000g,h; Fig. 4).

Earlier studies on the Manitoba snakes interpreted she-maleness as a tactic to confuse other males within a mating ball, thereby providing a mating advantage to the she-male (Mason and Crews, 1985). The more recent data falsified this hypothesis by showing that she-males rarely attract courtship if a female is present and that she-males typically cease courting in the presence of larger he-males (Shine et al., 2000g, 2003d). Thus, she-maleness has evolved via natural selection, not sexual selection.

### D. How Much Time and Effort to Devote to Courtship Versus Feeding and Thermoregulation?

Although a given den contains approximately equal numbers of each sex, females disperse soon after emerging whereas males remain at the den (Shine et al., 2006b). Thus, the operational sex ratio typically is massively biased toward males. Competition for newly emerging females thus is intense, with some females submerged under hundreds of amorous males

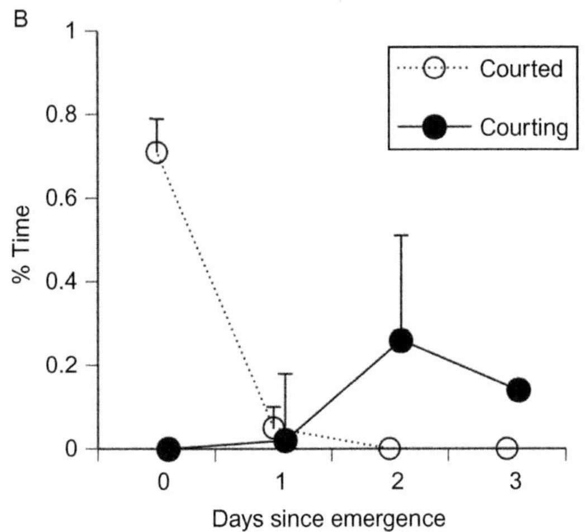


FIG. 4. When male gartersnakes emerge from their 8-month hibernation period, they are cold, slow, and covered in mud. They accelerate their recovery by producing female-like skin lipids, which induce courtship by other males. The figure shows (A) the mud-covered "she-male" (center) is being courted by two (nonmuddy) he-males and is also courting (chin rubbing) himself (photo by Tracy Langkilde) and (B) shifts in male behavior as a function of time since emergence from overwintering. Female mimicry is evident only for a day or two, after which time the male ceases to attract courtship and begins courting females. Graph (B) modified from Shine et al. (2000g). The graph shows mean values and associated standard errors. (For color version of this figure, the reader is referred to the online version of this chapter.)

even before the female has fully emerged from underground (Fig. 2A). Activities such as feeding and thermoregulation (basking) are incompatible with courtship, so how does a male allocate his time among these competing activities? Courtship is the priority. Males bask when females are not available, especially early in the morning, and may move *en masse* into low shrubs if rainfall cools the ground surface to below air temperature (Shine et al., 2005d). However, these small snakes heat and cool so quickly (due to high surface area to volume ratios) that a higher body temperature prior to courtship does not enhance male mating success (based on trials that manipulated male temperatures to test this idea: Shine et al., 2000e). In the field, males have been found mating with body temperatures $<10\,^{\circ}\mathrm{C}$ (Shine et al., 2000e). Feeding also is a low priority. High snake densities at the den eliminate almost all prey items so that males rarely if ever encounter a potential meal at the den. Standardized trials show that anorexia in courting males not only is partially facultative (i.e., due to a lack of encounter with potential prey) but also reflects an endogenous behavioral shift whereby appetite is initially low and increases over the period from emergence through to dispersal (O'Donnell et al., 2004b; Shine et al., 2003c).

## E. Where to Court?

Although writhing hordes of males at the den are a spectacular sight (and are a classic subject for photographers), courtship and mating occur in small groups away from the den (mostly, in clearings among the aspen) as well as beside the den entrance itself (Fig. 2). Densities of snakes are much lower in these clearings, but the area is greater and, thus, the absolute number of animals courting in the aspen clearings around the den may exceed that at the den (Shine et al., 2001a). Radiotelemetry and mark–recapture studies show that males often move between these two situations (Shine et al., 2001a), depending on their body size and condition (Shine et al., 2006b) and that they adopt different strategies for mate location and courtship in the two types of areas (Shine et al., 2005b; Fig. 5). Larger males and those with more energy reserves tend to focus their courtship in the den, where there are not only more females but also more competition from rival males, whereas smaller, lighter-bodied males focus their activities in aspen clearings surrounding the den (Shine et al., 2005b). Unlike females (that disperse in inconsistent directions if returned to the den), males that court in the aspen clearings return to the same area if displaced (Shine et al., 2005b).

## F. How to Find a Mate?

Although research on snake reproduction has emphasized the role of sex-specific skin lipids as sex-identification cues (e.g., see review by Mason and Parker, 2010), male red-sided gartersnakes flexibly use a wide range of other cues as well. Because most females are courted as they emerge from

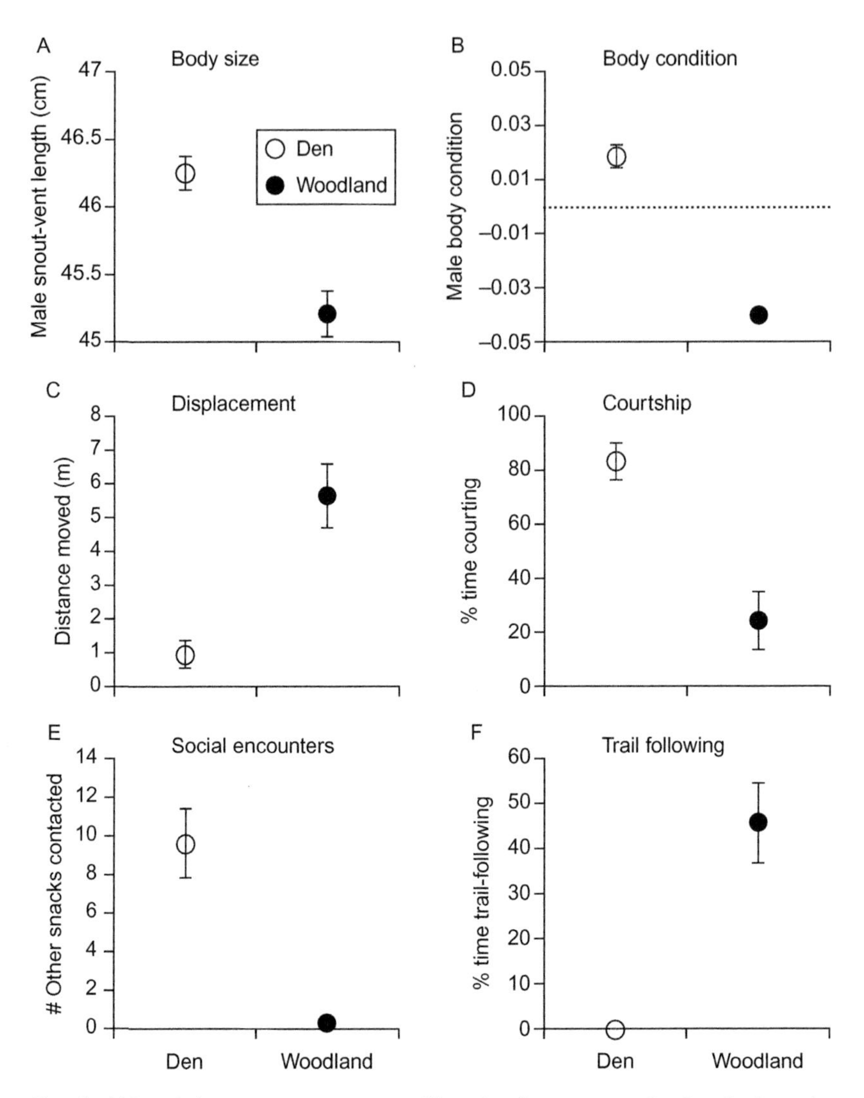


FIG. 5. Although the two areas are separated by only a few meters, males that obtain matings with newly emerged females at the dens (where snake densities are high) use different tactics than males focusing on dispersing females in the surrounding aspen (where snake densities are low). Our sampling of animals from these two areas showed that males that focus on dispersing females tend to be smaller (A) and in lower body condition (B) than males that are active in the den area. Timed focal observations showed that the woodland males move about much more than do den males (C) and spend less of their time courting for much less of the time (D) because of low rates of encounter with other snakes (E). Instead of courting, the aspen males spend most of their time following the pheromonal trails left by dispersing females (F). Figure modified from Shine et al. (2005b). The graphs show mean values and associated standard errors.

underground, they tend to be colder than males, and to be covered in mud. Males target courtship based on these cues, as well as on skin lipids (Shine and Mason, 2001b; Fig. 6). Additionally, vigorous courtship by other males offers a clear cue to female presence, and these behaviors attract additional courting males (Shine and Mason, 2001b; Shine et al., 2005b). Females courted by two males simultaneously are more likely to mate than are females courted by only a single male, so the male's probability of mating success is not necessarily reduced by the presence of a rival (Shine and Mason, 2005a).

In the den, males locate females mostly by direct contact (tongue flicking, for pheromones), whereas in the aspen clearings, males move about to locate and follow the pheromonal trails left by females (Shine et al., 2005b). Males in the aspen clearings also use visual cues (rapid movement, mimicked experimentally by twitching a rope) that are largely ignored by males in the frenetic den environment (Shine et al., 2005b). Woodland males spend much of their time in a distinctive head-up ("periscope") posture, visually scanning for females. Trail-following males rely primarily upon skin lipids, but if the trail is difficult to follow (e.g., crosses a dusty road), males switch to visual cues, elevating their heads and forebodies (the

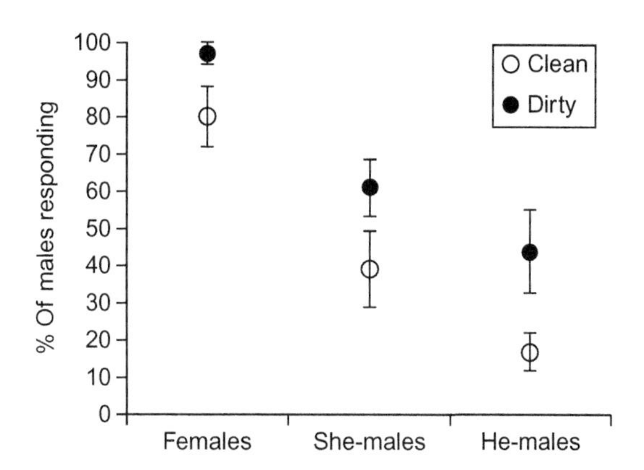


FIG. 6. Although pheromonal cues from skin lipids are the primary sex-identifying mechanism for gartersnakes, males use a wide range of cues to locate potential mates. For example, most snakes that have recently emerged from underground (and, thus, are covered in mud) are females—and hence, mud-covered snakes often attract courtship responses (tongue flicking) from males, regardless of pheromonal cues. This graph shows the proportion of free-ranging males that responded by courtship to snakes that we had experimentally covered in mud versus washed clean. We presented the target snakes to free-ranging males in the den area and simply recorded the proportion of arriving males that commenced courting the target snake. Muddy covering elicited courtship not only to females and female-mimicking males (she-males) but also to "normal" males (he-males). Figure modified from Shine and Mason (2001b).

"periscope" posture) and moving rapidly, without tongue flicking, toward any snake they see in the direction previously taken by the female (Shine et al., 2005f,g). Because males cannot detect female skin lipids by airborne olfaction, visual cues (which do not identify snake sex but work over longer distances) may offer more useful information to a male that has lost the pheromonal trail (Shine et al., 2005g). Trail-following males that lose the trail switch to visual cues; if no such cues are available, the male embarks on a series of highly organized sweeps around the last point at which the trail was detected (Shine et al., 2005g).

### G. Which Female to Court?

Among most animal species, females tend to be more selective than males in terms of mate choice, putatively reflecting the female's higher investment in her progeny (Andersson, 1994). Nonetheless, male mate choice occurs in many taxa, including red-sided gartersnakes. The brief mating season and high energy cost of courtship mean that males pay an opportunity cost if they invest time and energy into courting the "wrong" female: for example, one that is unlikely to mate, or that will produce only a small litter if she does so. Males have finite resources in terms of not only energy but also sperm (which are produced in the preceding summer, so not replaced during the mating season) and mating-plug material (secreted by the sexual segment of the kidneys; plug mass decreases in successive matings by the same male: Shine et al., 2000d). For all these reasons, then, male red-sided gartersnakes evaluate multiple females, often tongue flicking one female before moving on to court another.

The likely reproductive benefit of a mating for a male will be highest if he mates with a large female, in good body condition (these traits are predictors of her fecundity: Gregory, 1977, 2011). Experiments confirm that males can detect variation both in female size and in condition and that they allocate more intense courtship to chemical and visual cues from larger fatter females (Shine et al., 2001c, 2003a,e; Fig. 7A). Males also are more willing to follow a scent trail laid down by a larger rather than smaller female (LeMaster and Mason, 2001a, 2002). Experimental trials in which skin lipids were added and removed show that the discrimination of female size and condition by males is based on visual as well as chemical (pheromonal) cues from the female (Shine et al., 2003a; Fig. 7B, C). Tests varying the absolute amount of lipids on paper towels showed that the male preference for larger rather than smaller females was driven by the quality not quantity of lipids (Shine et al., 2003e). The probability of obtaining a mating will be sharply reduced if the female's cloaca is occluded by a plug from a previous mating. Males evaluate this aspect also; they can detect copulatory

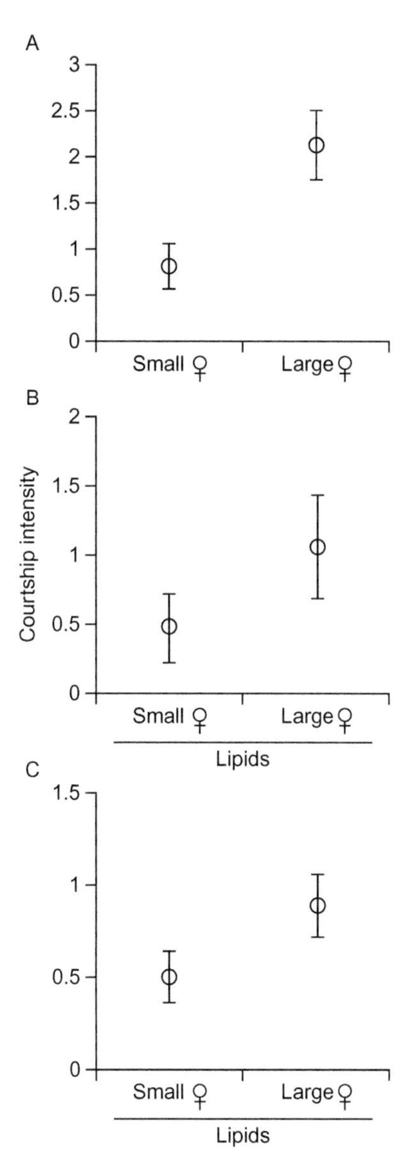


Fig. 7. Male gartersnakes exhibit mate choice with respect to several phenotypic traits of females. Because courting male gartersnakes at the Manitoba dens largely ignore human observers, we could quantify courtship intensity by presenting free-ranging male snakes with stimuli such as live female snakes, or sheets of paper containing skin lipids that had been nondestructively sampled from the bodies of those females. These tests showed that male snakes preferred to court (A) larger rather than smaller females if both were laid out in front of them and that the proximate cues allowing males to discriminate among females include the female's skin lipids, which predict a female's (B) body length (males prefer lipids from longer rather than shorter females), and (C) body condition (males prefer lipids from fatter rather than thinner females, even if the two females are identical in body length). Figure modified from Shine and Mason (2001b). The graphs show mean values and associated standard errors.

fluids (but perhaps not the plug itself) and cease courting when they encounter these airborne scents (Shine and Mason, 2012a; Shine et al., 2000d). In trail-following experiments, males respond more intensely to trails of unmated females than of mated females (O'Donnell et al., 2004a).

A male's tactics also depend upon the context of the encounter, notably the intensity of competition from rival males. Larger body size enhances male mating success not only because it enhances competitive ability in tail-wrestling bouts against smaller rivals but also because it facilitates forcible copulation (Shine and Mason, 2005a; Shine et al., 2001c). Thus, a small male is unlikely to succeed in courting a large female—and perhaps for this reason, small males allocate considerable effort to smaller females that are largely ignored by larger males. Hence, mating is size assortative both in the field and in standardized enclosure trials (Shine et al., 2001c). Newly emerged (weak) males also forego attempts at courtship in the presence of other males (Shine et al., 2000g).

Some (perhaps, all) matings result from coercion. A female snake's respiration is impeded by the caudocephalic waving of males draped along her body (Shine et al., 2003f). The resultant stress induces cloacal gaping, providing an opportunity for a male to insert his hemipenis (Shine et al., 2003f, 2004b, 2005e). Females that have recently emerged from the den after their long winter inactivity are weaker and slower, and easier to stress (Shine et al., 2003f,g, 2005e). Thus, males target their courtship at such females rather than stronger, faster postrecovery females (Shine et al., 2005e) and court most intensely at high temperatures when their performance superiority over females is maximized (Shine et al., 2005e). Males can distinguish newly emerged versus postrecovery females based on skin lipids, as well as other cues (Shine et al., 2005e).

A male's response to a female is not determined entirely by her own traits; males also adjust their courtship tactics to local conditions. A small female is likely to be vigorously courted by a male that has recently encountered mostly small females, but the same female will be ignored by a male that has encountered mostly larger females (Shine et al., 2006a). Presumably for this reason, males at the den (where large females are often encountered) are more size selective in their courtship than are males in the aspen clearings (where many of the females encountered are small: Shine et al., 2006a). Mate choice flexibility is evident in trailing as well as courting behaviors. For example, males are reluctant to trail follow a female if she has already been followed by other males, a response that may be adaptive or may simply reflect obscuring of the trail (Shine et al., 2005g).

Pheromonal cues provide information about a female's den of origin as well as her sex, size, and condition. Subtle differences in skin-lipid signatures among geographically separate den populations are reflected in mate choice: males court more vigorously to skin lipids of females from their own den than of females from other dens (LeMaster and Mason, 2003). Potential benefits to males of detecting female lipids, and benefits to females of escaping overly vigorous courtship, may result in a pheromonal "arms race" between the sexes. Such a system might produce relatively rapid local (den-specific) shifts in the nature of sex differences in pheromonal composition.

### H. Do Males Recognize Individual Females?

In some squamate reptiles, males recognize individual females and court more vigorously to a novel female than to one with which they have already mated ("the Coolidge Effect": Cooper, 1985; Cooper and Steele, 1997; Orrell and Jenssen, 2002; Steele and Cooper, 1997; Tokarz, 1992). The obverse effect also occurs, whereby male lizards selectively court or associate with their prior partner in preference to a novel female (Bull, 2000; Censky, 1995; Olsson and Shine, 1998). To test the factors responsible for the Coolidge Effect, we need studies on a wide range of taxa—including those with mating systems in which we would not expect (based on current theory) that such an effect would be evident. The huge numbers of snakes at Manitoba dens makes it unlikely that males could recognize individual females. We conducted experimental trials with red-sided gartersnakes at a communal den in Manitoba, to see whether previous exposure to a female (either courting or courting plus mating) modified male mate choice or courtship intensity. In keeping with prediction from theory (but contrary to an early anecdotal report), male gartersnakes did not modify their courtship behavior based upon their familiarity (or lack thereof) with a specific female (Shine and Mason, 2012b). At least in large courting aggregations, male snakes may maximize their fitness by basing mate choice upon generic attributes of the female (body size, condition, mated status) and the intensity of competition (numbers and sizes of rival males) rather than information derived from previous sexual encounters with a specific individual.

### I. What Courting Behaviors?

A male gartersnake encountering a solitary female has access to information on traits such as her body size, shape, and skin lipids, but in a crowded den environment, much of the female's body may be hidden beneath dozens (even hundreds) of frantically courting competitors. In such cases, a male may have very limited information on which to base his decision on whether or not to court the female, and if so where to position his body, and how much effort he should allocate to different components of his courtship repertoire (e.g., caudocephalic waving vs. chin rubbing or tail searching: Shine et al., 2004a).

Positioning himself on the female is not a trivial challenge if most of her body is hidden beneath other males. For example, how can he distinguish which end is her head, and which is her tail? Experimental manipulations suggest that males assess the female's orientation partly by her direction of movement, rather than by cues such as the location of her head or tail. Females rarely crawl backwards, and thus, the direction of movement offers a reliable cue to which end is her head (Shine et al., 2000f). The direction of scale overlap also offers important cues, perhaps because the distinctive chin-rubbing behavior of a courting male provides physical evidence about scale overlap, as well as providing tongue tip access to pheromones beneath overlapping scales as the male moves his head anteriorly (but not posteriorly) along the female's body (Shine et al., 2000f).

Once in an optimal position, how much effort should he allocate to courtship? On an ultimate level, the answer presumably lies in probable returns (successful matings) relative to costs (energy expenditure). As expected, males adjust their courtship intensity to factors that affect those benefits and costs. Larger and fatter (and hence more fecund) females attract more intense courtship (Shine et al., 2003e; Fig. 7), and smaller males reduce their courtship vigor if a larger rival is present (Shine and Mason, 2005a). Larger males are stronger, and it requires more force to dislodge their tails from the optimal locations close to the female's cloaca; thus, courtship by a small male is less likely to be successful if a larger male is present (Shine and Mason, 2001b, 2005a). Males that are collected in the crowded den environment are more sensitive to the numbers of rivals in this respect than are males collected in less crowded areas away from the den (Shine et al., 2005b).

For any given intensity of courtship, how should a male allocate his effort between behaviors that provide information (chin rubbing) versus those that elicit female cloacal gaping (caudocephalic waving) versus those that enable him to intromit his hemipenis before rival males (tail searching)? The answer depends upon the intensity of male–male competition. If he is alone with the female, he needs to prioritize caudocephalic waving; the sooner he can induce female receptivity, the less likely another male will arrive to join the mating ball before mating occurs. If he is one of many males courting the female, however, he allows other males to do most of the caudocephalic waving (which likely is energetically expensive, as well as making it difficult for him to remain in the optimal position on the female's body) and focuses instead on maintaining his own tailbase close to that of the female (Shine et al., 2003g). This flexible adjustment of tactics to the intensity of male–male rivalry is more important in the den than in the surrounding aspen clearings (where male densities are lower, and thus fewer males court each female), and hence, aspen males are less responsive to the number of rivals in this respect than are den males (Shine et al., 2005b).

### J. How to Induce Female Cooperation?

How does a male actually achieve intromission? Intuition suggests that forcible insemination would be impossible in snakes, because of the long and flexible body of the female (Devine, 1984). Nonetheless, our data show that male gartersnakes exploit aspects of female morphology, physiology, and antipredator responses to copulate without female cooperation. Rhythmic caudocephalic waves by courting males force anoxic air from the female's posterior saccular lung onto the respiratory surfaces of the anterior lung, inducing hypoxia (Shine et al., 2003f). Consistent with a shift to anaerobic respiration, mating females have higher lactate levels than unmated females and courted females also have higher respiratory rates (Shine et al., 2003f, 2004b). Respiratory rates increase still more if the female is actively stroked in a fashion designed to mimic caudocephalic waving by courting males (Shine et al., 2003f). Cloacal gaping is a widespread stress response in snakes and functions as an antipredator tactic, whereby the noxious contents of cloacal glands (plus feces) are expelled to discourage a predator (Greene, 1988). The immediately postemergence female snakes have no fecal material in their alimentary tract, so (as noted above) a gaping cloaca offers an opportunity for a male to insert his hemipenis (because the cloaca is the common orifice for reproductive and digestive functions). Experimental posterior-to-anterior stroking of a courted female's dorsal surface (mimicking caudocephalic waves) rapidly induces cloacal gaping and, thus, copulation (Shine et al., 2003f, 2004b, 2005e).

In keeping with the idea that males obtain matings by inducing anoxia, small males are capable of mating with larger females only if the latter animals are exhausted by prior exercise or prolonged courtship (Shine et al., 2003f, 2004b). Also, females that are thin are more quickly exhausted by courtship, and hence these females are more likely to mate (and to mate sooner) than are females in better body condition (Shine and Mason, 2005a). As a result, males presented with size-matched females selectively targeted courtship to females that were least able to resist (Shine et al., 2005e). Thus, cues that predict a female's ability to resist coercion may elicit more intense courtship responses of males—despite the fact that all else being equal, males respond more intensely to visual and chemical cues from larger fatter females (Shine et al., 2001c, 2003a). The end result is a complex matching of male courtship behavior to multiple phenotypic traits of the female, and of the situation in which she is encountered.

### K. Which Hemipenis to Use?

Male squamates have paired hemipenes, connected to independent systems (testis, sexual segment of the kidney, efferent ducts) on either side of the body (Greene, 1997; Zug et al., 2009). Thus, a given copulation may involve either the right hemipenis (and right testis and stored sperm that was produced by that testis, etc.) or the left. How does a male decide which hemipenis to use? Male red-sided gartersnakes alternate hemipenis use in successive matings, thereby maximizing the time available for replenishment of accessory copulatory fluids (including mating plug material). Spermiogenesis does not occur during the mating season so that sperm stores cannot be replenished at this time (Shine et al., 2000d). Both in the field and in enclosures, males preferentially use the right rather than the left hemipenis in mating, perhaps because the right-hand hemipenis and its associated structures are larger than are the corresponding organs on the left-hand side of the body (Shine et al., 2000c). For this reason, mating plugs deposited by the right hemipenis tend to be larger than those deposited by the left hemipenis (Shine et al., 2000d). This right-handedness preference disappears at low temperatures, perhaps because colder males are unable to flexibly adjust their positions and, thus, simply use whichever hemipenis they can most easily bring to bear (Shine et al., 2000c). Hemipenis use also is affected by anatomical asymmetry. Males that exhibit asymmetric ventral scales close to the cloaca (reflecting underlying asymmetries in rib number) preferentially use the hemipenis on the side of the body opposite to the missing rib, perhaps reflecting a reduced bodily flexibility (Shine et al., 2005c). Such anatomical asymmetries, in turn, may result from suboptimal thermoregulation during pregnancy by a gravid female; and thus, a female snake's thermal biology can influence the mating tactics (and success) of her adult sons (Shine et al., 2005c).

### L. How Much Mating Plug to Deposit?

Unlike snakes of most other species (including some gartersnakes: Ford, 1996), male red-sided gartersnakes deposit gelatinous plugs in the female's cloaca after transferring sperm (Shine et al., 2000d). The milky fluid rapidly hardens, occluding the cloaca and preventing remating for 2 or 3 days (experimental plug removal allows rapid remating: Shine et al., 2000d). The plug may serve as a postcopulatory mate-guarding device. Its primary importance probably occurs several hours after mating, not immediately, because the airborne scent of copulatory fluids discourages courtship by other males during and shortly after mating (Shine and Mason, 2012a; Shine et al., 2000d, 2004b). Larger males produce larger plugs, and larger females are allocated larger plugs. The matching of plug size to female body size may ensure effective sealing of the cloacal entrance (Shine et al., 2000d).

Mating plugs in many kinds of animals (e.g., insects, lizards, mammals) have been interpreted as adaptations to prevent sperm leakage rather than to prevent remating (Birkhead and Møller, 1998; Olsson and Madsen, 1998). To test this idea, we attached condoms to 14 recently mated female gartersnakes (of which 7 had the plug manually removed), such that the female's entire hindbody (from just anterior to the cloaca) was encased. Counts of sperm in the condoms 12 h later showed fewer sperm in females with intact plugs (1 of 7 females had 7000 sperm/ml, and the rest had none) than in females from which the plug had been removed (3 of 7 females had detectable sperm, with counts from 0 to 468,000 sperm/ml, mean = 174,342). Sample sizes are too small, and variance too high, to warrant statistical analysis of these data. The pattern tends to support the "barrier to sperm leakage" hypothesis, but it is noteworthy that four of seven females did not leak any sperm despite the removal of their mating plug. Why did so little sperm leak out? Trials in which we injected saline into the upper oviducts of dissected females showed that the fluid ran straight through the oviduct and out of the cloaca in seven unmated females but was impossible to force beyond a muscular contraction a few centimeters above the cloacal chamber in 13 mated animals (R. Shine, unpublished data). Thus, as in European vipers (Nilson and Andren, 1982) and South American rattlesnakes (Almeida-Santos and Salomão, 1997), copulatory fluids apparently induce oviductal muscular contraction that prevents sperm leakage (and presumably, ingress of sperm from later matings). Such postcopulatory impediments to remating are transitory, based on the high frequency of multiple paternity in litters (e.g., Garner et al., 2002; McCracken et al., 1999; Schwartz et al., 1989). Nonetheless, the plug (and oviductal contraction) may last long enough to allow many females to obtain sperm from only a single male before escaping from the high concentrations of snakes around the den.

### M. How Many Matings?

In experimental trials, males readily mate on several successive days (Shine et al., 2000d), but we do not know how often this happens in the wild. Plug size decreases in successive matings, and multiply-mated males eventually produce no plugs even after a prolonged copulation (Shine et al.,

2000d). Larger body size strongly enhances mating success in staged encounters, both when males are alone with females and when they are part of larger groups (Shine et al., 2000b, 2005a). Presumably, then, large males may often obtain multiple matings during a single breeding season, whereas smaller males fail to mate at all. In contrast to the directional sexual selection on male body size, relative tail length is under stabilizing selection. That is, males with tails that are about average length for their body size (based on residual scores from the general linear regression of tail length against snout-vent length) obtain more matings (Shine et al., 1999). Another determinant of mating success is vigor of courtship; heavier-bodied males (those with more energy stores) court more vigorously and, thus, obtain more matings (Shine et al., 2004a).

### N. When to Disperse from the Den?

Some male snakes stay at the dens for at least 2 weeks, whereas others depart the same day that they emerge (Shine and Mason, 2005b; Shine et al., 2006b). At a den 30 km away from the ones we studied, Gregory (2011) recorded some males staying up to 5 weeks. Mark–recapture studies show that males that emerge from hibernation in good body condition, and that lose condition only slowly thereafter, remain for longer at the den before dispersing than do thinner-bodied conspecifics, and those that lose condition rapidly (Gregory, 2011; Shine and Mason, 2005b). The range in body condition is considerable, with a 50-cm male weighing anywhere from 30 to 50 g (Shine and Mason, 2005b). Comparisons between courting and noncourting males confirm that courting imposes high energy costs and results in a rapid decrease in body condition (Shine and Mason, 2005b). Interestingly, dissections of snakes killed in mass mortality events at the den suggest that the energy reserves used over the mating period do not come from the usual energy-storage components of the body (abdominal fat bodies, liver) but instead are sequestered within the male's muscles (Shine and Mason, 2005b).

## V. Behavioral Tactics of Female Gartersnakes

### A. Where to Overwinter?

The large body sizes of adult female gartersnakes force them to use the (relatively scarce) sites with crevices wide enough and deep enough to allow snakes to escape lethally low soil-surface temperatures. Thus, these large snakes have little choice in terms of den sites.

### B. When to Emerge in Springtime?

Female gartersnakes face an unenviable trade-off. The activity season is only 4 months long, and early emergence maximizes the time available for a female to disperse to her summer range, feed, produce large ova, ovulate, gestate, bear her young, and return to the den prior to lethal autumn frosts (Gregory, 2009). However, early emergence also means that she will experience cold weather and massive harassment from amorous male conspecifics; operational sex ratios are highest early in the season, because males tend to emerge prior to females (Shine et al., 2006b). A female's body size affects her phenology, with larger females emerging first (Shine et al., 2006b). Larger size may provide thermal benefits (increased operative temperature: see below), perhaps shielding these larger snakes from the effects of cold weather in early spring. On a proximate level, thermal minima underground may trigger emergence (Lutterschmidt et al., 2006).

### C. How to Accelerate Recovery from Overwintering?

Like males, females are weak and slow after emerging from the den. High levels of harassment from courting males make it virtually impossible for a female to bask until she has dispersed many meters away, and only the largest females can disperse this far without mating (Shine et al., 2003f, 2005e,f, 2006b). Retarding of female dispersal by incessant courtship has been demonstrated experimentally, whereby unmated females painted with a courtship-discouraging pheromone (squalene) dispersed more rapidly from the den (Shine et al., 2005f). In nature, copulatory fluids have the same effect (Shine et al., 2000d, 2004b). Radiotelemetric monitoring showed that many females rest in aspen clearings at least 10 m from the main den for a few days, recuperating from their long period of winter inactivity (Shine et al., 2001a). By the time they are ready to move out to their summer range, they are stronger and faster (Shine et al., 2004b). Like she-males, unmated females in these aspen clearings may benefit from the warming, massage and protection against predators afforded by small groups of courting males (Shine et al., 2004b).

### D. How Much Time and Effort to Devote to Courtship Versus Feeding and Thermoregulation?

Female gartersnakes do not solicit courtship; instead, they devote much effort to avoiding it. Recently emerged females feed readily if given a chance, but rarely do so in nature because of the absence of prey items at the dens (Shine et al., 2003c). Intense harassment by males also may render feeding impossible. Interestingly, the cost–benefit balance for thermoregulation (such as basking) may depend upon a snake's body size. Our unpublished data on body temperatures show a strong thermal benefit to larger body size. Larger snakes maintained higher temperatures, both in the field (for 73 field-collected animals, snout-vent length vs. cloacal temperature, $r=0.24$, $P<0.01$) and in standardized trials where we anesthetized females to eliminate variance due to posture and activity (from 32 female snakes [snout-vent length 38.6–64.8 cm, 18–85 g mass] with cloacal thermistors, snout-vent length vs. cloacal temperature, $r=0.42$, $P<0.001$). The difference in mean hourly body temperature between the smallest and the largest snakes varied with time of day (maximum differential: 7.5 °C at midday). Size-dependent changes in convective boundary layer thickness may explain this pattern (F. Seebacher, personal communication). Regardless of the mechanism that is responsible, any such thermal effect may differentially advantage the (larger) female snakes, hence expediting their recovery from the long winter period of inactivity.

### E. Where to be Courted?

Most female snakes at the den have little control over the location of courtship; some are besieged by hundreds of frantically courting males even before they have fully emerged from underground (Figs. 2 and 8). Many females mate within a few minutes of emergence (Shine et al., 2004b, 2005e, 2006b).

### F. How to Find a Mate?

For a female gartersnake in a communal den, the challenge is to avoid males (and thus, avoid the energy costs and risks associated with prolonged courtship) not to attract them.

### G. Which Male to Mate with?

Many authors have pointed out the difficulty of detecting indirect female choice in a system with intense male–male competition (e.g., Andersson, 1994; Eberhard, 2002; Eberhard and Cordero, 2003). In theory, female gartersnakes could exert mate choice by intensifying male–male rivalry, or by producing specific pheromonal cues (on the skin surface, or as substrate-bound trails that males can follow) that attract specific types of males. For example, the trend for lipids of larger females to attract courtship from larger rather than smaller males (Shine et al., 2003a) could be interpreted in this way. In practice, this seems unlikely. For example, it is difficult to

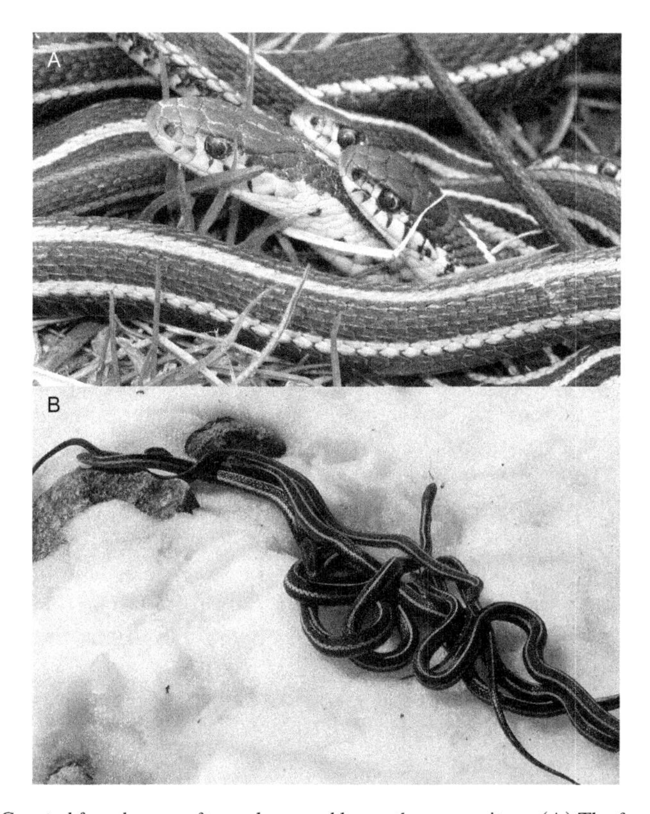


FIG. 8. Courted females are often submerged beneath many suitors. (A) The female (larger snake, center of photograph) often is unable to escape from the males, and eventually gapes her cloaca as a stress response, thereby enabling one of the males to achieve a copulation. (B) A small female gartersnake (head to left of photograph) attempting to disperse from the den the morning after a snowfall, when the number of males courting her (and thus, impeding her escape attempts) is minimized by cold conditions. Photographs by Jon Webb. (For color version of this figure, the reader is referred to the online version of this chapter.)

imagine how a female snake courted simultaneously by hundreds of suitors can influence which one intromits successfully; and the behaviors of females result in a generic avoidance of all males, not in selective mating (Shine et al., 2005f).

One consequence of this situation is that mate choice in this system is by males, not females. Males make subtle and sophisticated decisions about which female to court, whereas there is no evidence that females make any such evaluations of mate quality. That difference relates to "opportunity costs" of prolonged courtship for males, given limited time and nonreplenishing sperm (see above).

### H. How Many Matings?

The approximately equal sex ratios of adult snakes at the den (Shine et al., 2006b) suggest that on average, the numbers of matings per male must be about the same as those per female. Unlike the case in males, a female's body size may not affect the number of times that she mates: larger females not only attract more courtship but also are more capable of evading courting males and escaping from the den (Shine et al., 2004b). These two influences will tend to cancel out any overall effect of female body size on mating frequency. Because the mating plug precludes remating for a day or two, many females likely leave the den after mating only once (Shine et al., 2004b, 2006b). However, genetic studies show that multiple paternity is common, so females either may mate more than once at the den or may mate later in the season as they disperse to their summer ranges (Garner et al., 2002; McCracken et al., 1999). Occasional autumn matings also can produce viable progeny (Blanchard, 1943).

The high but variable frequency of multiple paternity in gartersnakes suggests opportunities for postcopulatory mate choice, whereby females selectively use sperm from some males (but not others) to fertilize their ova (as documented in lizards: Olsson et al., 1996). However, this topic has never been investigated in gartersnakes.

### I. What Courting Behaviors?

Like male conspecifics, female gartersnakes exhibit specific behaviors seen only in the context of courtship. However, the females' behaviors function to evade suitors not to facilitate mating. Although these behaviors may be more successful at repelling some males than others (perhaps based on a male's body size or strength), they constitute generalized male avoidance tactics rather than a form of mate choice (Shine et al., 2005f). Female avoidance of male aggregations is likely to confer significant fitness benefits; courtship can be so intense that females suffocate (Shine et al., 2001b).

Except in cold weather or for the largest females, it is difficult for a female snake to escape from the den area without mating (Shine et al., 2004b; Fig. 2A). Up in the aspen clearings, however, where courting group sizes are much smaller (Fig. 2B), females use a range of behavioral tactics to avoid males. First, females avoid areas where they are vigorously courted, or that contain abundant scent of male conspecifics (Shine et al., 2004b). Second, because rapid movement attracts courting males (see above), prolonged immobility is often effective. Third, if a female in the aspen clearings is courted, she will suddenly dash away from her suitors in a straight-line move of 1–2 m, then make an equally sudden turn to one side and again rely on immobility. After the courting males rush past, the female can resume her attempt to disperse from the den (Shine et al., 2004b).

Another tactic involves the female spinning on a longitudinal axis (1–6 times) to displace a male's attempt at intromission. Videotapes show that females almost always rotate their bodies rapidly in this fashion when males attempt copulation (axial rotation accompanied 82 of 83 cases of hemipenis insertion: Shine et al., 2003f), but that in most cases, the male remains attached (presumably due to the spines on his everted hemipenes). Female axial rotations that terminated copulation have been recorded in another gartersnake species, *Thamnophis marcianus* (Perry-Richardson et al., 1990), but the ability of female red-sided gartersnakes to terminate copulations in this (or any other) way has not been demonstrated.

A final—and at first sight puzzling—female behavior involves the tendency of juvenile female snakes to copulate: why mate if you cannot produce offspring? (Shine et al., 2004b). There are two possible explanations for this behavior. First, small females may be unable to resist courtship-induced hypoxia and, thus, mate simply because they cannot control their cloacal-gaping response to stress. Second, the copulatory fluids deposited during mating allow females to disperse from the den with little harassment, thus providing a direct benefit to female acceptance of mating (Shine et al., 2000d,i, 2001b, 2004b, 2005f). Both of these explanations may be valid. Experimental tests show that snakes likely to experience significant costs from being copulated (she-males) are very reluctant to gape their cloacas in response to hypoxia (Shine et al., 2003f). The fact that juvenile females are not as reluctant in this respect is consistent with the idea that copulation entails benefits as well as costs for them.

### J. When to Disperse from the Den?

Unlike males, females gain no benefit from remaining near the den for long periods. Most females disperse soon after emergence, sometimes waiting for a day or two in aspen clearings close to the den (where male numbers, and thus harassment, are lower: Shine et al., 2005b) to recover their strength and speed before undertaking the arduous and dangerous trek to their summer ranges (Shine et al., 2001a). Although larger females attract more courters, they are also better able to escape from groups of courting males (Shine et al., 2004b). In the absence of copulatory fluid scents (which discourage courtship: see above), females can escape the den by emerging in cold weather when few males are active (Fig. 8B). Cold conditions thus induce a burst of emergence and dispersal by small females (those most at risk from male harassment). Unfortunately, such conditions also increase a dispersing snake's vulnerability to predation by crows. As a result, small female gartersnakes are disproportionately represented among crow-kill victims (Shine et al., 2001b, 2006b).

## VI. Emergent Themes

### A. Male-Male Rivalry Takes Many Forms and Affects Many Traits

Rivalry between male snakes is a major feature of the mating system at the Manitoba dens. The most overt rivalry involves physical struggles for copulatory opportunities within a writhing ball of snakes, but competition takes many other forms also. Sexual selection likely works on a diversity of traits such as male body sizes (Shine et al., 2000b), relative tail lengths (Shine et al., 1999), sperm and mating plug allocation (Shine et al., 2000d), hemipenis alternation (Shine et al., 2000c), and the tactics employed when following a discontinuous pheromonal trail (Shine et al., 2005g). There may also be intense selection on a male's ability to detect and respond to specific cues that predict a female's presence, her probability of mating, her likely fecundity, and the probable intensity of competition from rival males. Such cues may emanate from the female herself, from the responses of other males to that female, or from substrate-bound or visual cues that predict female presence at a distance (Shine et al., 2005g). Perhaps the most remarkable result of male–male rivalry is female mimicry, whereby newly emerged males exploit the courtship behavior of their rivals to accelerate their own recovery from overwintering (Shine et al., 2000g,h, 2001d, 2003d).

One result of this complexity is that aspects of the mating system impose selection (natural and/or sexual) on a broad suite of traits. For example, males experience intrasexual selection on traits that enhance mating success through male–male rivalry, intersexual selection on traits that facilitate coercive mating, and natural selection on traits that enhance survival. The Manitoba dens thus impose a diverse array of selective processes on a diverse array of male traits. Even if we consider a single trait, such as body size in males, several simultaneous selective processes may be at work. At the Manitoba dens, larger body size enhances a male's mating success both because he obtains more matings and because he tends to mate with larger females (that, in turn, are likely to produce more offspring). Also, larger males are more successful both in outcompeting rivals in multi-male aggregations and in forcing copulations onto solitary females (Shine and Mason, 2005a). The proximate mechanisms whereby a male's body size enhances his reproductive success include increased vigor of courtship, enhanced ability to position his tailbase close to that of the female, and enhanced ability to resist attempts by other males to displace him (Shine and Mason, 2005a). However, body size does not affect a male's probability of finding a female, either in the den or in the aspen woodlands (Shine et al., 2005b). Even for intrasexual selection on a single trait, then, the actual causal links to male mating success may be complex. Intersexual selection (on male ability to obtain matings by coercion) may be intense also. Male gartersnakes that more effectively stress the female and/or can detect and target the females least able to withstand stress will have higher mating success (Shine et al., 2003f, 2005e). Thus, a trait such as male body size (which likely makes a male more capable of inducing anoxia in females) will be under both inter- and intrasexual selection.

Natural selection is important also, because snakes at the den are vulnerable to a wide range of threats. Mortality due to predator attack may impose selection on the timing of emergence and dispersal in both sexes, as well as for the adoption of female mimicry tactics by males (Langkilde et al., 2004; Shine et al., 2000a, 2001a, 2006b). Indeed, she-maleness in gartersnakes provides a clear example of an apparent shift in male reproductive tactics (the transitory adoption of female mimicry) being driven by natural selection (the danger of predation) rather than by reproductive costs and benefits *per se* (Shine et al., 2001d). As with sexual selection, body size is a trait under intense natural selection. For example, larger body size renders a male snake less vulnerable to mortality due to attack from crows, suffocation in a mating ball (Shine et al., 2001a; Fig. 9), or being run over by motor vehicles (Shine and Mason, 2004). Snakes that are suffocated within mating balls also tend to be individuals in poor body condition, and males are more vulnerable than females (Fig. 9). Other mortality sources, such as winterkill (freezing) and drowning at inundated dens, are random with respect to snake body size (Shine and Mason, 2004).

Some of the morphological variation that affects a snake's mating success is a result of environmental factors not genetics, and hence, even intense selection on such traits may not induce evolutionary change (or at least, will not induce change in those specific traits). For example, loss of the tail-tip is common (probably through parasitism and predation), and stub-tailed males are ineffective at performing the tail-wrestling maneuvers needed to obtain matings in multi-male aggregations (Shine et al., 1999). Similarly, asymmetries in the number of ribs on either side of the body, apparently a result of suboptimal maternal thermoregulation during embryogenesis, have long-term consequences for mating success of the sons born to such a female. Males with asymmetric rib numbers are less capable of obtaining matings in standardized trials (Shine et al., 2005c). In the field, males with rib number asymmetries departed the den sooner than did conspecific males without such asymmetries. The effects of asymmetry on mating success interact with

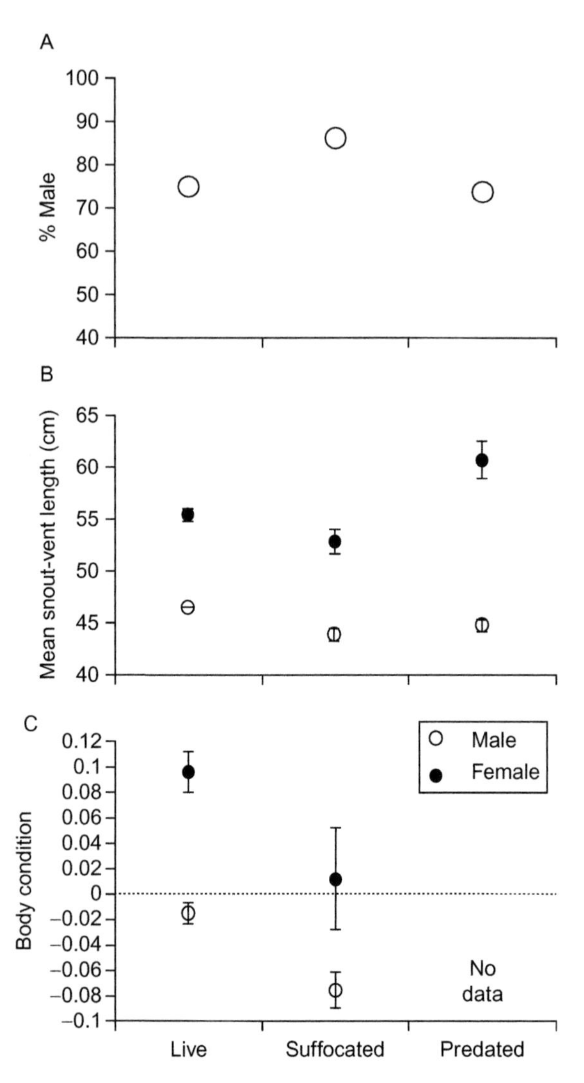


FIG. 9. The huge numbers of snakes at and near the Manitoba dens not only introduce novel risks to the snakes (such as the danger of suffocating within a large mating ball) but also enable researchers to quantify correlates of such mortality. Our sampling of live snakes, compared to snakes found dead either due to suffocation within balls or due to predator (crow) attack, shows that a snake's vulnerability is affected by its sex, its body length, and its body condition. Graphs show mean values and associated standard deviations. Modified from Shine et al. (2001b).

those of male body size, such that only the largest asymmetric males are likely to mate (at least at the den; we know very little about matings that occur after snakes leave the den area: Shine et al., 2005c).

Female gartersnakes at the dens also experience selection. Female–female rivalry appears to be absent, but female–male rivalry (sexual conflict) is intense. Traditionally, researchers have interpreted female attempts to evade males as tactics to enhance mate quality, rather than to escape harassment (e.g., Eberhard, 2002; Eberhard and Cordero, 2003). This hypothesis is not consistent with our data on the Manitoba gartersnakes. For example, females disperse without mating if given the opportunity to do so, and they avoid courtship regardless of the number of males or their phenotypes (Shine et al., 2005f). Contrary to the orthodoxy of "choosy females and indiscriminate males," then, mate choice is exhibited by male but not by female gartersnakes. Male choosiness reflects high opportunity costs: in the brief mating period, and with limited energy reserves, any time that a male spends courting one female detracts from his ability to court other females. The decrease in size of the mating plug in successive matings by the same male (Shine et al., 2000d) suggests that males need to allocate their resources carefully, to the most productive partner. The apparent lack of precopulatory mate choice in females reflects a female's inability to assess male traits that predict enhanced viability of her offspring. Even if she could make such an assessment, she has little control over which male mates with her.

Female snakes are subject to natural as well as sexual selection, with a strong link between the two processes. For example, females can evade courting males by dispersing in cold weather but must then run the gauntlet of predatory crows (Shine et al., 2006b). A trend for predation-induced mortality to be concentrated on unmated rather than mated female snakes at the Manitoba dens has been interpreted as evidence that mating reduces a snake's risk of predation (O'Donnell and Mason, 2007); however, that pattern also might arise if mortality is concentrated on snakes immediately as they emerge (while they are still cold and slow), before they have had time to mate.

## B. Snakes Obtain, Integrate, and Utilize Many Types of Information

Historically, scientific research on alternative sensory modalities has tended to follow independent lines, perhaps because the training and methodology needed for researchers to evaluate vision (for example) are very different from those needed to investigate chemical cues (Partan and Marler, 2005). Gartersnakes pay no attention to such distinctions between sensory modalities, instead utilizing and integrating any information that is available. For example, a courting male gartersnake assesses not only chemical cues (skin lipids; copulatory pheromones) of a potential partner but also her body temperature and appearance (e.g., her body length, her body condition, whether or not she is covered in mud, whether or not other males are courting her, and in which direction she is moving). Even within a single modality such as chemical cues, some are detected by olfaction (e.g., copulatory pheromone) and others by vomerolfaction (e.g., skin lipids: Mason and Parker, 2010; Shine and Mason, 2012a). Understanding the proximate and functional integration of such information remains a major challenge for future work.

Similarly, male snakes make surprisingly sophisticated evaluations of mate quality. Previous literature has focused on a snake's ability to identify the sex and species of another snake, but male gartersnakes also derive information on physical attributes of a female (length and condition) that predict her fecundity, and cues that predict her availability for mating (numbers and sizes of already courting males, copulatory pheromones, her ability to resist harassment). This complex discrimination coupled with population subdivision (based around den use) translates into significant interpopulation variation in pheromonal composition of females, and responses by males (LeMaster and Mason, 2003). With a few tongue flicks to the body of another snake, a male gartersnake can reliably identify the other snake's species, den of origin, sex, body length, body condition, mating history, and time since emergence from overwintering (e.g., Shine et al., 2003a, 2004d).

### C. Mating Tactics are Flexibly Adjusted to Local Conditions

The Manitoba gartersnakes adjust their tactics relative to temporally variable factors such as (a) their own energy reserves, (b) their own body size and strength relative to same sex rivals, (c) their own traits relative to individuals of the other sex, (d) the traits of a potential partner relative to other available partners they have recently encountered, (e) temporally variable environmental factors (e.g., temperature), and (f) spatially variable environmental factors (e.g., in the aspen clearings vs. the den).

Some of these flexible responses are driven by variable benefits (e.g., courting a larger rather than smaller female, in the absence rather than presence of rival males), whereas others are driven by costs (e.g., resistance to cloacal gaping in she-males). A balance between alternative costs is critical in some cases (e.g., small females dispersing in cold weather avoid costs of male harassment but must accept potential mortality costs of crow predation; Fig. 9). In yet others, a snakes' flexible decisions are driven by the balance between alternative benefits (e.g., a male ignores females smaller than most of those he has recently encountered).

The end result is a highly dynamic mating system, where males and females with different phenotypic traits emerge and interact at different times of the spring mating season; where a given male will shift over both short and long timescales in terms of whether he courts or is courted; and after recovering from overwintering, which females he courts, how intensely he courts them, where he courts them, what tactics he uses to locate, recognize and court them, and how he responds to female traits and to the presence of rival males.

### D. Mating Tactics Influence Spatial Patterns of Gene Flow

Because genetic exchange occurs primarily at overwintering dens, the mating system of red-sided gartersnakes reduces rates of gene flow among local demes. If each den population acts as a separate gene pool, stochastic events (founder effects, genetic drift) or local adaptation might result in spatial divergence in gene frequencies (Shine et al., 2003b). For this to occur, most snakes must return to the same den year after year; mark–recapture data support this assumption (Shine et al., 2001a). However, young-of-the-year snakes do not use the communal dens and presumably overwinter in smaller aggregations in the surrounding landscape (Shine et al., 2006b). As they grow older, the recruiting snakes likely rely on pheromonal trails left by den-bound adults to locate local dens (LeMaster et al., 2001). That process might increase rates of genetic exchange among adjacent dens, despite a preference for pheromonal cues from the snake's own den population (LeMaster and Mason, 2003). The avoidance of dens by young-of-the-year snakes plausibly reflects the dangers of being courted; thus, different aspects of the snakes' mating system have opposing effects on rates of spatial divergence in gene frequencies.

The relatively recent (postglacial) colonization of the Manitoba prairies by snakes has resulted in a complex mosaic of genetic relatedness within and between species (Placyk et al., 2007). Some Manitoba dens contain western plains gartersnakes, *Thamnophis radix haydeni*, as well as red-sided gartersnakes, *T. sirtalis parietalis* (Shine et al., 2004d). The two species are morphologically similar, and taxa are sympatric over a broad area of North America. Nonetheless, the reproductive isolation between the two taxa breaks down to some degree in the northern extreme of their joint ranges in Manitoba (Shine et al., 2004d). That breakdown may reflect the effects of a brief summer season and restricted denning habitat; both of these factors reduce the usual temporal and spatial interspecific differences of mating activity. Males court females (and female skin lipids) of the "wrong" species, albeit less vigorously than they court females of their own species; males of the two species show different responses of courtship vigor to ambient temperature. Male *T. radix* court vigorously only under warm conditions, whereas the courtship vigor of male *T. sirtalis* is unaffected by temperature (Shine et al., 2004d). This divergence means that a newly emerging (cold) female *T. radix* likely will be courted mostly by heterospecific males, and the same would be true of a female *T. sirtalis* on an unusually hot day. In practice, however, rates of hybridization are low.

### E. How Generally Applicable Are the Results from Manitoba Gartersnakes?

Model systems are a two-edged sword. On the one hand, studies on such a system can generate detailed insights—partly because of the ease of investigation, and partly because subsequent studies can build on the foundation provided by earlier work. On the other hand, that increased depth of understanding comes at a cost: our knowledge is based on a small and nonrandom sample of the biological lineage in which we are interested. The problem is a very general one—in every lineage, some taxa are easier to study than are others. However, the situation is particularly acute with snakes, because so many species are difficult to study, whereas red-sided gartersnakes are almost absurdly easy to investigate. There is no simple solution to this problem. Reviews of snake biology regularly bemoan the inordinate focus on research on the Manitoba dens and rightly point out that these large-den systems are not typical even for red-sided gartersnakes, let alone for snakes in general (e.g., Dubey et al., 2009; Seigel, 1996). Nonetheless, no other system has yielded such rich insights into the behavioral tactics of reproducing snakes—including, critically, experimental tests of relevant hypotheses.

The Manitoba work is part of a more general reassessment of reptile behavior, with other studies also revealing unsuspected sophistication and complexity in a broad range of traits. For example, the classical view of snakes as "nonsocial" organisms has come under increasing attack in recent years, with reports of kin recognition in rattlesnakes (Clark, 2004), "social" groups in seasnakes (Shine et al., 2005h), sex-specific defense of food resources in kukri snakes (Huang et al., 2011), and complex social systems in psammophiines (de Haan, 1984, 2003). A burgeoning literature on complex sociality in the snakes' sister group, the lizards, provides many examples of remarkably complex kin-based systems that involve long-term monogamy (e.g., Bull, 2000), kin recognition (Bull et al., 2001), and parental defense of offspring (O'Connor and Shine, 2004). Clearly, squamate reptiles are not simple robotic organisms (Greene, 1997). Studies on the Manitoba snakes confirm that squamates are capable of more subtle and flexible tactics than have been envisaged in classical views of reptile behavior.

The sophisticated reproductive tactics of red-sided gartersnakes cannot have evolved *de novo* in the short period since the retreating glaciers allowed snakes to occupy this area. Instead, that sophistication must be an elaboration of preexisting capacities in snakes. Although we lack the comparative data to evaluate whether or not other snake species exhibit the kinds of complex situation-dependent behavioral responses characteristic of Manitoba gartersnakes, similarities are likely to be more important than differences. For example:

1. A male's ability to use chemoreception to rapidly evaluate the size and reproductive condition of a female, as well as her sex and species, likely is a very general facet of snake biology (Andren, 1982, 1986; Blanchard and Blanchard, 1941; Ford, 1986; Noble, 1937; see review by Mason and Parker, 2010).

2. Males locate females using substrate-deposited pheromonal trails in most snakes, including arboreal as well as terrestrial species (e.g., brown tree snake *Boiga irregularis*—Greene et al., 2001; rattlesnake *Crotalus viridis*—Duvall and Schuett, 1997; whipsnake *Hierophis viridiflavus*—Fornasiero et al., 2007).

3. A link between female body size and fecundity is well-nigh universal among snakes (Shine, 2003), and a consequent male preference for courting larger rather than smaller females has been documented in taxa as divergent as European grass snakes (*Natrix natrix*—Luiselli, 1996) and Fijian sea kraits (*Laticauda colubrina*—Shetty and Shine, 2002).

4. Similarly, mate choice based on multiple sensory modalities has been documented or inferred in a diverse array of snakes ranging from hydrophiine seasnakes (*Emydocephalus annulatus*—Shine, 2005) to anacondas (*Eunectes murinus*—Rivas and Burghardt, 2001).

5. Larger males outcompete their smaller rivals in physical struggles for mating opportunities in viperids (e.g., *Vipera berus*—Madsen and Shine, 1993; *Agkistrodon contortrix*—Schuett, 1997; Schuett and Gillingham, 1989), pythons (e.g., *Morelia kinghorni*—Fearn et al., 2005), and colubrids (*Elaphe obsoleta*—Blouin-Demers et al., 2005; *Stegonotus cucullatus* – Dubey et al., 2009), but this pattern is not universal (e.g., the filesnake *Acrochordus arafurae* and the sea krait *L. colubrina*—Shetty and Shine, 2002; Shine, 1986).

6. Anecdotal records of male–male courtship (and even copulation) in other snake species suggest that female mimicry is not restricted to the Manitoba snakes (e.g., *Crotalus s. scutulatus*—Hardy, 1998).
7. Mate-searching male rattlesnakes (*C. viridis*) adopt complex and sophisticated spatial tactics to locate receptive females, as do the Manitoba gartersnakes (Duvall and Schuett, 1997; Duvall et al. 1993).
8. Reproduction is energetically expensive (e.g., Olsson et al., 1997), and conflicts between reproduction and feeding are widespread (e.g., Brischoux et al., 2011).
9. Mate choice by males rather than females may be typical of snakes (e.g., see Rivas and Burghardt, 2001; Shine, 2003).
10. In wide-ranging species with fragmented distributions, a preference for chemical cues from individuals of one's own population may be common (e.g., Fornasiero et al., 2011).

Another emerging theme of ecological research on snakes is the great flexibility of these animals in response to local conditions. For example, populations of gartersnakes living in adjacent habitats can have markedly different life history schedules (Bronikowski, 2000), and tropical filesnakes and pythons adjust their rates of growth and reproduction in response to annual (climatically driven) variation in prey resources (Madsen and Shine, 2000; Madsen et al., 2006). Patterns of sexual dimorphism and mating systems (e.g., presence or absence of male–male combat or mate guarding) can vary among populations within a single wide-ranging species (Huang et al., 2011; Luiselli, 1995; Pearson et al., 2002; Shine, 2003; Shine and Fitzgerald, 1995), and the importance of male–male rivalry for mating success can shift between years within a single population (Madsen and Shine, 1993). Even within the genus *Thamnophis*, gartersnake species (and even, populations within species) apparently vary in features such as whether or not males deposit mating plugs or exhibit caudocephalic waves (Ford, 1996; Perry-Richardson et al., 1990). Although some of these divergences have a genetic underpinning (e.g., Bronikowski, 2000), many are likely to reflect phenotypically plastic responses to unpredictable environmental variation. In consequence, the Manitoba gartersnakes' flexibility in behavioral traits is likely to be shared by other snake species, and future research is likely to reveal substantial variation in mating systems both through space and through time.

Overall, then, the gartersnakes of Manitoba do not tell us what other snakes actually do, but they tell us a great deal about what snakes are capable of doing. Fine-tuned responses to the attributes of conspecific animals may not be relevant to the mating systems of low-density populations of solitary snakes, where interactions with conspecifics may be brief and rare. By analogy, many of the behaviors exhibited by people living in large cities differ from those living in the countryside. Encounters with conspecifics play a larger role, and potentially have more influence on reproductive success, for individuals in large aggregations than for individuals living a more solitary existence. Nonetheless, the massive aggregations of snakes at the Manitoba dens are somewhat deceptive, in that a great deal of courtship and mating occurs in relatively small groups of animals, in the aspen clearings rather than at the den itself. In this respect, the interactions and adaptations revealed by this model system may not deviate as markedly from other snakes as would appear at first sight, when one is confronted by swarming masses of snakes at the den entrance (Shine et al., 2001a).

## VII. Summary

Although snakes are diverse and abundant in many parts of the world, they are difficult animals to study behaviorally. There is one remarkable exception, however: a population of red-sided gartersnakes (*T. sirtalis parietalis*) in the severely cold prairies of Manitoba, in central Canada. These small nonvenomous snakes overwinter in large communal hibernacula and court and mate after they emerge in spring, prior to dispersing to their summer ranges. The predictable availability of thousands of sex-crazed snakes, that readily tolerate observers and experimental manipulations, has facilitated detailed research into sexual tactics and conflict. Building on Robert T. Mason's pioneering studies of chemical ecology, our work has clarified communication modalities as well as the operation of natural selection and sexual selection in this system and has revealed an unsuspected subtlety and flexibility in mating tactics. For example, male snakes use information on a conspecific's skin lipids, body length, body condition, mating history, temperature, appearance, and behavior when selecting courtship targets. Males adjust their courtship intensity and tactics to attributes of the target snake and the intensity of competition from other males. The snakes also tailor their tactics to their own abilities; for example, males produce female-like skin lipids for a day or two after they emerge from hibernation, eliciting courtship from other males that serves to warm the "she-male" and protect it from predatory crows. This female mimicry works only when it is needed (i.e., when the she-male is cold, and before it recovers from overwintering). Females exhibit distinctive behaviors that function to evade courting males and facilitate rapid dispersal from the den.

Although the mating tactics of these gartersnakes have been influenced by system-specific attributes such as large aggregations and brief mating seasons, this "model system," nonetheless, has much to tell us about the reproductive behavior of other snake species. In particular, we should expect complex tactics, shifting in response to local abiotic and biotic circumstances; sophisticated male mate choice based on an integration of multiple sensory modalities; and a pervasive role for sexual conflict in mating system evolution in snakes. Snakes will always be hard to study behaviorally. The success of studies on the Manitoba systems should encourage us to look for equally tractable model systems in other places, based on snakes from other phylogenetic lineages. I look forward to a day when researchers exploit such opportunities to illuminate the mysterious world of snake sexuality so that our understanding of snake behavior does not depend so heavily upon the remarkable gartersnakes of the Canadian prairies.

## Acknowledgments

Many volunteers, colleagues, and students assisted with our research at the snake dens; I am particularly grateful to Melanie Elphick, Peter Harlow, Randy Krohmer, Amanda Lane, Mike LeMaster, Tracy Langkilde, Deb Lutterschmidt, Ignacio Moore, Ruth Nisbett, Dave O'Connor, Ryan O'Donnell, Mats Olsson, Ben Phillips, Dave Roberts, Mike Wall, Heather Waye, and Jonno Webb. Al, Jerry, and Susan Johnson made Chatfield a happy place to rest between forays to the snake den. My participation in the work was funded by the Australian Research Council and the Australian Academy of Science. Helpful comments on the chapter were provided by Tracy Langkilde and Mats Olsson. My most important collaborator, without doubt, was Bob Mason. Bob generously supported the work and shared his remarkable knowledge of this system. In fond recollection of many exciting days at the dens, Newfie pool in Chatfield, fried mushrooms at the King Buck Inn on snowy days, and the overpowering scent of snake schweeze in the DNR house after a day spent measuring 2000 gartersnakes, I gratefully dedicate this review to Dr. Bob.

## References

Aleksiuk, M., 1971. Temperature-dependent shifts in the metabolism of a cool temperature reptile, *Thamnophis sirtalis parietalis*. Comp. Biochem. Physiol. 39A, 495–503.

Aleksiuk, M., Gregory, P.T., 1974. Regulation of seasonal mating behavior in *Thamnophis sirtalis parietalis*. Copeia 1974, 681–689.

Aleksiuk, M., Stewart, K.W., 1971. Seasonal changes in the body composition of the garter snake (*Thamnophis sirtalis parietalis*) at northern latitudes. Ecology 52, 485–490.

Almeida-Santos, S.M., Salomão, M.G., 1997. Long-term sperm storage in the female neotropical rattlesnake *Crotalus durissus terrificus* (Viperidae: Crotalinae). Jpn. J. Herpetol. 17, 46–52.

Andersson, M., 1994. Sexual Selection. Princeton University Press, Princeton, NJ.

Andren, C., 1982. The role of the vomeronasal organs in the reproductive behavior of the adder, *Vipera berus*. Copeia 1982, 148–157.

Andren, C., 1986. Courtship, mating and agonistic behaviour in a free-living population of adders, *Vipera berus* (L.). Amphib.-Reptil. 7, 353–383.

Birkhead, T.R., Møller, A.P., 1998. Sperm Competition and Sexual Selection. Academic Press, London.

Blanchard, F.C., 1943. A test of fecundity of the garter snake *Thamnophis sirtalis sirtalis* (Linnaeus) in the year following the year of insemination. Pap. Mich. Acad. Sci. Arts Lett. 28, 313–316.

Blanchard, F.N., Blanchard, F.C., 1941. Mating of the garter snake *Thamnophis sirtalis sirtalis* (Linnaeus). Pap. Mich. Acad. Sci. Arts Lett. 27, 215–234.

Blouin-Demers, G., Gibbs, H.L., Weatherhead, P.J., 2005. Genetic evidence for sexual selection in black ratsnakes (*Elaphe obsoleta*). Anim. Behav. 69, 225–234.

Brischoux, F., Bonnet, X., Shine, R., 2011. Conflicts between feeding and reproduction in amphibious snakes (sea kraits, *Laticauda* spp.). Austral Ecol. 36, 46–52.

Bronikowski, A.M., 2000. Experimental evidence for the adaptive evolution of growth rate in the garter snake *Thamnophis elegans*. Evolution 54, 1760–1767.

Brown, W.S., Parker, W.S., 1976. Movement ecology of *Coluber constrictor* near communal hibernacula. Copeia 1976, 225–242.

Bull, C.M., 2000. Monogamy in lizards. Behav. Process. 51, 7–20.

Bull, C.M., Griffin, C.L., Bonnett, M., Gardner, M.G., Copper, S.J.B., 2001. Discrimination between related and unrelated individuals in the Australian lizard, *Egernia striolata*. Behav. Ecol. Sociobiol. 50, 173–179.

Camazine, B., Garstka, W., Tokarz, R., Crews, D., 1981. Effects of castration and androgen replacement on male courtship behavior in the red-sided garter snake (*Thamnophis sirtalis parietalis*). Horm. Behav. 14, 358–372.

Censky, E.J., 1995. Mating strategy and reproductive success in the teiid lizard, *Ameiva plei*. Behaviour 132, 529–557.

Clark, R.W., 2004. Kin recognition in rattlesnakes. Proc. R. Soc. B 271, S243–S245.

Cooper, W.E., 1985. Female residency and courtship intensity in a territorial lizard, *Holbrookia propinqua*. Amphib.-Reptil. 6, 63–69.

Cooper, W.E., Steele, L.J., 1997. Pheromonal discrimination of sex by male and female leopard geckos (*Eublepharis macularius*). J. Chem. Ecol. 23, 2967–2977.

Crews, D., 1983. Control of male sexual behavior in the Canadian red-sided garter snake. In: Balthazart, J., Prove, E., Gillies, R. (Eds.), Hormones and Behaviour in Higher Vertebrates. Springer-Verlag, Berlin, pp. 398–406, 443–473.

Crews, D., 1984. Gamete production, sex hormone secretion, and mating behavior uncoupled. Horm. Behav. 18, 22–28.

Crews, D., 1985. Effects of early sex steroid hormone treatment on courtship behavior and sexual attractivity in the red-sided garter snake, *Thamnophis sirtalis parietalis*. Physiol. Behav. 35, 569–575.

Crews, D., Garstka, W.R., 1982. Ecological physiology of a garter snake. Sci. Am. 247, 158–168.

Cwynar, L.C., MacDonald, G.M., 1987. Geographical variation of lodgepole pine in relation to population history. Am. Nat. 129, 463–469.

de Haan, C.C., 1984. Dimorphisme et comportement sexuel chez *Malpolon monspessulanus*: considerations sur la denomination subspecifique *insignitus*. Bull. Soc. Herp. Fr. 30, 19–26.

de Haan, C.C., 2003. Extrabuccal infralabial secretion outlets in *Dromophis*, *Mimophis* and *Psammophis* species (Serpentes, Colubridae, Psammophiini). A probable substitute for 'self-rubbing' and cloacal scent gland functions, and a cue for a taxonomic account. C. R. Biol. 326, 275–286.

Devine, M.C., 1975. Copulatory plugs, restricted mating opportunities and reproductive competition among male garter snakes. Nature 267, 345–346.
Devine, M.C., 1984. Potential for sperm competition in reptiles: behavioral and physiological consequences. In: Smith, R.L. (Ed.), Sperm Competition and the Evolution of Animal Mating Systems. Academic Press, Orlando, FL, pp. 509–521.
Dubey, S., Brown, G.P., Madsen, T., Shine, R., 2009. Sexual selection favours large body size in males of a tropical snake (*Stegonotus cucullatus*, Colubridae). Anim. Behav. 77, 177–182.
Duvall, D., Schuett, G.W., 1997. Straight-line movement and competitive mate-searching in prairie rattlesnakes, *Crotalus viridis viridis*. Anim. Behav. 54, 329–334.
Duvall, D., Schuett, G.W., Arnold, S.A., 1993. Ecology and evolution of snake mating systems. In: Seigel, R.A., Collins, J.T. (Eds.), Snakes: Ecology and Behavior. McGraw-Hill, New York, pp. 165–200.
Eberhard, W.G., 2002. The function of female resistance behavior: intromission by male coercion vs female cooperation in sepsid flies (Diptera: Sepsidae). Rev. Biol. Trop. 50, 485–505.
Eberhard, W.G., Cordero, C., 2003. Sexual conflict and female choice. Trends Ecol. Evol. 18, 438–439.
Fearn, S., Schwarzkopf, L., Shine, R., 2005. Giant snakes in tropical forests: a field study of the Australian scrub python, *Morelia kinghorni*. Wildl. Res. 32, 193–201.
Ford, N.B., 1986. The role of pheromone trails in the sociobiology of snakes. In: Duvall, D., Müller-Schwarze, D., Silverstein, R.M. (Eds.), Chemical Signals in Vertebrates. Vol. 4: Ecology, Evolution and Comparative Biology. Plenum, New York, pp. 261–278.
Ford, N.B., 1996. Behavior of garter snakes. In: Rossman, D.A., Ford, N.B., Seigel, R.A. (Eds.), The Garter Snakes: Evolution and Ecology. University of Oklahoma Press, Norman, OK, pp. 90–116.
Fornasiero, S., Bresciani, E., Dendi, F., Zuffi, M.A.L., 2007. Pheromone trailing in male European whip snake, *Hierophis viridiflavus*. Amphib.-Reptil. 28, 555–559.
Fornasiero, S., Dendi, F., Bresciani, E., Cecchinelli, E., Zuffi, M.A.L., 2011. The scent of others: chemical recognition in two distinct populations of the European whip snake, *Hierophis viridiflavus*. Amphib..-Reptil. 32, 39–47.
Garner, T.W.J., Gregory, P.T., McCracken, G.F., Burghardt, G.M., Koop, B.F., McLain, S.E., et al., 2002. Geographic variation of multiple paternity in the common garter snake (*Thamnophis sirtalis*). Copeia 2002, 15–23.
Gradstein, F., Ogg, J., Smith, A., 2004. A Geologic Time Scale. Cambridge University Press, Cambridge, UK.
Greene, H.W., 1988. Antipredator mechanisms in reptiles. Gans, C., Huey, R.B. (Eds.), In: Biology of the Reptilia, vol. 16. Alan R. Liss, New York, pp. 1–152.
Greene, H.W., 1997. Snakes: The Evolution of Mystery in Nature. University of California Press, Berkeley, CA.
Greene, M.J., Stark, S.L., Mason, R.T., 2001. Pheromone trailing behavior of the brown tree snake, *Boiga irregularis*. J. Chem. Ecol. 27, 2193–2201.
Gregory, P.T., 1974. Patterns of spring emergence of the red-sided garter snake (*Thamnophis sirtalis parietalis*) in the Interlake region of Manitoba. Can. J. Zool. 52, 1063–1069.
Gregory, P.T., 1977. Life-history parameters of the red-sided garter snake (*Thamnophis sirtalis parietalis*) in an extreme environment, the Interlake region of Manitoba. Natl. Mus. Can. Publ. Zool. 13, 1–44.
Gregory, P.T., 1984. Communal denning in snakes. In: Seigel, R.A., Hunt, L.E., Knight, J.L., Malaret, L., Zuschlag, N.L. (Eds.), Vertebrate Ecology and Systematics: A Tribute to Henry S. Fitch. Museum of Natural History, University of Kansas, Lawrence, KS, pp. 57–75.

Gregory, P.T., 2009. Northern lights and seasonal sex: the reproductive ecology of cool-climate snakes. Herpetologica 65, 1–13.

Gregory, P.T., 2011. Temporal dynamics of relative-mass variation of red-sided garter snakes (*Thamnophis sirtalis parietalis*) at a communal hibernaculum in Manitoba. Ecoscience 18, 1–8.

Gregory, P.T., Stewart, K.W., 1975. Long-distance dispersal and feeding strategy of the red-sided garter snake (*Thamnophis sirtalis parietalis*) in the Interlake of Manitoba. Can. J. Zool. 53, 238–245.

Hardy Sr., D.L., 1998. Male-male copulation in captive Mojave Rattlesnakes (*Crotalus s. scutulatus*): its possible significance in understanding the behaviour and physiology of crotaline copulation. Bull. Chic. Herp. Soc. 33, 258–262.

Hawley, A.W.L., Aleksiuk, M., 1975. Thermal regulation of spring mating behaviour in the red-sided garter snake (*Thamnophis sirtalis parietalis*). Can. J. Zool. 53, 768–776.

Hewitt, G.M., 2004. Genetic consequences of climatic oscillations in the Quaternary. Philos. Trans. R. Soc. Lond. B 359, 183–195.

Huang, G.-Z., Zhang, J.-J., Wang, D., Mason, R.T., Halpern, M., 2006. Female sex pheromone induces membrane responses in vomeronasal sensory neurons of male snakes. Chem. Senses 31, 521–529.

Huang, W.-S., Greene, H.W., Chang, T.-J., Shine, R., 2011. Territorial behavior in Taiwanese kukrisnakes (*Oligodon formosanus*). Proc. Natl. Acad. Sci. USA 108, 7455–7459.

Huey, R.B., Pianka, E.R., Schoener, T.W., 1983. Lizard Ecology: Studies of a Model Organism. Harvard University Press, Cambridge, MA.

Janzen, F.J., Krenz, J.G., Haselkorn, T.S., Brodie Jr., E.D., Brodie, E.D.I.I.I., 2002. Molecular phylogeography of common garter snakes (*Thamnophis sirtalis*) in western North America: implications for regional historical forces. Mol. Ecol. 11, 1739–1751.

Joger, U., Fritz, U., Guicking, D., Kalyabina-Hauf, S., Nagy, Z.T., Wink, M., 2007. Phylogeography of western Palaearctic reptiles—spatial and temporal speciation patterns. Zool. Anz. 246, 293–313.

Joy, J.E., Crews, D., 1985. Social dynamics of group courtship behavior in male red-sided garter snakes (*Thamnophis sirtalis parietalis*). J. Comp. Psychol. 99, 145–149.

Joy, J.E., Crews, D., 1988. Male mating success in red-sided gartersnakes: size is not important. Anim. Behav. 36, 1839–1841.

Kinghorn, J.R., 1929. The Snakes of Australia. Angus & Robertson, Sydney.

Klauber, L.M., 1956. Rattlesnakes, Their Habits, Life Histories and Influence on Mankind. University of California Press, Berkeley, CA.

Langkilde, T., Shine, R., Mason, R.T., 2004. Predatory attacks to the head versus body modify behavioral responses of garter snakes. Ethology 110, 937–947.

Larsen, K., 1987. Movements and behavior of migratory garter snakes, *Thamnophis sirtalis*. Can. J. Zool. 65, 2241–2247.

Lawson, P.A., 1989. Orientation abilities and mechanisms in a northern population of the common garter snake (*Thamnophis sirtalis*). Musk-Ox 37, 110–115.

LeMaster, M.P., Mason, R.T., 2001a. Evidence for a female sex pheromone mediating male trailing behavior in the red-sided garter snake, *Thamnophis sirtalis parietalis*. Chemoecology 11, 149–152.

LeMaster, M.P., Mason, R.T., 2001b. Annual and seasonal variation in the female sexual attractiveness pheromone of the red-sided garter snake (*Thamnophis sirtalis parietalis*). In: Marchlewska-Koj, A., Lepri, J., Müller-Schwarze, D. (Eds.), Chemical Signals in Vertebrates, vol. 9. Kluwer Academic/Plenum Publishers, New York, pp. 369–374.

LeMaster, M.P., Mason, R.T., 2002. Variation in a female sexual attractiveness pheromone controls male mate choice in garter snakes. J. Chem. Ecol. 28, 1269–1285.

LeMaster, M.P., Mason, R.T., 2003. Pheromonally mediated sexual isolation among denning populations of red-sided garter snakes, *Thamnophis sirtalis parietalis*. J. Chem. Ecol. 29, 1027–1043.

LeMaster, M.P., Moore, I.T., Mason, R.T., 2001. Conspecific trailing behavior of red-sided garter snakes, *Thamnophis sirtalis parietalis*, in the natural environment. Anim. Behav. 61, 827–833.

LeMaster, M.P., Stefani, A., Shine, R., Mason, R.T., 2007. Cross-dressing in chemical cues: exploring 'she-maleness' in newly-emerged male garter snakes. In: Hurst, J.L., Beynon, R.J., Roberts, S.C., Wyatt, T.D. (Eds.), Chemical Signals in Vertebrates, vol. 11. Springer, New York, pp. 223–230.

Luiselli, L., 1995. The mating strategy of the European adder, *Vipera berus*. Acta Oecol. 16, 375–388.

Luiselli, L., 1996. Individual success in mating balls of the grass snake, *Natrix natrix*: size is important. J. Zool. Lond. 239, 731–740.

Lutterschmidt, D.I., Mason, R.T., 2005. A serotonin receptor antagonist, but not melatonin, modulates hormonal responses to capture stress in two populations of garter snakes (*Thamnophis sirtalis parietalis* and *Thamnophis sirtalis concinnus*). Gen. Comp. Endocrinol. 141, 259–270.

Lutterschmidt, D.I., Mason, R.T., 2008. Geographic variation in timekeeping systems of three populations of garter snakes (*Thamnophis sirtalis*) in a common garden. Physiol. Biochem. Zool. 81, 810–825.

Lutterschmidt, D.I., Mason, R.T., 2009. Endocrine mechanisms mediating temperature-induced reproductive behavior in red-sided garter snakes (*Thamnophis sirtalis parietalis*). J. Exp. Biol. 212, 3108–3118.

Lutterschmidt, D.I., LeMaster, M.P., Mason, R.T., 2004. Effects of melatonin on the behavioral and hormonal responses of red-sided garter snakes (*Thamnophis sirtalis parietalis*) to exogenous corticosterone. Horm. Behav. 46, 692–702.

Lutterschmidt, D.I., LeMaster, M.P., Mason, R.T., 2006. Minimal overwintering temperatures of red-sided garter snakes (*Thamnophis sirtalis parietalis*): a possible cue for emergence? Can. J. Zool. 84, 771–777.

Macmillan, S., 1995. Restoration of an extirpated red-sided garter snake *Thamnophis sirtalis parietalis* population in the Interlake region of Manitoba, Canada. Biol. Conserv. 72, 13–16.

Madsen, T., Shine, R., 1993. Temporal variability in sexual selection on reproductive tactics and body size in male snakes. Am. Nat. 141, 167–171.

Madsen, T., Shine, R., 2000. Rain, fish and snakes: climatically-driven population dynamics of Arafura filesnakes in tropical Australia. Oecologia 124, 208–215.

Madsen, T., Shine, R., Loman, J., Håkansson, T., 1992. Why do female adders copulate so frequently? Nature 355, 440–441.

Madsen, T., Ujvari, B., Shine, R., Olsson, M., 2006. Rain, rats and pythons: climate-driven population dynamics of predators and prey in tropical Australia. Austral Ecol. 31, 30–37.

Mason, R.T., 1992. Reptilian pheromones. In: Gans, C., Crews, D. (Eds.), Biology of the Reptilia, vol. 18. Hormones, Brain, and Behavior. University of Chicago Press, Chicago, IL, pp. 114–228.

Mason, R.T., 1993. Chemical ecology of the red-sided garter snake, *Thamnophis sirtalis parietalis*. Brain Behav. Evol. 41, 261–268.

Mason, R.T., 1994. Hormonal and pheromonal correlates of reproductive behavior in garter snakes. In: Davies, K. (Ed.), Perspectives in Comparative Endocrinology. Research Journals Division of the National Research Council of Canada, Ottawa, pp. 427–432.

Mason, R.T., Crews, D., 1985. Female mimicry in garter snakes. Nature 316, 59–60.

Mason, R.T., Parker, M.R., 2010. Social behaviour and pheromonal communication in reptiles. J. Comp. Physiol. 196A, 729–749.
Mason, R.T., Chinn, J.W., Crews, D., 1987. Sex and seasonal differences in the skin lipids of garter snakes. Comp. Biochem. Physiol. 87B, 999–1003.
Mason, R.T., Fales, H.M., Jones, T.H., Pannell, L.K., Chinn, J.W., Crews, D., 1989. Sex pheromones in snakes. Science 245, 290–293.
Mason, R.T., Jones, T.H., Fales, H.M., Pannell, L.K., Crews, D., 1990. Characterized synthesis and behavioral responses to sex attractiveness pheromones of red-sided garter snakes (*Thamnophis sirtalis parietalis*). J. Chem. Ecol. 16, 2353–2369.
McCracken, G.F., Burghardt, G.M., Houts, S.E., 1999. Microsatellite markers and multiple paternity in the garter snake *Thamnophis sirtalis*. Mol. Ecol. 8, 1475–1479.
Mendonca, M.T., Crews, D., 1989. Effects of fall mating on ovarian development in the red-sided garter snake. Am. J. Physiol. 257, R1548–R1550.
Mendonca, M.T., Crews, D., 1990. Mating-induced ovarian recrudescence in the red-sided garter snake. J. Comp. Physiol. 166A, 629–632.
Moore, I.T., Mason, R.T., 2001. Behavioral and hormonal responses to corticosterone in the male red-sided garter snake, *Thamnophis sirtalis parietalis*. Physiol. Behav. 72, 669–674.
Moore, I.T., LeMaster, M.P., Mason, R.T., 2000. Behavioural and hormonal responses to capture stress in the male red-sided garter snake, *Thamnophis sirtalis parietalis*. Anim. Behav. 59, 529–534.
Moore, I.T., Greene, M.J., Mason, R.T., 2001. Environmental and seasonal adaptations of the adrenocortical and gonadal responses to capture stress in two populations of the male garter snake, *Thamnophis sirtalis*. J. Exp. Zool. 289, 99–108.
Nilson, G., Andren, C., 1982. Function of renal sex secretion and male hierarchy in the adder, *Vipera berus*, during reproduction. Horm. Behav. 16, 404–413.
Noble, G.K., 1937. The sense organs involved in the courtship of *Storeria*, *Thamnophis*, and other snakes. Bull. Am. Mus. Nat. Hist. 73, 673–725.
O'Connor, D., Shine, R., 2004. Parental care protects against infanticide in the lizard *Egernia saxatilis* (Scincidae). Anim. Behav. 68, 1361–1369.
O'Donnell, R.P., Mason, R.T., 2007. Mating is correlated with a reduced risk of predation in female red-sided garter snakes, *Thamnophis sirtalis parietalis*. Am. Midl. Nat. 157, 235–238.
O'Donnell, R.P., Ford, N.B., Shine, R., Mason, R.T., 2004a. Male red-sided garter snakes (*Thamnophis sirtalis parietalis*) determine female reproductive status from pheromone trails. Anim. Behav. 68, 677–683.
O'Donnell, R., Shine, R., Mason, R.T., 2004b. Seasonal anorexia in the male red-sided garter snake, *Thamnophis sirtalis parietalis*. Behav. Ecol. Sociobiol. 56, 413–419.
Olsson, M., Madsen, T., 1995. Female choice on male quantitative traits in lizards—why is it so rare? Behav. Ecol. Sociobiol. 36, 179–184.
Olsson, M., Madsen, T., 1998. Sexual selection and sperm competition in reptiles. In: Birkhead, T.R., Møller, A.P. (Eds.), Sperm Competition and Sexual Selection. Academic Press, London, pp. 503–578.
Olsson, M., Shine, R., 1998. Chemosensory mate recognition may facilitate prolonged mate guarding by male snow skinks, *Niveoscincus microlepidotus*. Behav. Ecol. Sociobiol. 43, 359–363.
Olsson, M., Shine, R., Gullberg, A., Madsen, T., Tegelström, H., 1996. Female lizards control paternity of their offspring by selective use of sperm. Nature 383, 585.
Olsson, M., Madsen, T., Shine, R., 1997. Is sperm really so cheap? Costs of reproduction in male adders, *Vipera berus*. Proc. R. Soc. B 264, 455–459.
Orrell, K.S., Jenssen, T.A., 2002. Male mate choice by the lizard *Anolis carolinensis*: a preference for novel females. Anim. Behav. 63, 1091–1102.

Parker, M.R., Mason, R.T., 2009. Low temperature dormancy affects both the quantity and quality of the female sexual attractiveness pheromone in red-sided garter snakes. J. Chem. Ecol. 35, 1234–1241.

Partan, S.R., Marler, P., 2005. Issues in the classification of multimodal communication signals. Am. Nat. 166, 231–245.

Pearson, D., Shine, R., Williams, A., 2002. Geographic variation in sexual size dimorphism within a single snake species (*Morelia spilota*, Pythonidae). Oecologia 131, 418–426.

Perry-Richardson, J.J., Schofield, C.W., Ford, N.B., 1990. Courtship of the garter snake, *Thamnophis marcianus*, with a description of a female behavior for interruption of coitus. J. Herpetol. 24, 76–78.

Pfrender, M., Mason, R.T., Wilmslow, J.T., Shine, R., 2001. *Thamnophis sirtalis parietalis* (red-sided gartersnake). Male-male copulation. Herpetol. Rev. 32, 52.

Placyk Jr., J.S., Burghardt, G.M., Small, R.L., King, R.B., Casper, G.S., Robinson, J.W., 2007. Post-glacial recolonization of the Great Lakes region by the common garter snake (*Thamnophis sirtalis*) inferred from mtDNA sequences. Mol. Phylogenet. Evol. 43, 452–467.

Rivas, J., Burghardt, G.M., 2001. Understanding sexual size dimorphism in snakes: wearing the snake's shoes. Anim. Behav. 62, F1–F6.

Rossman, D.A., Ford, N.B., Seigel, R.A., 1996. The Garter Snakes, Evolution and Ecology. University of Oklahoma Press, Norman, OK.

Schuett, G.W., 1997. Body size and agonistic experience affect dominance and mating success in male copperheads. Anim. Behav. 54, 213–224.

Schuett, G.W., Gillingham, J.C., 1989. Male-male agonistic behaviour of the copperhead, *Agkistrodon contortrix*. Amphib.-Reptil. 10, 243–266.

Schwartz, J.M., McCracken, G.F., Burghardt, G.M., 1989. Multiple paternity in wild populations of the garter snake, *Thamnophis sirtalis*. Behav. Ecol. Sociobiol. 25, 269–273.

Seigel, R.A., 1996. Ecology and conservation of garter snakes. In: Rossman, D.A., Ford, N.B., Seigel, R.A. (Eds.), The Garter Snakes: Evolution and Ecology. University of Oklahoma Press, Norman, OK, pp. 55–89.

Shetty, S., Shine, R., 2002. The mating system of yellow-lipped sea kraits (*Laticauda colubrina*, Laticaudinae). Herpetologica 58, 170–180.

Shine, R., 1986. Sexual differences in morphology and niche utilization in an aquatic snake, *Acrochordus arafurae*. Oecologia 69, 260–267.

Shine, R., 2003. Reproductive strategies in snakes. Proc. R. Soc. B 270, 995–1004.

Shine, R., 2005. All at sea: aquatic life modifies mate-recognition modalities in sea snakes (*Emydocephalus annulatus*, Hydrophiidae). Behav. Ecol. Sociobiol. 57, 591–598.

Shine, R., Bonnet, X., 2000. Snakes: a new "model organism" in ecological research? Trends Ecol. Evol. 15, 221–222.

Shine, R., Fitzgerald, M., 1995. Variation in mating systems and sexual size dimorphism between populations of the Australian python *Morelia spilota* (Serpentes: Pythonidae). Oecologia 103, 490–498.

Shine, R., Mason, R.T., 2001a. Serpentine cross-dressers. Nat. Hist. 110, 56–61.

Shine, R., Mason, R.T., 2001b. Courting male garter snakes use multiple cues to identify potential mates. Behav. Ecol. Sociobiol. 49, 465–473.

Shine, R., Mason, R.T., 2004. Patterns of mortality in a cold-climate population of garter snakes (*Thamnophis sirtalis parietalis*). Biol. Conserv. 120, 205–214.

Shine, R., Mason, R.T., 2005a. Does large body size in males evolve to facilitate forced insemination? A study on garter snakes. Evolution 59, 2426–2432.

Shine, R., Mason, R.T., 2005b. Do a male garter snake's energy stores limit his reproductive effort? Can. J. Zool. 83, 1265–1270.

Shine, R., Mason, R.T., 2012a. An airborne sex pheromone in snakes. Biol. Lett. 8, 183–185.

Shine, R., Mason, R.T., 2012b. Familiarity with a female does not affect a male's courtship intensity in garter snakes (*Thamnophis sirtalis parietalis*). Curr. Zool. in press.

Shine, R., Shetty, S., 2001. The influence of natural selection and sexual selection on the tails of sea-snakes (*Laticauda colubrina*). Biol. J. Linn. Soc. 74, 121–129.

Shine, R., Sun, L., 2002. Arboreal ambush site selection by pit-vipers (*Gloydius shedaoensis*). Anim. Behav. 63, 565–576.

Shine, R., Olsson, M.M., Moore, I.T., LeMaster, M.P., Mason, R.T., 1999. Why do male snakes have longer tails than females? Proc. R. Soc. B 266, 2147–2151.

Shine, R., Olsson, M.M., LeMaster, M.P., Moore, I.T., Mason, R.T., 2000a. Effects of sex, body size, temperature and location on the antipredator tactics of free-ranging gartersnakes (*Thamnophis sirtalis*, Colubridae). Behav. Ecol. 11, 239–245.

Shine, R., Olsson, M.M., Moore, I.T., LeMaster, M.P., Mason, R.T., 2000b. Body size enhances mating success in male garter snakes. Anim. Behav. 59, F4–F11.

Shine, R., Olsson, M.M., Moore, I.T., LeMaster, M.P., Mason, R.T., 2000c. Are snakes right-handed? Asymmetry in hemipenis size and usage in garter snakes (*Thamnophis sirtalis*). Behav. Ecol. 11, 411–415.

Shine, R., Olsson, M.M., Mason, R.T., 2000d. Chastity belts in gartersnakes: the functional significance of mating plugs. Biol. J. Linn. Soc. 70, 377–390.

Shine, R., Harlow, P.S., Elphick, M.J., Olsson, M.M., Mason, R.T., 2000e. Conflicts between courtship and thermoregulation: the thermal ecology of amorous male garter snakes (*Thamnophis sirtalis parietalis*). Physiol. Biochem. Zool. 73, 508–516.

Shine, R., O'Connor, D., Mason, R.T., 2000f. The problem with courting a cylindrical object: how does an amorous male snake determine which end is which? Behaviour 137, 727–739.

Shine, R., Harlow, P., LeMaster, M.P., Moore, I.T., Mason, R.T., 2000g. The transvestite serpent: why do male garter snakes court (some) other males? Anim. Behav. 59, 349–359.

Shine, R., O'Connor, D., Mason, R.T., 2000h. Female mimicry in gartersnakes: behavioural tactics of "she-males" and the males that court them. Can. J. Zool. 78, 1391–1396.

Shine, R., O'Connor, D., Mason, R.T., 2000i. Sexual conflict in the snake den. Behav. Ecol. Sociobiol. 48, 392–401.

Shine, R., Elphick, M.J., Harlow, P.S., Moore, I.T., LeMaster, M.P., Mason, R.T., 2001a. Movements, mating and dispersal of red-sided gartersnakes from a communal den in Manitoba. Copeia 2001, 82–91.

Shine, R., LeMaster, M.P., Moore, I.T., Olsson, M.M., Mason, R.T., 2001b. Bumpus in the snake den: effects of sex, size and body condition on mortality in red-sided garter snakes. Evolution 55, 598–604.

Shine, R., O'Connor, D., LeMaster, M.P., Mason, R.T., 2001c. Pick on someone your own size: ontogenetic shifts in mate choice by male garter snakes result in size-assortative mating. Anim. Behav. 61, 1133–1141.

Shine, R., Phillips, B., Waye, H., LeMaster, M., Mason, R.T., 2001d. Benefits of female mimicry in snakes. Nature 414, 267.

Shine, R., Reed, R.N., Shetty, S., LeMaster, M., Mason, R.T., 2002. Reproductive isolating mechanisms between two sympatric sibling species of sea-snakes. Evolution 56, 1655–1662.

Shine, R., Phillips, B., Waye, H., LeMaster, M., Mason, R.T., 2003a. The lexicon of love: what cues cause size-assortative courtship by male garter snakes? Behav. Ecol. Sociobiol. 53, 234–237.

Shine, R., Phillips, B., Waye, H., Mason, R.T., 2003b. Small-scale geographic variation in antipredator tactics of garter snakes. Herpetologica 59, 333–339.

Shine, R., Phillips, B., Waye, H., Mason, R.T., 2003c. Behavioral shifts associated with reproduction in garter snakes. Behav. Ecol. 14, 251–256.

Shine, R., Langkilde, T., Mason, R.T., 2003d. Confusion within mating balls of garter snakes: does misdirected courtship impose selection on male tactics? Anim. Behav. 66, 1011–1017.
Shine, R., Phillips, B., Waye, H., LeMaster, M., Mason, R.T., 2003e. Chemosensory cues allow courting male garter snakes to assess body length and body condition of potential mates. Behav. Ecol. Sociobiol. 54, 162–166.
Shine, R., Langkilde, T., Mason, R.T., 2003f. Cryptic forcible insemination: male snakes exploit female physiology, anatomy and behavior to obtain coercive matings. Am. Nat. 162, 653–667.
Shine, R., Langkilde, T., Mason, R.T., 2003g. The opportunistic serpent: male garter snakes adjust courtship tactics to mating opportunities. Behaviour 140, 1509–1526.
Shine, R., Langkilde, T., Mason, R.T., 2004a. Courtship tactics in garter snakes: how do a male's morphology and behavior influence his mating success? Anim. Behav. 67, 477–483.
Shine, R., Phillips, B., Langkilde, T., Lutterschmidt, D., Waye, H., Mason, R.T., 2004b. Mechanisms and consequences of sexual conflict in garter snakes (*Thamnophis sirtalis*, Colubridae). Behav. Ecol. 15, 654–660.
Shine, R., LeMaster, M., Wall, M., Langkilde, T., Mason, R.T., 2004c. Why did the snake cross the road? Effects of roads on movement and mate-location by garter snakes (*Thamnophis sirtalis parietalis*). Ecol. Soc. 9, 9. http://www.ecologyandsociety.org/vol9/iss1/art9.
Shine, R., Phillips, B., Waye, H., LeMaster, M., Mason, R.T., 2004d. Species isolating mechanisms in a mating system with male mate choice (garter snakes, *Thamnophis*). Can. J. Zool. 82, 1091–1099.
Shine, R., Bonnet, X., Elphick, M., Barrott, E., 2004e. A novel foraging mode in snakes: browsing by the sea snake *Emydocephalus annulatus* (Serpentes, Hydrophiidae). Funct. Ecol. 18, 16–24.
Shine, R., O'Donnell, R.P., Langkilde, T., Wall, M.D., Mason, R.T., 2005a. Snakes in search of sex: the relationship between mate-locating ability and mating success in male garter snakes. Anim. Behav. 69, 1251–1258.
Shine, R., Langkilde, T., Wall, M., Mason, R.T., 2005b. Alternative male mating tactics in garter snakes. Anim. Behav. 70, 387–396.
Shine, R., Langkilde, T., Wall, M., Mason, R.T., 2005c. The fitness correlates of scalation asymmetry in garter snakes (*Thamnophis sirtalis parietalis*). Funct. Ecol. 19, 306–314.
Shine, R., Wall, M., Langkilde, T., Mason, R.T., 2005d. Scaling the heights: thermally-driven arboreality in garter snakes. J. Therm. Biol. 30, 179–185.
Shine, R., Wall, M., Langkilde, T., Mason, R.T., 2005e. Battle of the sexes: forcibly-inseminating male garter snakes target courtship to more vulnerable females. Anim. Behav. 70, 1133–1140.
Shine, R., Wall, M., Langkilde, T., Mason, R.T., 2005f. Do female garter snakes evade males to avoid harassment or to enhance mate quality? Am. Nat. 165, 660–668.
Shine, R., Webb, J.K., Lane, A., Mason, R.T., 2005g. Mate-location tactics in garter snakes: effects of rival males, interrupted trails, and non-pheromonal cues. Funct. Ecol. 19, 1017–1024.
Shine, R., Shine, T., Shine, J.M., Shine, B.G., 2005h. Synchrony in capture dates suggests cryptic social organization in sea snakes (*Emydocephalus annulatus*, Hydrophiidae). Austral Ecol. 30, 805–811.
Shine, R., Webb, J.K., Lane, A., Mason, R.T., 2006a. Flexible mate choice: a male snake's preference for larger females is modified by the sizes of females that he encounters. Anim. Behav. 71, 203–209.
Shine, R., Langkilde, T., Wall, M., Mason, R.T., 2006b. Temporal dynamics of emergence and dispersal of garter snakes from a communal den in Manitoba. Wildl. Res. 33, 103–111.

Shine, R., Langkilde, T., Mason, R.T., 2012. Facultative pheromonal mimicry in snakes: "she-males" attract courtship only when it is useful. Behav. Ecol. Sociobiol. in press. DOI: 10.1007/s00265-012-1317-4.

Steele, L.J., Cooper Jr., W.E., 1997. Pheromonal discrimination between conspecific individuals by male and female leopard geckos (*Eublepharis macularius*). Herpetologica 53, 476–485.

Sun, L., Shine, R., Zhao, D., Tang, Z., 2001. Biotic and abiotic influences on activity patterns of insular pit-vipers (*Gloydius shedaoensis*, Viperidae) from north-eastern China. Biol. Conserv. 97, 387–398.

Tokarz, R.R., 1992. Male mating preference for unfamiliar females in the lizard, *Anolis sagrei*. Anim. Behav. 44, 843–849.

Viitanen, P., 1967. Hibernation and seasonal movements of the viper, *Vipera berus berus* (L.), in southern Finland. Ann. Zool. Fennici 4, 472–546.

Whittier, J.M., Mason, R.T., Crews, D., 1985. Mating in the red-sided garter snake, *Thamnophis sirtalis parietalis*: differential effects on male and female sexual behavior. Behav. Ecol. Sociobiol. 16, 257–261.

Whittier, J.M., Mason, R.T., Crews, D., 1987a. Plasma steroid hormone levels of female red-sided garter snakes, *Thamnophis sirtalis parietalis*: relationship to mating and gestation. Gen. Comp. Endocrinol. 67, 33–43.

Whittier, J.M., Mason, R.T., Crews, D., Licht, P., 1987b. Role of light and temperature in the regulation of reproduction in the red-sided garter snake, *Thamnophis sirtalis parietalis*. Can. J. Zool. 65, 2090–2096.

Zug, G.R., Vitt, L.J., Caldwell, J.P., 2009. Herpetology: An Introductory Biology of Amphibians and Reptiles. Academic Press, New York.

# The Evolution of Animal Nuptial Gifts

SARA LEWIS and ADAM SOUTH

DEPARTMENT OF BIOLOGY, TUFTS UNIVERSITY, MEDFORD, MASSACHUSETTS, USA

Rich gifts wax poor when givers prove unkind
William Shakespeare, *Hamlet, Prince of Denmark* (3.1.101)

## I. INTRODUCTION

### A. WHAT ARE NUPTIAL GIFTS?

Nuptial arrangements in many human cultures include gift-giving traditions (Cronk and Dunham, 2007; Mehdi, 2003), and this behavior plays an important role in the mating systems of other creatures as well (Boggs, 1995; Fabre, 1918; Gwynne, 2008; Lack, 1940; Thornhill, 1976; Vahed, 1998, 2007; Zeh and Smith, 1985). In species widely distributed across the animal kingdom, males transfer many different non-gametic materials to females during courtship and mating. Such materials can include lipids, carbohydrates, proteins, peptides, amino acids, uric acid, minerals, water, antipredator defensive compounds, anti-aphrodisiac pheromones, and neuroendocrine modulators of recipient physiology. These nuptial gifts are an important aspect of reproductive behavior and animal mating systems (Andersson, 1994; Thornhill and Alcock, 1983). However, when compared to more conspicuous sexually selected traits such as male weaponry or ornamentation, such gifts have received relatively little attention from behavioral, ecological, and evolutionary research. Nuptial gifts heighten male reproductive investment, thus limiting male mating rates and altering courtship sex roles and sexual size dimorphism (Boggs, 1995; Gwynne and Simmons, 1990; Leimar et al., 1994). Selection acts on both gift-givers and receivers to shape nuptial gift structure and biochemical composition, as well as gift-giving behaviors. Not only do nuptial gifts form the basis for dynamic coevolutionary interactions between the sexes, but they also link



0065-3454/12 $35.00
DOI: 10.1016/B978-0-12-394288-3.00002-2

together male and female resource budgets (Boggs, 1990). Because they are thus strategically poised at the intersection of nutritional ecology, sexual selection, and life-history evolution (Boggs, 2009), understanding the evolutionary origins and maintenance of nuptial gifts is of fundamental importance.

Animal nuptial gifts come in multitudinous forms (Fig. 1), including food offerings, various male body parts, hemolymph, salivary gland secretions, seminal fluid, spermatophores (sperm-containing packages manufactured by male reproductive glands), and love darts. Many birds engage in courtship feeding, during which males provide prey to their own pair-bond partner or to extra-pair females (Lack, 1940; Mougeot et al., 2006). Scorpionfly males offer females dead insects or secretions from their

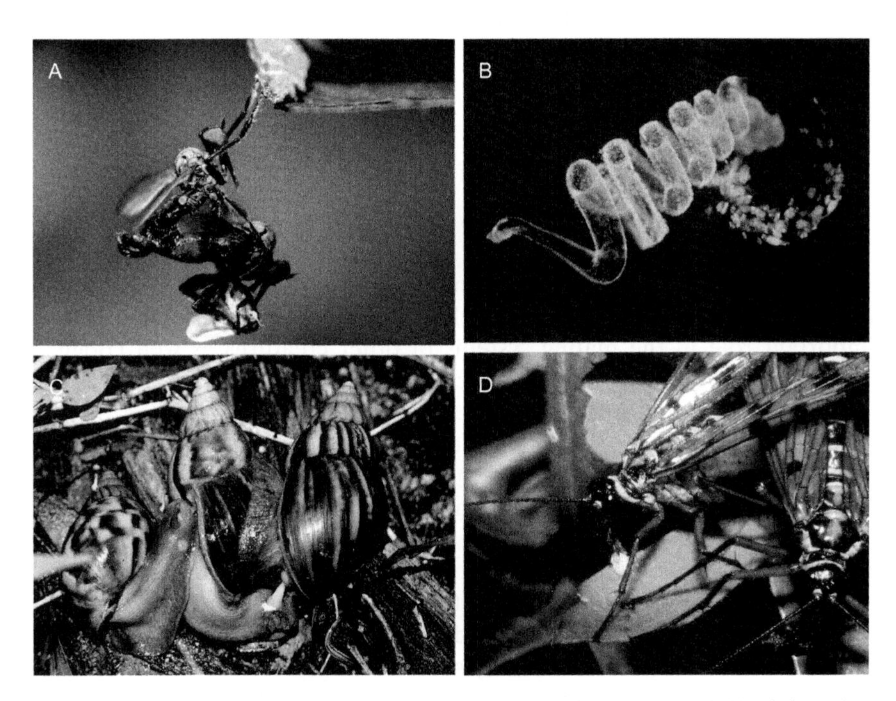


Fig. 1. A sampler illustrating the extraordinary diversity of animal nuptial gifts: (A) During mating, a female dance fly (Diptera: Empididae) feeds upon a dead insect provided by her mate (Photo by Rob Knell). (B) Sperm rings are released from a spermatophore manufactured by male accessory glands in *Photinus* fireflies (Coleoptera: Lampyridae). (C) During mating, hermaphroditic land snails (Gastropoda: Achatinidae) shoot their partner with a love dart that delivers mucus gland secretions (Photo by James Koh). (D) A male scorpionfly (Mecoptera: Panorpodidae) secretes a white salivary mass that will be consumed by a female during mating (Photo by Arp Kruithof). (For color version of this figure, the reader is referred to the online version of this chapter.)

enlarged, sexually dimorphic salivary glands (Liu and Hua, 2010; Thornhill, 1981). In some ground crickets, females imbibe hemolymph from a specialized spur located on their mate's hindleg (Gwynne, 1997; Piascik et al., 2010). In numerous animals (including salamanders, molluscs, crustaceans, annelids, leeches, and most insects), males transfer biochemically diverse spermatophores to females during mating (Mann, 1984). Nuptial gifts are not limited to animals with separate sexes, as during copulation many hermaphrodites inject chemicals that induce a physiological response in their partner (Koene and Schulenburg, 2005; Koene and Ter Maat, 2001; Michiels and Koene, 2006; Schilthuizen, 2005). Neither is gift-giving an exclusively male behavior: in heteropteran Zeus bugs, males feed upon glandular secretions provided by the female (Arnqvist et al., 2003).

Clearly, if we intend to move beyond merely describing these traits to begin formulating and answering questions about how animal nuptial gifts have evolved, we will need to start with a carefully articulated, coherent definition that encompasses this remarkable diversity. In this review, we begin by proposing such a definition, and then offer a conceptual framework for systematically classifying nuptial gifts. We go on to discuss some ecological conditions and life-history traits that might favor the evolution of nutritive nuptial gifts, that is, those that contribute to female resource budgets. From the male perspective, gift-giving behavior will usually (but not always) provide a net benefit. We analyze these potential benefits by describing how nuptial gifts can increase male reproductive fitness over multiple selection episodes that take place before, during, and after copulation. As a case study, we describe previous work on the katydid *Requena verticalis* that has elucidated how gift-giving males benefit from this behavior. Rigorously testing the many scenarios that have been proposed about nuptial gift evolution requires a comparative phylogenetic approach, and we discuss results from three insect groups where such an approach has been applied: crickets and katydids, fireflies, and *Drosophila* fruitflies. We also summarize work on rates of evolutionary change in an important constituent of *Drosophila* nuptial gifts known as seminal fluid proteins. Finally, we suggest several directions for future research that promise to deepen our understanding of nuptial gift evolution.

## B. Toward a Broader View of Nuptial Gifts

Before considering how nuptial gifts might have evolved, it is essential to clarify some relevant terminology. Previous studies have most often relied on *ad hoc* definitions of nuptial gifts, an approach reminiscent of the infamous "I know it when I see it" definition of hard-core pornography that Justice Potter Stewart used in his written opinion on the US Supreme

Court case Jacobellis v. Ohio (1964). The *Oxford English Dictionary* (1989) provides a legal definition of gift as "the transference of property or a thing by one person to another, voluntarily." Further, in colloquial English, the term "gift" generally implies some benefit for the recipient. However, because coevolutionary interactions between the sexes can continually alter costs and benefits for both givers and receivers, we contend that a broader view is essential for understanding the evolution of animal nuptial gifts. Within the scientific community, some researchers have limited the scope of nuptial gifts to encompass only "nutritive" gifts, that is, those that contain male-derived substances used by females to sustain metabolic activities (e.g., Andersson, 1994; Boggs, 1995; Gwynne, 2008; Thornhill, 1976). Others have excluded from consideration any gifts that are not contained within a consolidated package (e.g., Thornhill and Alcock, 1983). Again, we suggest that such restrictions may hinder progress toward the ultimate goal of understanding nuptial gift evolution.

As an alternative to this disjointed approach, we advocate the following definition (modified from South et al., 2011b; Lewis et al., 2011): *Nuptial gifts are materials beyond the obligatory gametes that are transferred from one sex to another during courtship or mating.* Importantly, this definition makes no assumptions concerning either: (1) how the gift currently affects fitness; thus, at certain times during its evolutionary trajectory, a gift might be beneficial, neutral, or detrimental to either sex, or (2) the presence or absence of gift-wrapping: thus, we include soluble proteins and other materials that are transmitted in seminal fluid or mucus as gifts, albeit unpackaged. In articulating this broad definition, we hope to unify what have previously been disconnected lines of investigation. For example, the protein content of insect spermatophores is often used as a measure of gift quality (Bissoondath and Wiklund, 1996; Cratsley et al., 2003; Wedell, 1994). This reasonable inference is based on female vitellogenesis being protein-limited (Wheeler, 1996), coupled with evidence that male-derived amino acids are incorporated into female eggs and soma (e.g., Boggs and Gilbert, 1979; Rooney and Lewis, 1999). However, male seminal fluid in *Drosophila melanogaster* contains soluble proteins secreted by reproductive accessory glands and the male ejaculatory duct (Chapman, 2008). Many of these reproductive proteins have been identified and demonstrated to alter female reproduction by stimulating ovulation and oviposition, increasing sperm storage, and lengthening females' latency to remate (Avila et al., 2011; Wolfner, 2009). However, because these proteins are transmitted in seminal fluid and are not encapsulated within a discrete package, traditionally they have not been considered as nuptial gifts (but see Markow, 2002; Simmons and Parker, 1989; Vahed, 1998). Yet it is becoming clear that male spermatophores contain many of the same protein classes (Andres et al., 2006, 2008; Braswell et al., 2006; Sonenshine et al., 2011; South et al., 2011a) and these components may produce similar effects on females. It could perhaps be argued that including seminal fluid makes our nuptial gift definition overly broad. However, because various constituents of seminal fluid have been shown to exert diverse effects on both male and female fitness (Gillott, 2003; Leopold, 1976; Poiani, 2006), such inclusion seems appropriate. Thus, we argue that drawing an arbitrary distinction between seminal products encased within a discrete package versus unpackaged seminal products transferred in a liquid ejaculate may have inadvertently obscured basic similarities in gift composition and function, as well as similarities in the evolutionary origin and maintenance of male reproductive accessory glands, the main gift-producing structures.

A key point is that this broad perspective on nuptial gifts allows for possible changes over evolutionary time in how gifts will affect the recipient's net fitness. While some degree of cooperation is required for sexual reproduction to occur, males and females have distinct reproductive interests (Arnqvist and Rowe, 2005; Parker, 1979; Trivers, 1972). As a result, coevolutionary interactions between the sexes will cause nuptial gifts to evolve dynamically in a manner that alters the cost/benefit ratio of nuptial gifts for each sex. Thus, a nuptial gift that originates because it provides mutual fitness benefits to both sexes may evolve into a gift that reduces the recipient's net fitness, and *vice versa*.

In summary, even though some may find fault with our definition, there is an undeniable need for a more systematic approach to defining what exactly constitutes a nuptial gift. Furthermore, a broader definition such as the one we propose here will allow us to better track the changes in nuptial gift costs and benefits that are certain to occur over evolutionary time.

## C. Classifying Nuptial Gift Diversity

In any contest, insects would surely emerge as the undisputed champions of gift diversity. For comprehensive insight into this fascinating diversity, readers are referred to excellent reviews by Boggs (1995) and Vahed (1998). Here, we highlight just a few notable patterns observed among insects before proposing a classification scheme that will encompass animal nuptial gifts.

First, gifts are conspicuously diverse, not only between different insect groups, but also within particular clades. For example, spermatophores are ubiquitous within the insect order Lepidoptera, yet they are absent in the Diptera and occur only sporadically within the Coleoptera (Davey,

1960; Mann, 1984). Within the beetle family Lampyridae (fireflies), some males pass elaborate spermatophores while firefly males of other species transfer free (unpackaged) ejaculates (Lewis and Cratsley, 2008; South et al., 2011b). Beyond spermatophores, orthopteran nuptial gifts have flowered into an especially impressive display of diversity (described in Section IV.A).

A second notable pattern is that some groups show surprising plasticity in their gift-giving behavior. For example, male *Panorpa* scorpionflies (Mecoptera: Panorpidae) pursue alternative mating tactics using different gift types (Sauer et al., 1998; Thornhill, 1981). In *P. cognata*, gift-giving behavior depends on a male's nutritional state: well-fed males secrete salivary masses that females consume during copulation, while low-nutrition males instead offer females a dead arthropod (Engqvist, 2007b). Similarly, in several empidid dance flies (Diptera: Empididae), males optionally offer females either a dead prey insect or inedible tokens such as silk balloons or seed tufts (Preston-Mafham, 1999; Vahed, 2007).

Here we propose a taxonomy for animal gifts that we hope will facilitate mapping the landscape of nuptial gift evolution (for other classification schemes see Gwynne, 2008; Simmons and Parker, 1989; Vahed, 1998). Table I presents four nuptial gift categories, with examples of relevant structures and behaviors from various taxa. One key distinction is based on the method of gift production. Thus, we distinguish between *endogenous gifts* that are manufactured by males themselves and *exogenous gifts* that consist of externally procured food items such as seeds or prey that males gather and then transfer to females. Another important distinction is based on how gifts are absorbed by the recipient. Gwynne (2008) distinguished *oral gifts* that are taken in through the female digestive system (e.g., food items, spermatophylaces, hindwing secretions), from gifts we term *genital gifts* that are absorbed through the female reproductive tract: this includes both unpackaged secretions from male reproductive glands (conveyed in liquid seminal fluid) as well as those encased in discrete packages (spermatophores). We propose here another category consisting of *transdermal gifts* that are injected through the skin into the partner's body (e.g., snail love darts, intradermally implanted squid spermatophores, hypodermic insemination in leeches and bedbugs). Although nuptial gifts are often commingled together into a single category (e.g., Arnqvist and Nilsson, 2000), we believe the distinctions drawn here will prove useful as a basis for future studies of the evolution of nuptial gift structure and composition. The primary reason for proposing this classification scheme is because, as discussed below, very different predictions can be made about how various gift types might affect fitness of both sexes (see also Simmons and Parker, 1989 and Section III).

TABLE I

A Classification Scheme For Nuptial Gifts

| Gift production | Gift absorption | Nuptial gift examples | Taxonomic group and references |
|---|---|---|---|
| Endogenous | Oral | Hemolymph from tibial spurs | Ground crickets (Piascik et al., 2010) |
| | | Spermatophylax | Katydids and crickets (Gwynne, 1997) |
| | | Salivary secretions | *Panorpa* scorpionflies (Engqvist, 2007a) |
| | | Anal secretions | *Drosophila nebulosa* (Steele, 1986) |
| | | Metanotal secretions | Tree crickets (Brown, 1997; Bussiére et al., 2005) |
| | | Male body (sexual cannibalism) | Red-backed spider, mantids (Elgar and Schneider, 2004) |
| Endogenous | Genital | Spermatophores | Salamanders, lepidopterans, molluscs, copepods, crabs, spiders (Mann, 1984) |
| | | Seminal fluid proteins | *Drosophila* spp. (Chapman, 2008; Markow, 2002; Wolfner, 2007) |
| Endogenous | Transdermal | Love darts | Land snails (Koene and Schulenburg, 2005) |
| | | Setal gland injection | Earthworms (Koene et al., 2005) |
| | | Intradermal spermatophore implantation | Squid (Hoving and Laptikhovsky, 2007), leeches (Mann, 1984) |
| | | Hemocoelic injection of seminal fluid | Bedbugs (Stutt and Siva-Jothy, 2001) |
| Exogenous | Oral | Courtship feeding | Birds: kestrels, shrikes (Lack, 1940; Mougeot et al., 2006) |
| | | Seeds | Lygaeid bugs (Carayon, 1964) |
| | | Insect prey | Hangingflies, scorpionflies (Thornhill, 1981), empidid flies (Cumming, 1994), *Pisaura* spiders (Austad and Thornhill, 1986) |

### 1. *Exogenous Oral Gifts*

These consist of food items that males capture or collect, so these are most likely to contain nutritive materials (defined as substances that contribute to female metabolic reserves). Thus, most exogenous oral gifts are predicted to deliver net fitness benefits to females, measured as increased lifetime fecundity. From the male perspective, these gifts are generally predicted to increase male fitness across several selection episodes (reviewed in Gwynne, 2008; Vahed, 1998, 2007). First, because they can be assessed (visually or by gustation) prior to mating, exogenous oral gifts should affect a male's ability to attract and successfully mate with females. Second, because females remain stationary while feeding, food gifts may make it easier for males to initiate copulation. Third, because females feed on these gifts while copulating, such gifts are expected to increase both copulation duration and the quantity of sperm transferred.

### 2. *Endogenous Oral Gifts*

This category includes diverse materials that are secreted by male salivary, reproductive, and other glands, as well as parts or the whole of the male's body; these materials are then consumed by females before, during, or after copulation (reviewed by Boggs, 1995; Elgar and Schneider, 2004; Vahed, 1998). Thus, in *Oecanthus nigricans* tree crickets (Orthoptera), females feed upon proteinaceous secretions produced by dorsally located male glands, while females of some true flies (Diptera) and scorpionflies (Mecoptera) consume male salivary secretions, and female *Allonemobius* ground crickets drink hemolymph from male hindleg spurs (Bidochka and Snedden, 1985). Females in many katydids and crickets (Orthoptera) consume a spermatophylax, a gelatinous portion of the spermatophore produced by male reproductive glands. Many mantids and orb-weaving spiders engage in sexual cannibalism, where females kill and consume males either before or after insemination (Elgar and Schneider, 2004); in both cases, the male body represents an endogenous oral gift under our definition, even when it is given involuntarily (i.e., gifts can have a negative effect on male fitness).

Since they derive from such diverse sources, endogenous oral gifts are likely to have quite varied effects on females. Some endogenous oral gifts, such as hemolymph or male body parts, may closely resemble exogenous gifts of prey or other food items in contributing to females' nutrient budgets (Boggs, 1995; Gwynne, 2008). Rather than replicating whatever nutritional mixtures are available in the diet, however, glandular gifts have the potential to provide more targeted dietary supplements. These specialized glandular gifts might supply nutrients which are otherwise absent or limited in female diets, such as macronutrients (proteins, lipids, carbohydrates), micronutrients (sodium, zinc), or defensive compounds (cantharidin, pyrrolizidine alkaloids, cyanogenic glycosides). In cockroaches (Dictyoptera: Blattidae), males provide endogenous oral gifts that constitute an important nitrogen source for females and their eggs (reviewed by Boggs, 1995; Vahed, 1998). In many cockroaches, males accumulate uric acid in their accessory glands before packaging it into their spermatophore; after mating, females expel and eat the spermatophore. In other roaches, females feed directly on uric acid as it is secreted from male glands.

On the other hand, sexual conflict theory predicts the evolution of male glandular gifts that benefit males even though they may adversely affect female net fitness (Arnqvist and Nilsson, 2000; Arnqvist and Rowe, 2005; Rice, 1998). Through reciprocal sexual coevolution, an escalating arms race might then ensue in which females evolve the ability to metabolize or otherwise counteract manipulative male substances (Arnqvist and Nilsson, 2000; Eberhard, 1996). However, it has been pointed out (Gwynne, 2008) that such oral gifts might be less likely to contain manipulative substances because those would be subject to degradation while passing through the digestive tract. Thus, the category of endogenous oral gifts is diverse and includes nuptial gifts that may have positive, negative, or no effects on female fitness.

From the male perspective, when endogenous oral gifts (such as secreted salivary masses) can be inspected by females, they could resemble exogenous gifts that increase male mating success, copulation duration, and possibly sperm quantity transferred during copulation. For example in spiders, sexual cannibalism that takes place after insemination can benefit the sacrificed male by prolonging copulation duration, thus increasing sperm transfer and male paternity share, in addition to increasing female fecundity and offspring survival (Andrade, 1996; Elgar and Schneider, 2004; Herberstein et al., 2011; Welke and Schneider, 2012). For orally ingested glandular gifts, such as the orthopteran spermatophylax, males may be selected to incorporate phagostimulants that increase their gifts' gustatory appeal for females (Sakaluk, 2000; Vahed, 2007). Selection may also alter male gift composition to slow female consumption rates if this allows more time for males to transfer sperm. For example, in many crickets and katydids (Orthoptera: Ensifera), the male spermatophylax has a sticky, gelatinous consistency that prevents rapid ingestion by females (Vahed, 2007).

### 3. *Endogenous Genital Gifts*

This category includes materials that are produced by secretory tissue in the male reproductive tract, transferred in seminal fluid or spermatophores, and absorbed through the female genital tract. Although spermatophores may have originated to prevent sperm loss or desiccation (Davey, 1960; Khalifa, 1949), in many animals these structures have become vastly elaborated (Mann, 1984; Thornhill, 1976). Many ideas have been proposed about the evolutionary origin of elaborated male ejaculates such as spermatophores. Wickler (1985) proposed that spermatophores originated as a way for males to prevent females from digesting sperm, as an adaptation secondary to intrasexual selection for greater sperm quantity. It has also been suggested that female choice, based on the quality or quantity of nonsperm ejaculate components, might have favored the elaboration of male ejaculates (Cordero, 1996). Arnqvist and Nilsson (2000) proposed that elaborated male ejaculates represent "manipulative and sinister superstimuli" that evolved through sexual conflict over female remating rates. However, given the wide taxonomic distribution and diversity of endogenous genital gifts, it is unrealistic to expect a single explanation for their evolution. Rather, even a brief overview of gift constituents indicates that endogenous genital gifts have probably had multiple evolutionary origins and diverse trajectories.

Like orally ingested glandular gifts, the products of male reproductive glands can also supply nutrients that are absent or limited within female diets. The geometric framework in nutritional ecology may provide a useful perspective for thinking about the evolution of nutritive nuptial gifts. This framework is based on locating an organism's nutritional requirements and dietary choices within a multidimensional resource space (Raubenheimer, 2011; Raubenheimer et al., 2009). Importantly, rather than replicating nutritional mixtures that are available in the female diet, nuptial gifts could provide a vector that targets the female-specific requirements for vitellogenesis (Boggs, 1990). Thus, selection may shape male glandular products to augment females' resources by providing them with entirely different nutritional mixtures compared to those gained through feeding. Empirical studies of numerous Orthoptera, Lepidoptera, and Coleoptera have demonstrated that diverse substances derived from endogenous genital gifts are incorporated into female somatic tissue and eggs; these substances include amino acids, zinc, phosphorus, and sodium transferred in male spermatophores (reviewed by Boggs, 1995; Vahed, 1998). For example, many lepidopteran males engage in puddling behavior on damp soil, dung, or carrion where they obtain sodium, which is a scarce nutrient for most folivores (Molleman, 2010). Males accumulate this element in their reproductive glands and transfer sodium-rich spermatophores during mating; in the moth *Gluphisia septentrionis*, a single spermatophore contains $>50\%$ of the male's total body sodium content (Smedley and Eisner, 1996). Females pass sodium along to their eggs, and in the skipper, *Thymelicus lineola*, such gifts enhance larval survivorship (Pivnick and McNeil, 1987, but see

Molleman et al., 2004). In addition, reproductive glands can serve as a reservoir for defensive compounds that males derive from dietary sources, and these compounds are later transferred to females within spermatophores or seminal fluid (reviewed by Vahed, 1998). Thus, endogenous genital gifts can contain defensive compounds that protect the female or her eggs against predators or microbial attack; such gifts include cantharidin in *Neopyrochroa flabellata* beetles (Eisner et al., 1996), pyrrolizidine alkaloids in *Utetheisa ornatrix* moths (Eisner and Meinwald, 1995), cyanogenic glycosides in several *Heliconius* butterflies (Cardoso and Gilbert, 2007), and vicilin-derived peptides in *Callosobruchus maculatus* cowpea beetles (Alexandre et al., 2011).

On the other hand, some components of endogenous genital gifts may reduce female fitness. In some male insects, reproductive accessory glands manufacture compounds that have diverse effects on female reproductive physiology and behavior (Eberhard, 1996; Gillott, 2003). In *D. melanogaster*, for example, seminal fluid proteins have been shown to heighten female oogenesis and oviposition, increase sperm storage and utilization, and to reduce female remating rates and life span (reviewed by Chapman, 2008; Chapman and Davies, 2004; Ravi Ram and Wolfner, 2007a,b; Wolfner, 2007). For most taxa, little is known concerning the nature of these secretions, although recent work has elucidated seminal fluid composition in *Aedes* mosquitoes (Sirot et al., 2008), *Gryllus* and *Allonemobius* crickets (Andres et al., 2006; Braswell et al., 2006), *Heliconius* butterflies (Walters and Harrison, 2010), *Tribolium* flour beetles (South et al., 2011a), and honeybees (Baer et al., 2009; Collins et al., 2006). In many species, male gifts contain anti-aphrodisiacs that reduce a female's likelihood of remating (*Tenebrio* beetles, Happ, 1969; *Pieris napi* butterflies, Andersson et al., 2004; *Heliconius* butterflies, Estrada et al., 2011). Selection on males to reduce sperm competition risk favors inclusion of such substances, yet anti-aphrodisiacs can lower female net fitness if they depress remating rates below some female optimum.

Thus, endogenous genital gifts are complex mixtures that have likely been shaped by multiple selective forces. While it has been argued that male ejaculate composition will be selected primarily to manipulate female reproductive physiology and should carry a net fitness cost borne by gift recipients (Arnqvist and Nilsson, 2000; Arnqvist and Rowe, 2005), it is clear that understanding the complex effects that male ejaculates have on females will require a broad and balanced perspective.

### 4. *Endogenous Transdermal Gifts*

These nuptial gifts include male seminal and glandular products that are transferred and absorbed outside the female's digestive or reproductive systems. This happens during extragenital insemination in bedbugs (Stutt and Siva-Jothy, 2001) and intradermal spermatophore implantation in deep-sea squid (Hoving and Laptikhovsky, 2007). In the bedbug, *Cimex lectularius*, male ejaculates include (in addition to sperm) antioxidants, micronutrients, and antibacterial compounds (Reinhardt et al., 2009). Hypodermic injection of seminal products is particularly widespread among simultaneous hermaphrodites such as leeches, sea slugs, and polyclad flatworms (Michiels and Koene, 2006). Another type of transdermal gift consists of allohormones, substances that induce a direct physiological response in the recipient (Koene and Ter Maat, 2001). These can be injected through the skin of a mating partner during copulation while sperm are being passed to the reproductive organs. This mode of delivery allows male products to bypass both digestive and reproductive tracts, where various gift components might get broken down. During copulation, hermaphroditic earthworms *Lumbricus terrestris* use their ventral copulatory setae to inject their partner with setal gland products that induce sperm uptake and storage (Koene et al., 2005). A similar benefit for male function occurs in *Helix aspersa* land snails, which penetrate their partners with a calcareous dart coated with allohormones produced by a mucus gland; these substances inhibit sperm digestion and enhance sperm storage by the recipient (Koene and Schulenburg, 2005; Schilthuizen, 2005). As in other endogenous gifts, selection on transdermal gift production may favor the inclusion of compounds that benefit males yet are detrimental to female fitness.

## II. Effects on Recipient Fitness

Empirical studies in numerous taxa have documented how male gifts affect several different female fitness components (including egg and clutch size, rate and timing of offspring production, longevity), as well as female net fitness (lifetime fecundity measured as the total number of eggs or offspring produced). Many studies have found that nuptial gifts can provide females with direct material benefits measured as an increase in the recipient's net fitness. Such evidence has been compiled and summarized by previous literature reviews for arthropod nuptial gifts (Boggs, 1995; Gwynne, 2008; Rooney and Lewis, 1999; Vahed, 1998, 2007) and for sexual cannibalism (Elgar and Schneider, 2004), as well as by some meta-analyses (Arnqvist and Nilsson, 2000; South and Lewis, 2011).

Rather than recapitulating these synopses here, we simply advocate that the term nuptial gift be used in its broadest sense, that is, independently of whether such materials currently exert a positive, a negative, or no effect on recipient net fitness (Fig. 2). Others have used narrower terminology, using

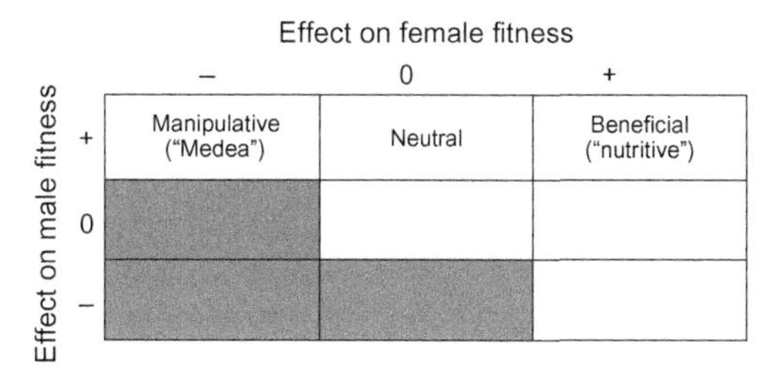


FIG. 2. Our nuptial gift definition encompasses a range of possibilities for how nuptial gifts might influence male and female net fitness (benefit minus cost). Gifts that provide a net fitness benefit to males can have negative, positive, or no effects on female fitness (top row). When gifts provide a net fitness benefit for females, they can be maintained whether or not males derive a fitness benefit; thus, sexual cannibalism would fall into the rightmost column (e.g., cases when the male is consumed before insemination would fall into the bottom right cell). The position of any gift is likely to shift over evolutionary time, as sexual interactions modify costs and benefits for each sex. (Gray areas indicate that evolutionary maintenance is unlikely as these gifts carry net fitness costs for one or both sexes.) (For color version of this figure, the reader is referred to the online version of this chapter.)

nuptial gifts to mean only nutritive gifts or those that increase female fitness (i.e., those falling within the upper-right cell of Fig. 2). On the other hand, because male and female reproductive interests are not perfectly aligned, sexual conflict may drive the evolution of nuptial gifts that provide fitness benefits to males while reducing female net fitness. Arnqvist and Nilsson (2000) and Arnqvist and Rowe (2005) suggested that the term "Medea gift" (named after a mythological Greek sorceress who used a beautifully embroidered, poisonous robe to murder a rival) should be used for any gifts that reduce female net fitness (i.e., those falling within the upper-left cell of Fig. 2). However, because coevolutionary interactions are expected to create dynamic changes over time in gifts' cost/benefit ratios for their recipients, we believe such restrictive terminology is counterproductive to the goal of understanding nuptial gift evolution. One example of this shifting balance of costs and benefits is seen in the bedbug *C. lectularius*. Although traumatic insemination through the abdominal wall causes wounding that reduces female life span (Stutt and Siva-Jothy, 2001), male ejaculates contain compounds that increase female net fitness via increased lifetime fecundity and oviposition rate, and delayed reproductive senescence (Reinhardt et al., 2009). One evolutionary scenario proposed by these authors is that male ejaculates were originally detrimental and that subsequent female counteradaptations evolved to neutralize, and eventually reverse, these harmful effects. Alternatively, they suggest, male ejaculates may have positively affected female net fitness when they originated. While distinguishing between these evolutionary trajectories must await future phylogenetic studies coupled with ancestral trait reconstruction, it is clear that a more holistic framework will be required to understand the evolution of nuptial gifts.

Possible evolutionary trajectories leading to manipulative male gifts have been presented in detail by others (e.g., Arnqvist and Rowe, 2005; Eberhard, 1996; Rice, 1998; Sakaluk, 2000). Here we focus on circumstances that might lead to the evolution of nutritive nuptial gifts, that is, those providing material benefits that directly increase fitness for the gift recipient.

Among the ways that nuptial gifts might increase a male's fitness relative to other males in the same population is by enhancing female fecundity relative to other females in the population. The enhanced fecundity hypothesis for paternal investment was proposed by Tallamy (1994), who suggested that male investment will evolve whenever males can provide materials whose availability constrains female reproductive output. While he mainly focused on postzygotic male investment (e.g., paternal brood care), Tallamy pointed out that this hypothesis should also apply to the evolution of nuptial gifts (i.e., prezygotic male investment). In addition, recent theoretical work has shown that depending on the degree of fecundity enhancement, male nuptial gifts can alter intersexual coevolutionary dynamics and lead to a stable evolutionary equilibrium with mutual fitness benefits for both sexes (Alonzo and Pizzari, 2010).

In general, female reproduction will be resource-constrained because of the higher gametic investment by this sex (Trivers, 1972); in oviparous organisms, all the nutritional resources required for embryogenesis must be contained within each egg. Female egg production is most often limited by protein availability (Wheeler, 1996). In insects, as in most oviparous animals, oocyte development is fueled mainly by vitellogenin, a female-specific glycolipoprotein; insect eggs also contain lipids and some carbohydrates in the form of glycogen (Klowden, 2007). Females need to obtain these macronutrients from larval feeding, from adult feeding, or from male nuptial gifts (Boggs, 1990). Thus, the enhanced fecundity hypothesis predicts that selection for nuptial gifts will be influenced by the availability and quality of specific nutritional resources needed for female reproduction. Resource availability will in turn depend on organismal life-history traits, as well as on temporal and habitat variation within a particular species. Below we consider some combinations of ecological conditions and life-history traits that are expected to favor the evolution of nutritive gifts that enhance female fecundity through male contributions to female resource budgets (Boggs, 1990, 1995). We do not discuss mating systems; while several studies have explored the relationship between nuptial gifts and polyandry (e.g., Karlsson, 1995; Karlsson et al., 1997), it is difficult to determine causal relationships between these two highly correlated features.

First we discuss some life-history features that are expected to lead to female-specific resource limitation. These include location of a species along the continuum between income and capital breeding, temporal dynamics of female oogenesis, and requirements for female dispersal and flight. Income breeders are those that fuel reproduction using current energetic income, while capital breeders support their reproduction with energy stores accumulated at an earlier life stage (Houston et al., 2007; Stearns, 1992). In purely capital breeders, male nuptial gifts could provide resources to supplement reserves that otherwise would be depleted over a female's reproductive life span. One such example is *Photinus ignitus* firefly beetles (reviewed by Lewis and Cratsley, 2008; Lewis et al., 2004), which are capital breeders that entirely lack adult feeding. Both sexes mate repeatedly over their 2-week adult life span. Males manufacture a complex spermatophore from several reproductive glands, and spermatophore-derived proteins are allocated to females' developing oocytes. Females that receive multiple spermatophores gain increased lifetime fecundity. In addition, a seasonal reversal in courtship roles occurs: late in the mating season when both sexes face depleted resource stores, females compete for access to gift-providing males and males selectively mate with more fecund females (Cratsley and Lewis, 2005).

Nuptial gift evolution may also depend on interspecific life-history differences in the temporal dynamics of female oogenesis (Boggs, 1990, 1995, 2009). In some insect taxa, adult females emerge with their entire complement of eggs already matured (e.g., mayflies), while in others (e.g., *P. ignitus* fireflies) females will continue to mature eggs throughout their reproductive lives (Jervis and Ferns, 2004; Jervis et al., 2005 compiled relevant data for many parasitoid wasps). When egg maturation is distributed over time, selection should favor male nuptial gifts that could enhance female reproductive output by replenishing resources.

A final life-history trait that may alter selection for nuptial gifts relates to female mobility. If females must fly in order to locate food, mates or suitable oviposition sites, to disperse, or to escape predators, then wing-loading constraints may restrict how many mature eggs a female can carry at any point in time. In addition, females face a trade-off between allocating resources to flight or to oogenesis (Boggs, 2009; Wheeler, 1996). Lewis and Cratsley (2008) presented a conceptual model proposing that because flightless (wingless) females can devote all their resources to egg production, selection for nuptial gifts will be relaxed due to limited scope for any further increases in females fecundity. A recent evolutionary trait analysis in fireflies supported this predicted intersexual correlation between female flight ability and male nuptial gifts (South et al., 2011b; see Section V.B below). Females of ancestral fireflies most likely were fully winged and received male nuptial gifts in the form of spermatophores. In several lineages, after females lost their flight ability (possibly driven by fecundity selection), males subsequently lost the ability to produce these nuptial gifts.

Selection for nuptial gifts should also be influenced by within-species variation in the availability and quality of specific resources required for female reproduction. When such resources are limited, females may increase their mating activity to gain access to nutritive nuptial gifts (Boggs, 1990; Gwynne, 1990). For example, in the pollen katydid *Kawanaphila nartee* (Orthoptera: Tettigonidae), scarcity of pollen (a protein-rich food source for both sexes) generates intersexual competition among females for access to endogenous oral gifts in the form of the male spermatophylax (Simmons and Bailey, 1990). In the pollen-feeding butterfly *Heliconius cydno*, pollen load varies among females and is negatively correlated with number of matings, and thus nuptial gifts, that females acquire (Boggs, 1990). In addition, many experimental studies have found that nuptial gifts provide larger fecundity increases when females are food-limited (Gwynne and Simmons, 1990; reviewed by Boggs, 1990; Gwynne, 1991, 2008). Finally, Leimar et al. (1994) provided comparative data from butterflies suggesting that variation in available resources (rather than average) will increase selection for nuptial gifts. Similarly, another comparative study across butterfly species found increased polyandry with greater variation in female body size, again indirectly suggesting that species with more variable larval food resources might experience increased selection for nuptial gifts (Karlsson, 1995).

Thus, several ecological conditions and life-history traits linked to female resource allocation are predicted to favor the evolution of fecundity-enhancing male gifts. Indeed, an entire suite of correlated life-history traits seems likely to select for fecundity-enhancing nuptial gifts. Based on the connections outlined here between nuptial gifts, life-history traits, and nutritional ecology, testing hypotheses about trait combinations that favor the evolution of nutritive nuptial gifts seems like an important and relatively unexplored research area. As Boggs (1995) pointed out nearly 20 years ago, we still need rigorous comparative phylogenetic studies focused on testing for evolutionary associations between nuptial gift presence (and type) and interspecific variation in resource conditions and life-history traits.

## III. Potential Gift Benefits for Males

Considerable evidence indicates that the collection and manufacture of nuptial gifts is costly for males (reviewed by Boggs, 1995). In addition, males have been shown to strategically allocate their gifts depending on female reproductive status or age (e.g., Simmons et al., 1993; Sirot et al.,

2011; Wedell, 1992). While colloquial usage views gifts as something given voluntarily, in the case of sexual cannibalism there may also be involuntary gift-giving that carries a net fitness cost for males (Elgar and Schneider, 2004, lower-right cell of Fig. 2). Yet as mentioned above, in some species cannibalized males gain posthumous benefits through increased paternity share and decreased likelihood of the female remating (Andrade, 1996; Herberstein et al., 2011: upper-right cell of Fig. 2). In most cases, however, the male structures that produce nuptial gifts and the various behaviors associated with gift-giving will only be maintained if they confer a net fitness benefit on males; that is, the gift-giving males must be able to sire more offspring compared to other males in the same population. These fitness advantages could accrue across different reproductive episodes, including through higher mating success, increased paternity share relative to other males mating with the same female, and/or enhanced female fecundity compared to other females in the population.

In determining what specific benefits a male might derive from his gift-giving behavior, much previous work has been caught up in a largely unproductive semantic debate. For several decades many attempts were made to distinguish between two particular hypotheses for the origin and maintenance of male gifts (Alexander and Borgia, 1979; Gwynne, 1984; Sakaluk, 1986; Simmons and Parker, 1989; Vahed, 1998). The mating effort hypothesis suggested that gifts function to ensure mating and sperm transfer, while the paternal investment hypothesis suggested that gifts function to increase the number or quality of the gift-giver's own offspring. However, attempts to sort nuptial gifts neatly into these two categories were unsuccessful, and a fatal terminological quagmire gradually developed (see also Gwynne, 2008; Simmons, 1995; Simmons and Parker, 1989; Vahed, 1998; Wickler, 1985). Among the reasons for this failure was that these two hypotheses represent two nonindependent gift functions (i.e., the latter depends on the former), and also that the paternity data necessary for empirical tests of the paternal investment hypothesis were lacking.

Moving forward, we suggest that a more constructive approach will be to think about nuptial gifts as selection targets during several sequential episodes that occur *before*, *during*, and *after* mating (Fig. 3). A similar approach was suggested by Gwynne (1997; his Table 6-1). For example, nuptial gifts may enhance a male's mating success by increasing his ability to attract females (episode 1) and to successfully copulate with them (episode 2). During copulation, nuptial gifts may improve a male's insemination success (episode 3, measured as whether or not any sperm transfer occurs), or increase the number of sperm transferred (episode 4). After mating, nuptial gifts may increase the viability and storage of male sperm within the female reproductive tract (episode 5). In competitive mating situations

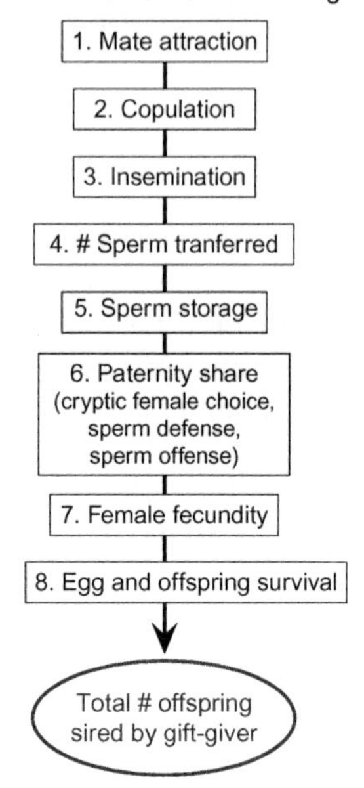


FIG. 3. Potential fitness benefits gained by males from nuptial gifts across sequential selection episodes. Nuptial gifts may increase mate attraction, copulation, or insemination success, quantity of sperm transferred or stored, and paternity share in competitive mating situations (gained through cryptic female choice, protecting paternity share when the female remates, and/or when the female has previously mated). Nuptial gifts may also provide fitness benefits to males by increasing overall female fecundity, egg or offspring survival, if on average such gifts increase the production and survival of the gift-giving male's own offspring. (For color version of this figure, the reader is referred to the online version of this chapter.)

(polyandry or polygamy), providing a nuptial gift may increase male paternity share (proportion of offspring sired by the gift-giving male) relative to other males mating with the same female (episode 6); this may occur through cryptic female choice favoring certain male or gift traits, or by increased sperm defense or offense. As discussed in Section II, nuptial gifts may also increase a male's fitness by enhancing overall female fecundity (episode 7), as well as egg and offspring survival (episode 8), relative to that of other females in the population. Numerous experimental or observational studies of nuptial gift function have demonstrated that larger or more nuptial gifts lead to higher male fitness during one or more of these sequential selection episodes; we provide just a few examples below.

Many oral nuptial gifts (both exogenous and endogenous) provide a benefit to males by attracting females and also by increasing the likelihood that females will copulate once they have been attracted. Among such gifts are the edible and inedible gifts offered by male empidid dance flies (Lebas and Hockham, 2005; Preston-Mafham, 1999), prey and salivary secretions provided by *Panorpa* scorpionflies (Engqvist, 2007a), and food regurgitated by *Drosophila subobscura* males (Steele, 1986). Some oral gifts, such as male hindwings and hemolymph in *Cyphoderris* hump-winged crickets (Eggert and Sakaluk, 1994), also facilitate successful insemination, as males can more readily accomplish sperm transfer when females hold still while feeding.

Oral nuptial gifts (both exogenous and endogenous) also function during mating to increase the quantity of sperm transferred to the female reproductive tract; these include the spermatophylax in several crickets and bushcrickets (Sakaluk, 1984; reviewed by Gwynne, 1997; Vahed, 1998, 2007), prey gifts in *Bittacus* hangingflies (Thornhill, 1976), and salivary secretions in *Panorpa* scorpionflies (Engqvist, 2007a).

Other nuptial gifts benefit males during postcopulatory episodes of selection, including enhancing female storage of sperm that will later fertilize eggs. Acp36DE is a seminal fluid protein (endogenous genital) transferred by *D. melanogaster* males that causes an increase in sperm numbers stored within the female reproductive tract (Qazi and Wolfner, 2003). Similarly, hermaphroditic *H. aspersa* garden snails penetrate their partner with a mucus-coated dart (endogenous transdermal gift) that increases sperm storage (Rogers and Chase, 2001).

Many endogenous gifts (both oral and genital) also include materials that inhibit the female's mating receptivity (thus reducing the risk of sperm competition) and/or increase her latency to remate (thus increasing sperm defense). Such receptivity-inhibiting materials include *Drosophila* sex peptide, unknown ejaculate components in *R. verticalis* katydids (Gwynne, 1986), nonfertilizing apyrene sperm in Lepidoptera (Wedell, 2005), and salivary secretions of *Panorpa* scorpionflies (Engqvist, 2007a). In addition, many spermatophores (endogenous genital) contain anti-aphrodisiacs that deter other males from approaching a mated female (e.g., Estrada et al., 2011; Happ, 1969).

Although we lack information on male paternity share for most gift-giving taxa, nuptial gifts can also influence what proportion of offspring produced by a multiply-mated female gets sired by the gift-giving male. For example, larger salivary secretions (endogenous oral gifts) offered by male *Panorpa* scorpionflies increase male paternity share by increasing copulation duration (Engqvist et al., 2007; Sauer et al., 1998). Also, larger spermatophores (endogenous genital gifts) increase the paternity share of male *Photinus greeni* fireflies (South and Lewis, 2012a). Many nuptial gifts (including exogenous oral, endogenous oral, and endogenous genital gifts) have been demonstrated to play a role in increasing female fecundity, either through nutritive contributions or allohormones that stimulate female ovulation or oviposition (see Arnqvist and Nilsson, 2000; Boggs, 1995; Eberhard, 1996; Gwynne, 2008; South and Lewis, 2011). A central concern is that these gifts will provide a fitness benefit to males only if they increase the total number of offspring sired by the gift-giving male, yet the requisite information on offspring paternity is often not gathered.

As described in Section I.C.3, endogenous genital gifts can contain chemical defenses that protect a female and/or her eggs against predators, thus increasing offspring survival (e.g., cantharidin in spermatophore of male *Neopyrochroa* beetles [Eisner et al., 1996], pyrrolizidine alkaloids in *U. ornatrix* moths [reviewed by Eisner and Meinwald, 1995]).

Thus, male costs incurred in manufacturing or procuring nuptial gifts are apparently outweighed by fitness benefits that can accrue during multiple selection episodes before, during, and after mating. Endogenous nuptial gifts are especially likely to contain complex mixtures that will operate across multiple selection episodes to increase male fitness. Also, nuptial gift composition and the associated male fitness benefits will shift dynamically over time due to coevolutionary interactions between the sexes.

## IV. A Case Study of Male Benefits: *Requena verticalis*

### A. Orthopteran Nuptial Gifts

Orthopteran insects (grasshoppers, crickets, and katydids) display a dazzling array of endogenously produced nuptial gifts (Fig. 4); these include male body parts and glandular secretions that females absorb orally, genitally, or in some cases, both. Female *Pteronemobius* and *Allonemobius* (Gryllidae: Nemobiinae) ground crickets receive an endogenous oral gift by chewing on a modified hindleg spur and drinking the male's hemolymph (Fedorka and Mousseau, 2003; Mays, 1971). Similarly, female *Cyphoderris* hump-winged crickets (Tettigonioidea: Haglidae) drink hemolymph after feeding on the male's fleshy hindwings (Dodson et al., 1983; Morris, 1979). In *Oecanthus* tree crickets (Gryllidae: Oecanthinae), females consume secretions produced by dorsal glands on the male's thorax (Brown, 1997); such metanotal gland feeding also occurs in many other orthopterans (see Vahed, 1998). Many species with endogenous oral nuptial gifts also transfer a genital gift during mating in the form of a spermatophore (Gwynne, 2001; Vahed, 1998). In most species within the suborder Ensifera (katydids, crickets, and wetas), males produce spermatophores that can comprise between 2% and 40% of their total body weight. These two-part structures are produced by two distinct accessory glands: the smooth glands produce the small, sperm-containing ampulla, while the rough glands produce the

FIG. 4. Diverse endogenous oral gifts are produced by males within the insect order Orthoptera (crickets, katydids, grasshoppers and their allies): (A) After mating, a female Mormon cricket (*Anabrus simplex*) consumes the gelatinous spermatophylax portion of the male's spermatophore (photo by Darryl Gwynne). (B) A female tree cricket (*Oecanthus quadripunctatus*) feeds on the secretions from a male's metanotal gland (photo by Kevin Judge). (C) A female hump-winged cricket (*Cyphoderris*) feeds on a male's hindwings (photo by David Funk). (For color version of this figure, the reader is referred to the online version of this chapter.)

larger, gelatinous spermatophylax (Gwynne, 1997, 2001). During copulation, the ampulla tube is inserted into the female's genital opening, while the remainder of the gift is deposited externally. When the male departs after coupling, the female ingests the spermatophylax while sperm and associated seminal fluid drain from the ampulla into the female's reproductive system. Once the female finishes consuming the spermatophylax, she removes and consumes whatever remains of the ampulla (Gwynne, 1984; Gwynne et al., 1984; Sakaluk, 1984).

The Australian katydid *R. verticalis* (family Tettigoniidae) has been extensively studied as a model system for understanding the costs and benefits of nuptial gifts for both sexes. Thus, this species provides an excellent case study for illustrating the episodes of selection framework presented in Section III. Like other tettigonids, *R. verticalis* males produce a two-part spermatophore that is approximately 15–20% of total male body weight (Davies and Dadour, 1989). The spermatophylax alone (without the sperm-containing ampulla) comprises 78% of total spermatophore weight (Bowen et al., 1984). The spermatophylax is composed of 13.5% protein (Bowen et al., 1984). In this species, both sexes mate multiple times and both courtship and mating are costly to males. The chirping acoustic signals that males use to attract females require an energetic investment that averages 3.2 kJ/h (Bailey et al., 1993) and spermatophore production requires 1.1 kJ (Simmons et al., 1992); together these two components make up approximately 70% of a male's daily energy budget (Simmons et al., 1992). Nuptial gift costs also limit male mating frequency. After mating, males require 2.5–5 days (depending on diet quality) to manufacture another spermatophore before they are able to mate again (Davies and Dadour, 1989; Gwynne, 1990). Furthermore, when male diet is restricted, males invest less energy into courtship signals but nuptial gift production remains constant (Simmons et al., 1992).

What fitness benefits might balance out these well-established costs of nuptial gifts for *R. verticalis* males? Below we expand on Gwynne's (1997) analysis of current gift function to examine the fitness benefits that males derive from spermatophylax production across multiple selection episodes that occur before, during, and after mating.

## B. Fitness Benefits to *R. verticalis* Males

### 1. *Male Insemination Success and Number of Sperm Transferred*

When *R. verticalis* females are deprived of a spermatophylax they will remove and eat the ampulla, effectively halting sperm transfer (Gwynne et al., 1984). Gwynne et al. (1984) demonstrated that while sperm drainage from the ampulla is completed within 3 h, females take ~5 h to eat the spermatophylax before moving on to consume the ampulla. In some other orthopteran species, in contrast, male spermatophylax size attains only the minimum necessary to ensure complete sperm drainage (e.g., the cricket *Gryllodes supplicans*; Sakaluk, 1984). Thus, the *R. verticalis* spermatophylax serves to protect male ejaculates by insuring insemination and maximizing the number of sperm transferred (Fig. 3, selection episodes 3 and 4).

### 2. *Paternity Share*

Male nuptial gifts in *R. verticalis* also affect male paternity share postmating (Fig. 3, episode 6). Laboratory studies indicate that *R. verticalis* generally show complete first-male sperm precedence; that is, the first male that mates with a virgin female will sire all of her offspring even when the female remates (Gwynne, 1988a; Simmons and Achmann, 2000). When females were given a longer intermating interval and allowed to oviposit between matings, second males gained ~20% paternity share (Gwynne and Snedden, 1995). In addition, first males that had greater spermatophore mass retained higher paternity share. Radiolabeling studies show that amino acids derived from second males become incorporated into eggs that were fertilized by the first male; thus, second male gifts are allocated to offspring sired by another male. However, such cuckoldry may happen infrequently under natural conditions, as field estimates of female polyandry suggest that females in nature remate less frequently than in the lab (Simmons et al., 2007).

The high degree of first-male paternity seen in *R. verticalis* suggests that males should be selected to preferentially mate with virgin females, but males appear incapable of discriminating females' mating status (Lynam et al., 1992; Simmons et al., 1993, 1994). This may represent sexual conflict, with selection acting on females to hide their mating status to obtain the benefits provided by additional spermatophores (Simmons et al., 1994). However, Simmons et al. (1994) did find that males are able to discriminate among potential mates based on female age. By preferentially mating with younger females, males may increase their chance of mating with virgins and may thus gain higher paternity share.

Sperm competition theory predicts that males should strategically allocate their ejaculates depending on female mating status (Simmons, 2001); when mating with a previously mated female, males should maximize sperm number in the ampulla (to increase their sperm offense ability), but minimize spermatophylax investment due to their the low probability of siring offspring. Instead, Simmons et al. (1993) found that when mating with young females, *R. verticalis* males transfer identical spermatophores regardless of female mating status; however, males transfer spermatophores with 50% more sperm and 25% less spermatophylax material when mating with older compared to younger females. Thus, *R. verticalis* males appear to strategically allocate their ejaculates when mating with older females to increase their sperm offense ability, and thus their potential paternity share.

Nuptial gifts produced by *R. verticalis* also affect male postcopulatory fitness by increasing sperm defense. Given the high cost of producing nuptial gifts, males should be selected to increase female latency to remate as a mechanism of reducing sperm competition. By experimentally manipulating ampulla attachment times, Gwynne (1986) demonstrated that the ampulla contains receptivity-reducing substances that act in a dose-dependent fashion, normally rendering females non-receptive for approximately 4 days. Substances in the ampulla also appear to negatively affect female longevity. Wedell et al. (2008) found that when females received the contents of three male ampullas (each without a spermatophylax), they had significantly shorter life spans, and this negative effect was not counteracted by spermatophylax consumption.

### 3. *Female Fecundity and Egg/Offspring Survival*

Finally, nuptial gifts can also increase male fitness through effects on female fecundity and the survival of offspring sired by the gift-giving male (Fig. 3, selection episodes 7 and 8). Radiolabeling experiments demonstrated that male protein derived from the *R. verticalis* spermatophore is incorporated into the female's eggs (Bowen et al., 1984; Gwynne, 1988a). Furthermore, females that consume more spermatophylaces produce more and heavier eggs (Gwynne, 1984), and offspring from larger eggs had greater overwintering survival (Gwynne, 1988b). If receiving spermatophylax nutrients directly benefits female fitness, nutrient-limited females would be expected to seek out matings to obtain additional nuptial gifts. Indeed, female *R. verticalis* females kept on a low-quality diet remate more often than females kept on a high-quality diet (Gwynne, 1990).

By applying this framework in *R. verticalis*, we see that nuptial gifts increase male fitness across several episodes of selection, ultimately increasing the number of offspring sired by the gift-giving male. A complete spermatophore (ampulla + spermatophylax) is necessary for insemination to occur (i.e., ejaculate protection), as otherwise the ampulla will be removed and eaten before sperm transfer. Presence of a spermatophylax increases the duration of ampulla attachment, and spermatophylax size exceeds that required for complete sperm transfer. Unidentified substances present in the male ampulla act to reduce female receptivity to additional matings, helping to ensure a male's paternity share relative to his rivals. Additionally, males discriminate against older females that have likely already mated as a mechanism to reduce incidence of cuckoldry. Spermatophylax consumption increases female fecundity and has the potential to enhance fitness by increasing the number of offspring sired by the gift-giving male. Paternity success of second mating males increases if they mate with a female after she has had an opportunity to oviposit. Finally, spermatophylax consumption increases egg size, which enhances survival of a male's offspring. Thus, this work on *R. verticalis* clearly illustrates how costly nuptial gifts might provide males with demonstrable fitness benefits measured across several sequential episodes of selection.

## V. Phylogenetic Insights into the Evolution of Nuptial Gifts

Despite the key role that nuptial gifts play in the reproductive ecology of so many animals, surprisingly few studies have rigorously examined the evolution of nuptial gifts using a comparative phylogenetic approach. To thoroughly test the various evolutionary scenarios that have been proposed for nuptial gifts, it will be essential to map gifts and other relevant traits onto robust phylogenies developed for particular taxa. Using this approach will provide insight into the evolutionary sequence of gift trait transitions and will also allow tests of correlated evolution between nuptial gifts, life-history, and ecological traits. To date, however relatively few studies have applied these methods. Here we review work from three insect taxa where a comparative phylogenetic approach has provided insight into nuptial gift evolution: 1) endogenous oral gifts within katydids and crickets (Ensifera: Orthoptera), 2) correlated evolution of wingless females and male nuptial gifts in fireflies (Lampyridae: Coleoptera), and 3) patterns of male ejaculate incorporation, as well as rates of seminal fluid protein evolution, in the genus *Drosophila* (Drosophilidae: Diptera).

### A. Endogenous Oral Gifts in Katydids and Crickets

The first comparative phylogenetic study of nuptial gift evolution was presented by Gwynne (1995, 1997, 2001), who examined the origins and elaboration of edible glandular gifts within the orthopteran suborder Ensifera (katydids, crickets and their allies). Gwynne's (1995) phylogenetic reconstruction was based upon morphological characters and suggested that the ancestral trait in this group was an exposed spermatophore (essentially a naked sperm-containing ampulla) that was deposited externally on the female genitalia (Fig. 5). Female consumption of this unprotected spermatophore was hypothesized to be ancestral for all ensiferans, followed in the superfamily Tettigonioidea by the origin of the spermatophylax as an edible addition to the spermatophore. In the evolutionary branch leading to the family Gryllidae (true crickets), there were numerous origins of diverse glandular gifts consumed by females before and after mating, along with male spermatophylaces.

A more detailed analysis (Gwynne, 1995, 1997) showed a total of 11 origins of males producing endogenous oral gifts within the Ensifera; these included 3 origins of a spermatophylax, 4 origins of metanotal glands, 1 tibial gland, and 3 others instances where females feed on other male body parts. This work also indicated several independent spermatophylax losses or size reductions; these occurred in some wetas (Stenopelmatidae; loss in *Deinacrida*, size reduction in *Hemideina*) and katydids (Tettigoniidae; loss in *Tympanophora*, *Decticita*, size reduction in *Neoconocephalus*). Interestingly, such losses were often associated with origins of other endogenous gifts, such as secretions from metanotal or tibial glands. Within the Grylloidea, Gwynne's analysis reveals that there were also seven likely losses of nuptial gifts and three origins of postcopulatory mate guarding.

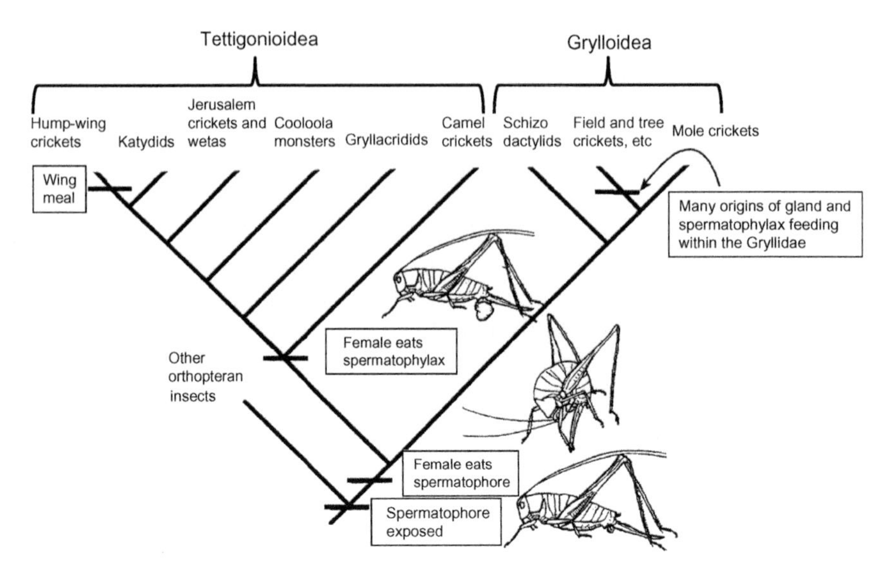


FIG. 5. Proposed evolution of spermatophylax nuptial gifts within the orthopteran suborder Ensifera (two superfamilies indicated), with gift-related traits mapped onto the most parisimonious tree based on morphological characters (figure modified from Gwynne, 2001).

This analysis supports an evolutionary scenario in which males first used a simple, externally attached ampulla to transfer their sperm (for review see Gwynne, 2001). Food limitation may have initially driven females to consume this proteinaceous package, leading to sexual conflict over ampulla attachment times. Selection on males to maximize sperm transfer could have lead to the origin of male reproductive glands that produced an additional spermatophore component, the edible spermatophylax. Thus, the spermatophylax likely originated as an ejaculate protection mechanism, prolonging ampulla attachment times and allowing sufficient time for sperm to fully drain from the ampulla into the female reproductive tract. Further elaboration of the spermatophylax might have occurred if male gifts increased the number of offspring sired by increasing female fecundity and/or offspring survival. As females increased their mating rates to obtain nutritional supplements from these oral gifts, male ejaculates (genital gifts) would have undergone selection to include compounds that suppress female receptivity to further matings, thus reducing sperm competition risk.

While this analysis provides considerable insight into the evolution of orthopteran nuptial gifts, additional work could increase taxon coverage and incorporate more detailed information on species' life-history and ecological traits. Further studies could also help elucidate what conditions led to the spermatophylax loss seen across several ensiferan lineages, and what factors underlie the explosion of nuptial gift diversity seen among modern day Orthoptera.

### B. Firefly Spermatophores: Coevolution with Female Flight

Recent work on fireflies (Coleoptera: Lampyridae) also shows the power of a comparative phylogenetic approach and offers new insights into how nuptial gift evolution is linked to other life-history traits (South et al., 2011b). This analysis also allowed reconstruction of ancestral character states as well as the sequence of evolutionary transitions, and demonstrated trait coevolution between the sexes.

Based upon Tallamy's (1994) enhanced fecundity hypothesis and Boggs' (1990) female allocation model, Lewis and Cratsley (2008) developed a conceptual model to explore the evolution of nuptial gifts in lampyrids as a function of female allocation trade-offs between flight and reproduction. Because nuptial gifts can link together male and female resource budgets, they have the potential to alter the allocation strategies used by both sexes. Thus, selection for nuptial gifts might depend on female reproductive allocation, which in turn depends on allocation to other activities, including flight. If females do not require flight, fecundity selection can act to maximize female reproductive allocation. In this case, because female reproductive output is already at its maximum ($E_{max}$ in Boggs, 1990), male nuptial gifts will have limited scope to further enhance female fecundity. On the other hand, when female reproductive allocation is constrained by the energetic and biomechanical demands of flight, nuptial gifts could provide larger proportional fecundity increases for females. Therefore, this model predicts that nuptial gifts would not be selected in species with flightless females. Fireflies present an opportunity to test this relationship, as they demonstrate variation in not only nuptial gift-giving, but also in female flight ability.

As fireflies are capital breeders and both sexes mate multiply, nuptial gifts can have major fitness consequences for both sexes (Lewis and Cratsley, 2008). Firefly nuptial gifts consist of spermatophores (endogenous genital gifts) that are manufactured by several accessory glands and transferred to females during mating (Lewis et al., 2004). Some female fireflies possess a specialized reproductive sac to receive and break down the spermatophore after sperm are released into the female spermatheca (van der Reijden et al., 1997). Radiolabeling experiments in *Photinus* fireflies have shown that spermatophore-derived proteins are incorporated into the female's developing oocytes (Rooney and Lewis, 1999), and male gifts benefit females by increasing their lifetime fecundity

(Rooney and Lewis, 2002) and longevity (South and Lewis, 2012b). Gift production is costly for males, as spermatophore size declines across successive matings in *Photinus* (Cratsley et al., 2003). Among the 2000 extant species of firefly worldwide, spermatophores are present in some, yet absent in others (Hayashi and Suzuki, 2003; Lewis et al., 2004; South et al., 2008, 2011b; van der Reijden et al., 1997). Those species that lack spermatophores show reduced male accessory glands and females do not have a spermatophore-receiving sac (Demary and Lewis, 2007; South et al., 2011b). What accounts for such interspecific variation in nuptial gifts?

Fireflies also exhibit extensive interspecific variation in life-history traits. In some fireflies, females have greatly reduced wings and as a result are flightless, while in other species both sexes have normal wings and can fly (Jeng, 2008). Hayashi and Suzuki (2003) first proposed that female wing reduction might be negatively associated with male nuptial gifts in Japanese fireflies. Thus, the existing variation in both spermatophore production and female flight within the Lampyridae provided an opportunity to test Lewis and Cratsley's (2008) model and to examine whether this life-history trait could help explain how nuptial gifts are distributed across fireflies.

South et al. (2011b) performed a phylogenetic analysis of the relationship between spermatophore production and female flightlessness within the Lampyridae (Fig. 6). These two traits were measured in 32 taxa and mapped onto a lampyrid molecular phylogeny constructed by Stanger-Hall et al. (2007). Ancestral state reconstruction revealed it was highly likely that firefly males originally produced spermatophores, but these nuptial gifts were subsequently lost in four separate lineages (Fig. 6, right). This reconstruction also revealed that ancestral fireflies had flight-capable females, and females then lost their flight ability at least five times (Fig. 6, left). Furthermore, this work revealed a remarkably congruent pattern between male nuptial gifts and female flight, with the correlated loss of both female flight and male gifts occurring in many lineages. This congruence (statistically confirmed by Pagel's test of correlated evolution) demonstrated coevolution between two traits expressed in different sexes. Finally, transitional probability analysis demonstrated that first females lost their flight ability, subsequently followed by male spermatophore loss.

Thus, female flight ability provides a compelling explanation for observed patterns of nuptial gifts in fireflies, but what selected for female flight loss in the first place? Based upon considerable evidence demonstrating that flightless females can allocate more to reproduction, the most likely explanation for female-specific flight loss is selection for increased fecundity. Thus, these results strongly support the conclusion that male nuptial gifts are co-adapted with patterns of female reproductive allocation, at least

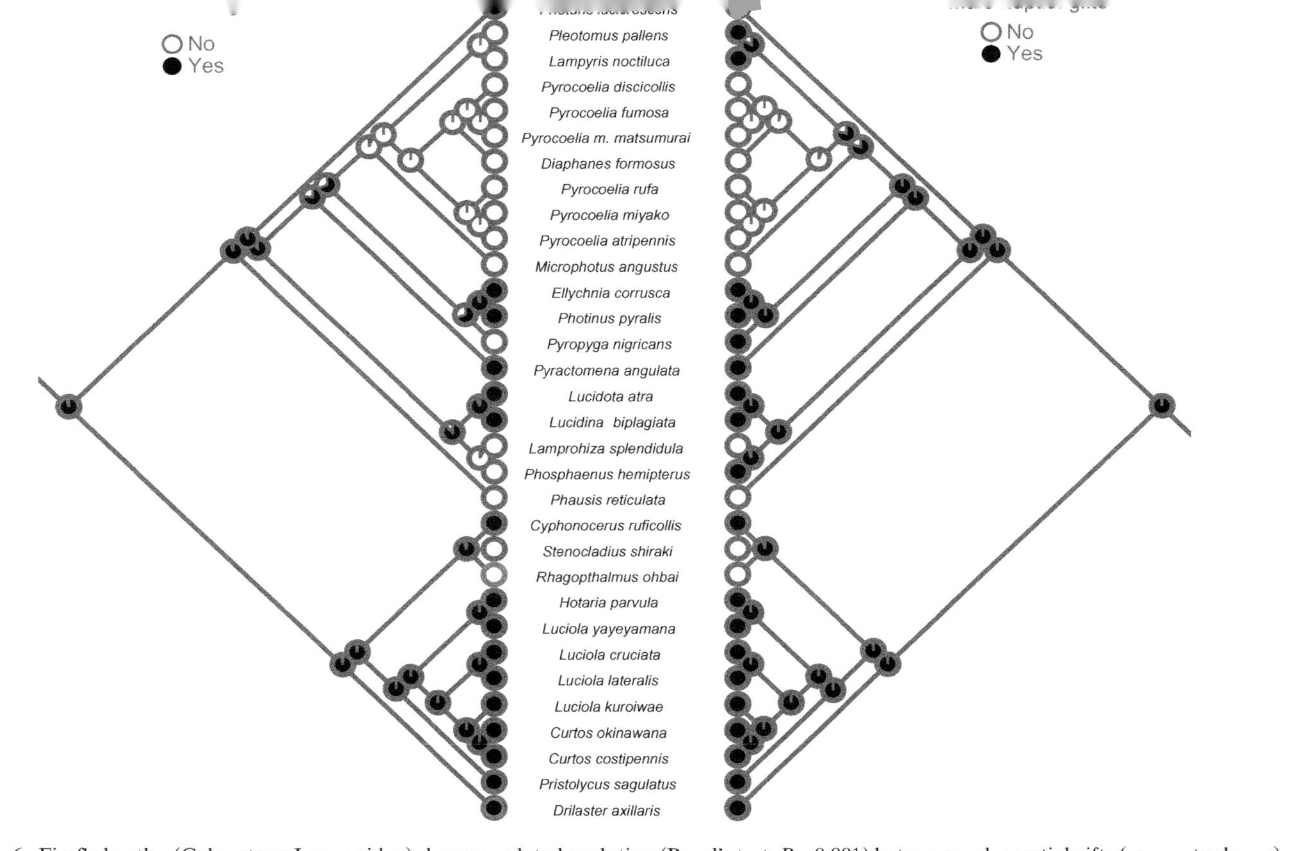


FIG. 6. Firefly beetles (Coleoptera: Lampyridae) show correlated evolution (Pagel's test, $P < 0.001$) between male nuptial gifts (spermatophores) and female flight ability (based on presence of functional wings). For 32 worldwide firefly species, these two traits were mapped onto a lampyrid molecular phylogeny based on 18S, 16S, and *cox1* DNA sequences (Stanger-Hall et al., 2007). For each trait, pie charts at each node indicate the proportional likelihood support for ancestral states (figure modified from South et al., 2011b). (For color version of this figure, the reader is referred to the online version of this chapter.)

in fireflies. These results could be broadly applicable to other capital breeders and could help explain patterns of nuptial gift evolution in other taxa. Further studies are needed to see whether variation in other ecological and life-history traits associated with resource allocation can provide additional insights into nuptial gift evolution.

### C. Female Incorporation of Ejaculate-Derived Proteins in *Drosophila* Fruitflies

In our taxonomy of nuptial gifts, the category of endogenous genital gifts explicitly includes seminal products that are transferred in a liquid ejaculate; this occurs in many Diptera, including *Drosophila* fruitflies. *Drosophila* species vary widely in several aspects of their mating systems, including female remating latency, male ejaculate composition, mating behavior, and the degree to which substances from male ejaculates are incorporated into female tissue (Markow, 2002; Markow and Ankney, 1984; Markow and O'Grady, 2005; Pitnick et al., 1997). Studies mapping reproductive traits onto a *Drosophila* phylogeny provide insight into the evolutionary history of these unpackaged nuptial gifts.

For 34 species of *Drosophila*, Pitnick et al. (1997) used radiolabeled amino acids to determine how much protein transferred in male ejaculates was incorporated into female ovarian or somatic tissue. *Drosophila* species showed dramatic variation in the degree to which females incorporated male-derived proteins (Fig. 7). Females in most species, including those in the *melanogaster* group, showed no incorporation into their ovarian tissue and about half showed no incorporation into somatic tissue. However, *Drosophila* species within the *subpalustris* group showed substantial incorporation into somatic tissue, and those within the *mojavensis* cluster showed substantial incorporation of male-derived protein into both somatic tissue and oocytes.

By mapping these data onto a molecular phylogeny, Pitnick et al. (1997) showed that incorporation of male-derived protein into female somatic tissue has independently evolved multiple times (Fig. 7). In the *mojavensis* cluster, high levels of ovarian incorporation were also seen to accompany high incorporation into somatic tissue. This phylogeny also reveals some degree of lability, as incorporation into both tissue types seems to have been subject to both gains and losses. Thus, this work provides evidence that multiple *Drosophila* groups have evolved male ejaculates that contribute to female somatic maintenance or reproduction.

As pointed out by Pitnick et al. (1997), these patterns of female ejaculate incorporation might be related to differences among species in their nutritional ecology, as the host resources exploited by *Drosophila* vary

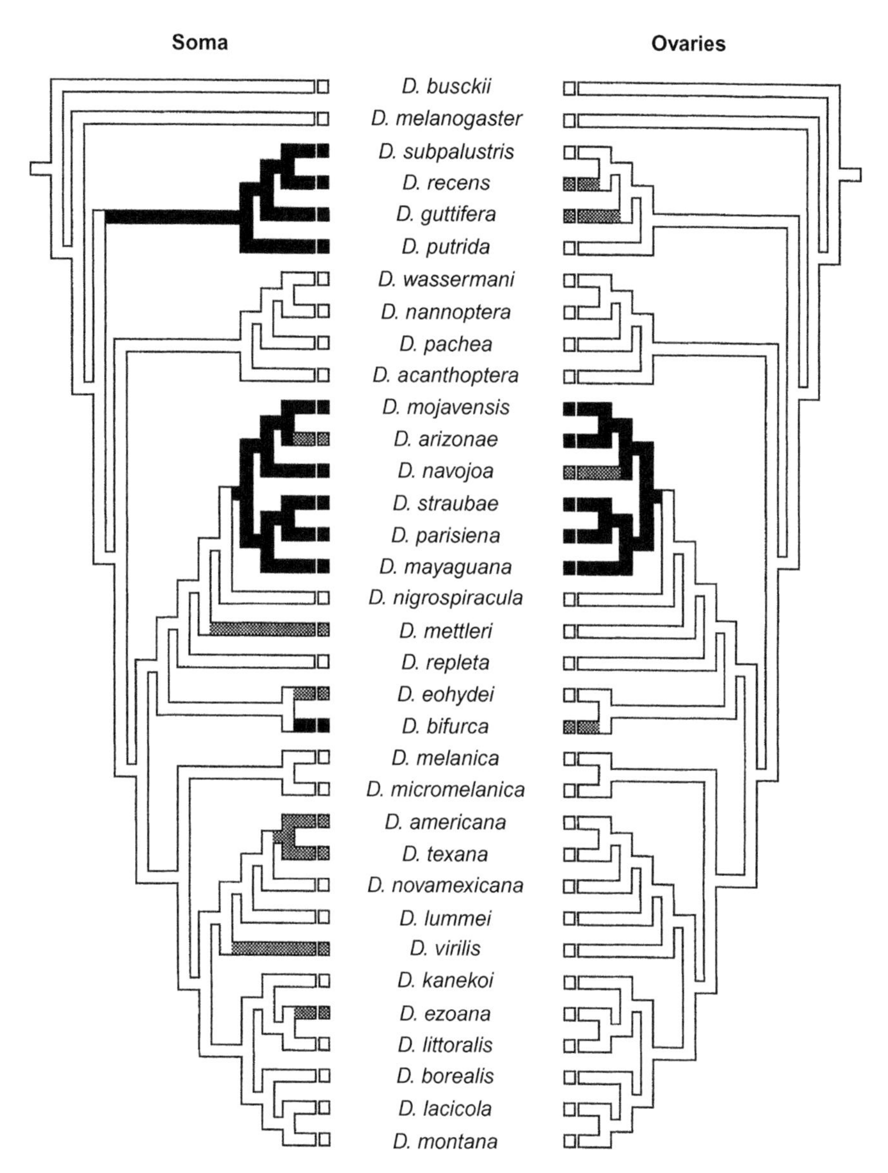


FIG. 7. Phylogenetic distribution of nutritive male ejaculates in 34 species of *Drosophila* fruitflies (Diptera: Drosophilidae). Shading indicates the degree to which females have incorporated $^{14}$C-labeled proteins derived from male ejaculates into their somatic tissue (shown on left) and ovaries (shown on right) 6–8 h after mating. White bars indicate no female incorporation (0–50 corrected DPM), gray bars indicate a small degree of female incorporation (51–100 DPM), and black bars indicate substantial incorporation of male-derived protein by females (>100). Figure from Pitnick et al. (1997).

widely in quality (Markow and O'Grady, 2008). Species in the *mojavensis* group (*D. mojavensis*, *D. navojoa*, *D. straubae*, *D. parisiena*, *D. mayaguana*, and *D. arizonae*) all breed and feed on necrotic cactus (see Markow and O'Grady, 2005, 2008), which contains lower levels of both nitrogen and phosphorus compared to the fruit hosts used by the *melanogaster* group (Markow et al., 1999). In addition, by manipulating nutritional content of the cactus host, Brazner et al. (1984) showed that *D. mojavensis* are likely to undergo frequent nutritional stress. Markow et al. (1990) showed that *D. mojavensis* females kept on low-quality diets experienced enhanced fecundity from the receipt of male ejaculate, and suggested that nutritive ejaculates are more likely to evolve when adults are subject to nutrient limitation. Consistent with the enhanced fecundity hypothesis (Tallamy, 1994), females in some cactophilic, and presumably nutrient-limited, *Drosophila* species show a high degree of male ejaculate incorporation (Pitnick et al., 1997). In addition, Markow et al. (2001) found that in *Drosophila nigrospiracula*, a cactophilic species subject to larval phosphorus limitation, mated females incorporate phosphorus derived from male ejaculates into oocytes. Therefore, variation in resource availability among *Drosophila* species may be one factor in the evolution of nutritive male ejaculates that contribute to female somatic maintenance and reproduction.

The presence of nutritive male ejaculates also shows strong phylogenetic correlations with other features of *Drosophila* mating systems (Markow, 2002). Across 21 *Drosophila* species, Markow (2002) found strong congruence between female mating frequency and exaggerated male ejaculates (these include nutritive male ejaculates). Additionally, Markow (2002) suggested that the evolution of nutritive male ejaculates may have been preceded by higher female remating rates. However, because so many reproductive traits covary with mating systems, additional work is needed to test the sequence of these evolutionary transitions. The extensive knowledge base for *Drosophila* concerning host use, life histories, and reproductive traits makes this a compelling system for examining specific factors that promote the evolution of endogenous genital gifts.

In summary, in this section we present previous work that has taken a comparative phylogenetic approach to describe and test hypotheses about nuptial gift evolution. To rigorously test the various evolutionary scenarios that have been proposed for nuptial gifts, it will be essential to map gifts and other relevant traits onto robust phylogenies developed for particular taxa. We emphasize how valuable it is to include life-history and ecological traits in such evolutionary analyses. This approach should provide insight into the evolutionary sequence of gift transitions and will also allow formal tests of correlated evolution between male gifts and other traits that can influence their evolutionary trajectory.

## D. Evolutionary Rates of *Drosophila* Seminal Proteins

*Drosophila* male ejaculates are certainly the most well-characterized of all endogenous nuptial gifts. Their nonsperm components comprise a complex cocktail of molecules produced by male accessory glands and secretory tissues in the male ejaculatory duct. While many different types of molecules are transferred within *Drosophila* male ejaculates, research has focused on seminal fluid proteins (SFPs). Once transferred, these SFPs engage in dynamic molecular interactions within the female reproductive tract, and this sexual interplay is likely to influence SFP evolution. Because these molecules have been so well-studied, research on SFP evolutionary rates can contribute to a broader understanding of nuptial gift evolution.

Nearly 150 different SFPs have been identified from the ejaculate of *D. melanogaster* males, and these proteins initiate many physiological and behavioral changes within mated females (reviewed by Avila et al., 2011). Significant changes in female gene expression are seen 1–3 h following the receipt of ejaculate and are maximized at 6 h postmating. Conformational changes of the female reproductive tract allow for sperm storage, and the oviduct shows increased innervation and enhanced formation of myofibrils (Adams and Wolfner, 2007; Kapelnikov et al., 2008). Specific SFPs are necessary for female sperm storage and release, while others improve sperm survival (Ravi Ram and Wolfner, 2007a; Xue and Noll, 2000). Male SFPs increase female egg production and ovulation (Heifetz et al., 2000; Ravi Ram and Wolfner, 2007a,b), initiate the formation of a mating plug (a gelatinous mass containing sperm; Bretman et al., 2010; Lung and Wolfner, 2001), and cause females to actively reject courting males. Female activity levels also increase following mating, with increased foraging (Carvalho et al., 2006) and 70% less sleep (Isaac et al., 2010), possibly leading to shorter life spans for mated females (Isaac et al., 2010; Wigby and Chapman, 2005).

Notably, rapid evolution of genes encoding male SFPs has been documented in *Drosophila* as well as in other taxa (Clark et al., 2006; Swanson and Vacquier, 2002; Vacquier, 1998). Comparisons between *D. melanogaster* and *D. simulans* demonstrated high rates of nonsynonymous nucleotide substitution in SFP genes compared to non-SFP genes (Swanson et al., 2001). Sequence comparisons between *D. melanogaster* and *D. pseudoobscura* of 52 SFP-encoding genes from male reproductive accessory glands detected only 58% conserved as true orthologs (Mueller et al., 2005). Such rapid and dynamic evolution of SFPs is likely due to postcopulatory sexual selection (Clark et al., 2006; Panhuis et al., 2006; Swanson and Vacquier, 2002). Sperm competition (Birkhead and Moller, 1998), cryptic female choice (Eberhard, 1996), and sexual conflict (Parker, 1979) may all contribute to a coevolutionary arms race between and within sexes over control of reproductive outcomes.

Comparisons between *Drosophila* species can be used to test the prediction that SFP evolution will proceed more rapidly when postcopulatory sexual selection is more intense. Mating systems and reproductive ecology differ dramatically between species in the *repleta* group and those in the *melanogaster* group. *D. repleta* males transfer a nutritive ejaculate, and females remate more frequently (Markow, 2002; Markow and Ankney, 1984; Pitnick et al., 1997). In addition, many *repleta* species show an insemination reaction, consisting of an opaque mass that develops within the female vagina after mating. This is thought to prevent females from remating, thus protecting the male's nutritional investment from cuckholdry by rival males (Markow and Ankney, 1984, 1988). Based on these differences in reproductive ecology, species in the *D. repleta* group appear subject to more intense postcopulatory sexual selection and thus are predicted to show faster rates of SFP evolution compared to *D. melanogaster*. Supporting this prediction, several studies have shown that SFP genes expressed by male accessory glands in the *repleta* group evolve more rapidly than those in the *D. melanogaster* group (Almeida and DeSalle, 2009; Wagstaff and Begun, 2005, 2007). In the *repleta* group, SFP genes also show high rates of gene duplication, which is suggested to facilitate adaptive protein evolution (Ohno, 1970; Walsh, 2003). Thus, *repleta* SFPs appear to be undergoing rapid evolution, potentially due to differences in their reproductive ecology.

Consistent with the prediction that sexual coevolution is responsible for rapid evolutionary changes in male gifts, some female reproductive proteins in the *repleta* group also show rapid adaptive evolution, and gene duplication has also been important in the evolution of these proteins (Kelleher et al., 2007). Of particular interest are several digestive proteases, which Kelleher et al. (2007) suggest might play a role in breaking down the mating-induced insemination reaction. Interestingly, male ejaculates in *D. mojavensis* contain protease inhibitors (Wagstaff and Begun, 2005), two of which have also experienced lineage-specific gene duplication events (Kelleher et al., 2009). The reproductive tract of female *D. arizonae* (a close sister species to *D. mojavensis*) shows exceptionally high proteolytic activity that is negatively regulated by mating (Kelleher and Pennington, 2009). Taken together, these results from different *repleta* species suggest active sexually antagonistic coevolution around the insemination reaction, with male protease inhibitors acting to prevent male ejaculate components getting broken down by female proteases.

Thus, rapid evolution of *Drosophila* nuptial gifts appears to be driven by a complex sexual interplay taking place at the molecular level. While some male-derived proteins are incorporated into female oocytes and somatic tissue, other SFPs may have evolved to counter defenses mounted by females to prevent male manipulation. Further exploration of these dynamic sexual interactions should provide many insights into the constantly shifting balance between the costs and benefits of nuptial gifts.

## VI. Conclusions and Future Directions

Animal nuptial gifts take multitudinous forms, and their evolutionary stories promise to be just as diverse. In this overview, we have tried to offer a fresh perspective on the evolution of animal nuptial gifts. We argue for a broader definition of nuptial gifts that can accommodate anticipated lability of nuptial gift structure and function arising from coevolutionary interactions both between and within the sexes. By systematically classifying nuptial gifts according to how they are produced (endogenous vs. exogenous) and how they are absorbed by the recipient (oral, genital, or transdermal), we hope to establish a robust framework for testing predictions about how gifts influence both male and female fitness. Rather than attempting to place potential benefits gained by gift-giving males into the falsely dichotomous categories of parental investment versus mating effort, we illustrate how nuptial gifts might enhance male fitness across multiple selection episodes that occur before, during, and after mating. Finally, we highlight some studies that have greatly advanced our understanding by using comparative phylogenetic methods to examine how nuptial gifts and associated life-history traits have changed over evolutionary time.

We hope this foundation will inspire future research efforts to enhance our understanding of nuptial gift evolution. Despite many advances, there remain several areas that clearly call out for more focused research efforts:

- We have detailed morphological descriptions of the glands which are responsible for manufacturing many endogenous gifts (e.g., Leopold, 1976; Liu and Hua, 2010). In many taxa, nuptial gifts are the combined productions of multiple glands, yet much work remains to fully characterize these glandular products. Transcriptome studies of gene expression within gift-manufacturing glands will provide insight into differences and similarities in their gene products and associated functions. For example, to what extent has convergent evolution occurred between those male reproductive glands that produce oral versus genital gifts, or between reproductive and salivary glands?
- In considering selection for nutritive nuptial gifts, the geometric framework developed for nutritional ecology (Raubenheimer, 2011; Raubenheimer et al., 2009) provides a powerful tool for testing whether male gifts evolved to support female reproduction. Does

selection shape male glandular products to provide novel nutritional mixtures that will supplement females' dietary resources, that is, do such gifts act as vectors that specifically target the requirements of vitellogenesis? We need more detailed biochemical analyses of different types of nuptial gifts to test many of the predictions laid out here.

- Most importantly, there is a compelling need for additional phylogenetic analyses of nuptial gift traits that can provide insight into the evolutionary origin and maintenance of nuptial gifts across different taxonomic groups. Continuing to examine evolutionary patterns within the Orthoptera will be especially interesting, because their nuptial gift types are so variable. Phylogenetic analysis would also be worthwhile in the Lepidoptera, where reconstructing ancestral character states could shed light on possible trajectories of spermatophore evolution. Finally, because nuptial gifts lie at the intersection of nutritional ecology, sexual selection, and life-history evolution, testing informed predictions concerning evolutionary associations between nuptial gifts and relevant ecological and life-history traits is of fundamental importance.

**Acknowledgments**

We are grateful to Jane Brockmann for inviting us to write this chapter, and to Nooria Al-Wathiqui, Robert Burns, and Natasha Tigreros for in-depth discussions of these ideas during the Spring 2011 Nuptial Gifts seminar at Tufts University. We also thank Carol Boggs, Jeremy Bono, Darryl Gwynne, Erin S. Kelleher, and Scott Sakaluk for their many insightful comments on earlier versions of this manuscript. This research was supported by NSF award I0B-0543738.

**References**

Adams, E.M., Wolfner, M.F., 2007. Seminal proteins but not sperm induce morphological changes in the *Drosophila melanogaster* female reproductive tract during sperm storage. J. Insect Physiol. 53, 319–331.

Alexander, R.D., Borgia, G., 1979. On the origin and basis of the male-female phenomenon. In: Blum, M.S., Blum, N.A. (Eds.), Sexual Selection and Reproductive Competition in the Insects. Academic Press, New York, pp. 417–440.

Alexandre, D., Linhares, R.T., Queiroz, B., Fontoura, L., Uchoa, A.F., Samuels, R.I., et al., 2011. Vicilin-derived peptides are transferred from males to females as seminal nuptial gift in the seed-feeding beetle *Callosobruchus maculatus*. J. Insect Physiol. 57, 801–808.

Almeida, F.C., DeSalle, R., 2009. Orthology, function and evolution of accessory gland proteins in the *Drosophila repleta* group. Genetics 181, 235–245.

Alonzo, S.H., Pizzari, T., 2010. Male fecundity stimulation: conflict and cooperation within and between the sexes. Am. Nat. 175, 174–185.

Andersson, M., 1994. Sexual Selection. Princeton University Press, New Jersey.

Andersson, J., Borg-Karlson, A., Wiklund, C., 2004. Sexual conflict and anti-aphrodisiac titre in a polyandrous butterfly: male ejaculate tailoring and absence of female control. Proc. Biol. Sci. 271, 1765–1770.
Andrade, M.C.B., 1996. Sexual selection for male sacrifice in the Australian redback spider. Science 271, 70–72.
Andres, J.A., Maroja, L.S., Bogdanowicz, S.M., Swanson, W.J., Harrison, R.G., 2006. Molecular evolution of seminal proteins in field crickets. Mol. Biol. Evol. 23, 1574–1584.
Andres, J.A., Maroja, L.S., Harrison, R.G., 2008. Searching for candidate speciation genes using a proteomic approach: seminal proteins in field crickets. Proc. R. Soc. Lond. B 275, 1975–1983.
Arnqvist, G., Nilsson, T., 2000. The evolution of polyandry: multiple mating and female fitness in insects. Anim. Behav. 60, 145–164.
Arnqvist, G., Rowe, L., 2005. Sexual Conflict. Princeton University Press, Princeton, NJ.
Arnqvist, G., Jones, T.M., Elgar, M.A., 2003. Reversal of sex roles in nuptial feeding. Nature 424, 387.
Austad, S.N., Thornhill, R., 1986. Female reproductive variation in a nuptial-feeding spider, *Pisaura mirabilis*. Bull. Cr. Arachnol. Soc. 7, 48–52.
Avila, F., Sirot, L.K., Laflamme, B.A., Rubinstein, C.D., Wolfner, M.F., 2011. Seminal fluid proteins: identification and function. Annu. Rev. Entomol. 56, 21–40.
Baer, B., Heazlewood, J.L., Taylor, N.L., Eubel, H., Millar, A.H., 2009. The seminal fluid proteome of the honeybee *Apis mellifera*. Proteomics 9, 2085–2097.
Bailey, W.J., Withers, P.C., Endersby, M., Gaull, K., 1993. The energetic costs of calling in the bushcricket *Requena verticalis* (Orthoptera: Tettigoniidae: Listroscelidinae). J. Exp. Biol. 178, 21–37.
Bidochka, M.J., Snedden, W.A., 1985. Effect of nuptial feeding on the mating behaviour of female ground crickets. Can. J. Zool. 63, 207–208.
Birkhead, T.R., Moller, A.P., 1998. Sperm Competition and Sexual Selection. Academic Press, London.
Bissoondath, C.J., Wiklund, C., 1996. Effect of male mating history and body size on ejaculate size and quality in two polyandrous butterflies, *Pieris napi* and *Pieris rapae* (Lepidoptera: Pieridae). Funct. Ecol. 10, 457–464.
Boggs, C.L., 1990. A general model of the role of male-donated nutrients in female insects' reproduction. Am. Nat. 136, 598–617.
Boggs, C.L., 1995. Male nuptial gifts: phenotypic consequences and evolutionary implications. In: Leather, S.R., Hardie, J. (Eds.), Insect Reproduction. CRC Press, Boca Raton, FL, pp. 215–242.
Boggs, C.L., 2009. Understanding insect life histories and senescence through a resource allocation lens. Funct. Ecol. 23, 27–37.
Boggs, C.L., Gilbert, L.E., 1979. Male contribution to egg production in butterflies: evidence for transfer of nutrients at mating. Science 206, 83–84.
Bowen, B.J., Codd, C.G., Gwynne, D.T., 1984. The katydid spermatophore (Orthoptera: Tettigoniidae); male investment and its fate in the mated female. Aust. J. Zool. 32, 23–31.
Braswell, W.E., Andres, J.A., Maroja, L.S., Harrison, R.G., Howard, D.J., Swanson, W.J., 2006. Identification and comparative analysis of accessory gland proteins in Orthoptera. Genome 49, 1069–1080.
Brazner, J., Aberdeen, V., Starmer, W.T., 1984. Host-plant shift and adult survival in the cactus breeding *Drosophila mojavensis*. Ecol. Entomol. 9, 375–381.
Bretman, A., Lawniczak, M.K., Boone, J., Chapman, T., 2010. A mating plug protein reduces early female remating in *Drosophila melanogaster*. J. Insect Physiol. 56, 107–113.
Brown, W.D., 1997. Courtship feeding in tree crickets increases insemination and female reproductive life span. Anim. Behav. 54, 1369–1382.

Bussiére, L.F., Basit, H.A., Gwynne, D.T., 2005. Preferred males are not always good providers: female choice and male investment in tree crickets. Behav. Ecol. 16, 223–231.

Carayon, J., 1964. Un cas d'offrande nuptiale chez les Heteropteres. C. R. Hebd. Acad. Sci. 259, 4815–4818.

Cardoso, M.Z., Gilbert, L.E., 2007. A male gift to its partner? Cyanogenic glycosides in the spermatophore of longwing butterflies (*Heliconius*). Naturwissenschaften 94, 39–42.

Carvalho, G.B., Kapahi, P., Anderson, D.J., Benzer, S., 2006. Allocrine modulation of feeding behavior by the sex peptide of *Drosophila*. Curr. Biol. 16, 692–696.

Chapman, T., 2008. The soup in my fly: evolution, form and function of seminal fluid proteins. PLoS Biol. 6, 1379–1382.

Chapman, T., Davies, S.J., 2004. Functions and analysis of the seminal fluid proteins of *Drosophila melanogaster* fruit flies. Peptides 25, 1477–1490.

Clark, N.L., Aagaard, J.E., Swanson, W.J., 2006. Evolution of reproductive proteins from animals and plants. Reproduction 131, 11–22.

Collins, A.M., Caperna, T.J., Williams, V., Garrett, W.M., Evans, J.D., 2006. Proteomic analyses of male contributions to honey bee sperm storage and mating. Insect Mol. Biol. 15, 541–549.

Cordero, C., 1996. On the evolutionary origins of nuptial seminal gifts in insects. J. Insect Behav. 9, 969–974.

Cratsley, C.K., Lewis, S.M., 2005. Seasonal variation in mate choice of *Photinus ignitus* fireflies. Ethology 111, 89–100.

Cratsley, C.K., Rooney, J., Lewis, S.M., 2003. Limits to nuptial gift production by male fireflies *Photinus ignitus*. J. Insect Behav. 16, 361–370.

Cronk, L., Dunham, B., 2007. Amounts spend on engagement rings reflect aspects of male and female mate quality. Hum. Nat. 18, 329–333.

Cumming, J.M., 1994. Sexual selection and the evolution of dance fly mating systems (Diptera: Empididae). Can. Entomol. 126, 907–920.

Davey, K.G., 1960. The evolution of spermatophores in insects. Proc. R. Entomol. Soc. Lond. 35, 107–113.

Davies, P.M., Dadour, I.R., 1989. A cost of mating by male *Requena verticalis* (Orthoptera: Tettigoniidae). Ecol. Entomol. 14, 467–469.

Demary, K.C., Lewis, S.M., 2007. Male reproductive allocation in fireflies (*Photinus* spp.). Invertebr. Biol. 126, 74–80.

Dodson, G.N., Morris, G.K., Gwynne, D.T., 1983. Mating behaviour of the primitive orthopteran genus *Cyphoderris* (Haglidae). In: Gwynne, D.T., Morris, G.K. (Eds.), Orthopteran Mating Systems: Sexual Competition in a Diverse Group of Insects. Westview Press, Boulder, CO, pp. 305–318.

Eberhard, W.G., 1996. Female Control: Sexual Selection by Cryptic Female Choice. Princeton University Press, Princeton, NJ.

Eggert, A.K., Sakaluk, S.K., 1994. Sexual cannibalism and its relation to male mating success in sagebrush crickets, *Cyphoderris strepitans* (Haglidae: Orthoptera). Anim. Behav. 47, 1171–1177.

Eisner, T., Meinwald, J., 1995. The chemistry of sexual selection. Proc. Natl. Acad. Sci. USA 92, 50–55.

Eisner, T., Smedley, S.R., Young, D.K., Eisner, M., Roach, B., Meinwald, J., 1996. Chemical basis of courtship in a beetle (*Neopyrochroa flabellata*): cantharidin as "nuptial gift" Proc. Natl. Acad. Sci. USA 93, 6499–6503.

Elgar, M.A., Schneider, J.M., 2004. Evolutionary significance of sexual cannibalism. Adv. Stud. Behav. 34, 135–163.

Engqvist, L., 2007a. Nuptial gift consumption influences female remating in a scorpionfly: male or female control of mating rate? Evol. Ecol. 21, 41–61.
Engqvist, L., 2007b. Sex, food and conflicts: nutrition dependent nuptial feeding and pre-mating struggles in scorpionflies. Behav. Ecol. Sociobiol. 61, 703–710.
Engqvist, L., Dekomien, G., Lippmann, T., Epplen, J.T., Sauer, K.P., 2007. Sperm transfer and paternity in the scorpionfly *Panorpa cognata*: large variance in traits favoured by post-copulatory episdoes of sexual selection. Evol. Ecol. 21, 801–816.
Estrada, C., Schulz, S., Yildizhan, S., Gilbert, L.E., 2011. Sexual selection drives the evolution of antiaphrodisiac pheromones in butterflies. Evolution 65, 2843–2854.
Fabre, J.H., 1918. The Life of the Grasshopper. Hodder and Stoughton, London.
Fedorka, K.M., Mousseau, T.A., 2003. Tibial spur feeding in ground crickets: larger males contribute larger gifts (Orthoptera: Gryllidae). Fla. Entomol. 85, 317–323.
Gillott, C., 2003. Male accessory gland secretions: modulators of female reproductive physiology and behavior. Annu. Rev. Entomol. 48, 163–184.
Gwynne, D.T., 1984. Courtship feeding increases female reproductive success in bushcrickets. Nature 307, 361–363.
Gwynne, D.T., 1986. Courtship feeding in katydids: investment in offspring or in obtaining fertilisations? Am. Nat. 128, 342–352.
Gwynne, D.T., 1988a. Courtship feeding in katydids benefits the mating male's offspring. Behav. Ecol. Sociobiol. 23, 373–377.
Gwynne, D.T., 1988b. Courtship feeding and the fitness of female katydids. Evolution 42, 545–555.
Gwynne, D.T., 1990. Testing parental investment and the control of sexual selection in katydids: the operational sex ratio. Am. Nat. 136, 474–484.
Gwynne, D.T., 1991. Sexual competition among females: what causes courtship-role reversal? Trends Ecol. Evol. 6, 118–121.
Gwynne, D.T., 1995. Phylogeny of the Ensifera (Orthoptera): a hypothesis supporting multiple origins of acoustical signalling, complex spermatophores and maternal care in crickets, katydids and weta. J. Orthop. Res. 4, 203–218.
Gwynne, D.T., 1997. The evolution of edible "sperm sacs" and other forms of courtship feeding in crickets, katydids and their kin (Orthoptera: Ensifera). In: Choe, J., Crespi, B. (Eds.), The Evolution of Mating Systems in Insects and Arachnids. Cambridge University Press, Cambridge, UK, pp. 110–129.
Gwynne, D.T., 2001. Katydids and Bushcrickets: Reproductive Behaviour and Evolution of the Tettigoniidae. Cornell University Press, Ithaca, NY.
Gwynne, D.T., 2008. Sexual conflict over nuptial gifts in insects. Annu. Rev. Entomol. 53, 83–101.
Gwynne, D.T., Simmons, L.W., 1990. Experimental reversal of courtship roles in an insect. Nature 346, 172–174.
Gwynne, D.T., Snedden, A.W., 1995. Paternity and female remating in *Requena verticalis* (Orthoptera: Tettigoniidae). Ecol. Entomol. 20, 191–194.
Gwynne, D.T., Bowen, B.J., Codd, C.G., 1984. The function of the katydid spermatophore and its role in fecundity and insemination. Aust. J. Zool. 32, 15–22.
Happ, G., 1969. Multiple sex pheromones of the mealworm beetle, *Tenebrio molitor* L. Nature 222, 180–181.
Hayashi, F., Suzuki, H., 2003. Fireflies with and without prespermatophores: evolutionary origins and life-history consequences. Entomol. Sci. 6, 3–10.
Heifetz, Y., Lung, O., Frongillo Jr., E.A., Wolfner, M.F., 2000. The *Drosophila* seminal fluid protein Acp26Aa stimulates release of oocytes by the ovary. Curr. Biol. 10, 99–102.

Herberstein, M.E., Schneider, J.M., Harmer, A.M.T., Gaskett, A.C., Robinson, K., Shaddick, K., et al., 2011. Sperm storage and copulation duration in a sexually cannibalistic spider. J. Ethol. 29, 9–15.

Houston, A.I., Stephens, P.A., Boyd, I.L., Harding, K.C., McNamara, J.H., 2007. Capital or income breeding? A theoretical model of female reproductive strategies. Behav. Ecol. 18, 241–250.

Hoving, H.J.T., Laptikhovsky, V., 2007. Getting under the skin: autonomous implantation of squid spermatophores. Biol. Bull. 212, 177–179.

Isaac, R.E., Li, C., Leedale, A.E., Shirras, A.D., 2010. *Drosophila* male sex peptide inhibits siesta sleep and promotes locomotor activity in the post-mated female. Proc. Biol. Sci. 277, 65–70.

Jacobellis v. Ohio, 1964. No. 378-184. Supreme Court of the United States, 22 June 1964.

Jeng, M.L., 2008. Comprehensive phylogenetics, systematics, and evolution of neoteny of Lampyridae (Insecta: Coleoptera). Unpublished PhD thesis, University of Kansas.

Jervis, M.A., Ferns, P.N., 2004. The timing of egg maturation in insects: ovigeny index and initial egg load as measures of fitness and of resource allocation. Oikos 107, 449–460.

Jervis, M.A., Boggs, C.L., Ferns, P.N., 2005. Egg maturation strategy and its associated trade-offs: a synthesis focusing on Lepidoptera. Ecol. Entomol. 30, 359–375.

Kapelnikov, A., Rivlin, P.K., Hoy, R.R., Heifetz, Y., 2008. Tissue remodeling: a mating-induced differentiation program for the *Drosophila* oviduct. BMC Dev. Biol. 8, 114.

Karlsson, B., 1995. Resource allocation and mating systems in butterflies. Evolution 49, 955–961.

Karlsson, B., Leimar, O., Wiklund, C., 1997. Unpredictable environments, nuptial gifts and the evolution of size dimorphism in insects: an experiment. Proc. Biol. Sci. 264, 475–479.

Kelleher, E.S., Pennington, J.E., 2009. Protease gene duplication and proteolytic activity in *Drosophila* female reproductive tracts. Mol. Biol. Evol. 26, 2125–2134.

Kelleher, E.S., Swanson, W.J., Markow, T.A., 2007. Gene duplication and adaptive evolution of digestive proteases in *Drosophila arizonae* female reproductive tracts. PLoS Genet. 3, 1541–1549.

Kelleher, E.S., Watts, T.D., LaFlamme, B.A., Haynes, P.A., Markow, T.A., 2009. Proteomic analysis of *Drosophila mojavensis* male accessory glands suggests novel classes of seminal fluid proteins. Insect Biochem. Mol. Biol. 39, 366–371.

Khalifa, A., 1949. Spermatophore production in Tricoptera and some other insects. Trans. R. Entomol. Soc. Lond. 100, 449–479.

Klowden, M.J., 2007. Physiological Systems in Insects. Academic Press, San Diego, CA.

Koene, J.M., Schulenburg, H., 2005. Shooting darts: co-evolution and counter-adaptation in hermaphroditic snails. BMC Evol. Biol. 5, 25.

Koene, J.M., Ter Maat, A., 2001. "Allohormones": a class of bioactive substances favoured by sexual selection. J. Comp. Physiol. A 187, 323–326.

Koene, J.M., Pfortner, T., Michiels, N.K., 2005. Piercing the partner's skin influences sperm uptake in the earthworm *Lumbricus terrestris*. Behav. Ecol. Sociobiol. 59, 243–249.

Lack, D., 1940. Courtship feeding in birds. Auk 57, 169–178.

Lebas, N.R., Hockham, L.R., 2005. An invasion of cheats: the evolution of worthless nuptial gifts. Curr. Biol. 15, 64–67.

Leimar, O., Karlsson, B., Wiklund, C., 1994. Unpredictable food and sexual size dimorphism in insects. Proc. R. Soc. Lond. B 258, 121–125.

Leopold, R.A., 1976. The role of male accessory glands in insect reproduction. Annu. Rev. Entomol. 21, 199–221.

Lewis, S.M., Cratsley, C.K., 2008. Flash signal evolution, mate choice, and predation in fireflies. Annu. Rev. Entomol. 53, 293–321.

Lewis, S.M., Cratsley, C.K., Rooney, J.A., 2004. Nuptial gifts and sexual selection in *Photinus* fireflies. Integr. Comp. Biol. 44, 234–237.
Lewis, S.M., South, A., Burns, R., Al-Wathiqui, N., 2011. Nuptial gifts. Curr. Biol. 21, R644–R645.
Liu, S., Hua, B., 2010. Histology and ultrastructure of the salivary glands and salivary pumps in the scorpionfly *Panorpa obtusa* (Mecoptera:Panorpidae). Acta Zool. 91, 457–465.
Lung, O., Wolfner, M.F., 2001. Identification and characterization of the major *Drosophila melanogaster* mating plug protein. Insect Biochem. Mol. Biol. 31, 543–551.
Lynam, A.J., Morris, S., Gwynne, D.T., 1992. Differential mating success of virgin female katydids *Requena verticalis* (Orthoptera: Tettigoniidae). J. Insect Behav. 5, 51–59.
Mann, T., 1984. Spermatophores: Development, Structure, Biochemical Attributes and Role in the Transfer of Spermatozoa. Springer, Berlin.
Markow, T.A., 2002. Female remating, operational sex ratio, and the arena of sexual selection in *Drosophila* species. Evolution 56, 1725–1734.
Markow, T.A., Ankney, P.F., 1984. *Drosophila* males contribute to oogenesis in a multiple mating species. Science 224, 302–303.
Markow, T.A., Ankney, P.T., 1988. Insemination reaction in *Drosophila*: found in species whose males contribute material to oocytes before fertilization. Evolution 42, 1097–1101.
Markow, T.A., O'Grady, P.M., 2005. Evolutionary genetics of reproductive behavior in *Drosophila*: connecting the dots. Annu. Rev. Genet. 39, 263–291.
Markow, T.A., O'Grady, P.M., 2008. Reproductive ecology of *Drosophila*. Funct. Ecol. 22, 747–759.
Markow, T.A., Gallagher, P.D., Krebs, R.A., 1990. Ejaculate derived nutritional contribution and female reproductive success in *Drosophila mojavensis* (Patterson and Crow). Funct. Ecol. 4, 67–73.
Markow, T.A., Raphael, B., Dobberfuhl, D., Breitmeyer, C.M., Elser, J.J., 1999. Elemental stoichiometry of *Drosophila* and their hosts. Funct. Ecol. 13, 78–84.
Markow, T.A., Coppola, A., Watts, T.D., 2001. How *Drosophila* males make eggs: it is elemental. Proc. Biol. Sci. 268, 1527–1532.
Mays, D.L., 1971. Mating behaviour of nemobiini crickets *Hygronemobius*, *Nemobius* and *Pteronemobius* (Orthoptera: Gryllidae). Fla. Entomol. 54, 113–126.
Mehdi, R., 2003. Danish law and the practice of *mahr* among Muslim Pakistanis in Denmark. Int. J. Sociol. Law 31, 115–129.
Michiels, N.K., Koene, J.M., 2006. Sexual selection favors harmful mating in hermaphrodites more than in gonochrists. Integr. Comp. Biol. 46, 473–480.
Molleman, F., 2010. Puddling: from natural history to understanding how it affects fitness. Entamol. Exp. Appl. 134, 107–113.
Molleman, F., Zwaan, B.J., Brakefield, P.M., 2004. The effect of male sodium diet and mating history on female reproduction in the puddling squinting bush brown *Bicyclus anynana* (Lepidoptera). Behav. Ecol. Sociobiol. 56, 404–411.
Morris, G.K., 1979. Mating systems, paternal investment and aggressive behaviour of acoustic orthoptera. Fla. Entomol. 62, 9–17.
Mougeot, F., Arroyo, B.E., Bretagnolle, V., 2006. Paternity assurance responses to first-year and adult male territorial intrusions in a courtship-feeding raptor. Anim. Behav. 71, 101–108.
Mueller, J.L., Ravi Ram, K., McGraw, L.A., Bloch Qazi, M.C., Siggia, E.D., Clark, A.G., et al., 2005. Cross-species comparison of *Drosophila* male accessory gland protein genes. Genetics 171, 131–143.
Ohno, S., 1970. Evolution by Gene Duplication. Springer-Verlag, Berlin/Heidelberg, Germany/New York.

Oxford English Dictionary, 1989. Online September 2011. Oxford University Press. (Accessed 16 September 2011). http://www.oed.com/view/Entry/78177?rskey=pENIVG&result=98.
Panhuis, T.M., Clark, N.L., Swanson, W.J., 2006. Rapid evolution of reproductive proteins in abalone and *Drosophila*. Philos. Trans. R. Soc. Lond. B Biol. Sci. 361, 261–268.
Parker, G.A., 1979. Sexual selection and sexual conflict. In: Blum, M.S., Blum, N.A. (Eds.), Sexual Selection and Reproductive Competition in Insects. Academic Press, London, pp. 123–166.
Piascik, E.K., Judge, K.A., Gwynne, D.T., 2010. Polyandry and tibial spur chewing in the Carolina ground cricket (*Eunemobius carolinus*). Can. J. Zool. 88, 988–994.
Pitnick, S., Spicer, G.S., Markow, T., 1997. Phylogenetic examination of female incorporation of ejaculate in *Drosophila*. Evolution 51, 833–845.
Pivnick, K.A., McNeil, J.N., 1987. Puddling in butterflies: sodium affects reproductive success in *Thymelicus lineola*. Physiol. Entomol. 12, 461–472.
Poiani, A., 2006. Complexity of seminal fluid: a review. Behav. Ecol. Sociobiol. 60, 289–310.
Preston-Mafham, K.G., 1999. Courtship and mating in *Empis* (*Xanthempis*) *trigramma* Meig., *E. tessellate* F. and *E.* (*Polyblepharis*) *opaca* F. (Diptera: Empididae) and the possible implications of 'cheating' behavior. J. Zool. 247, 239–246.
Qazi, M.C., Wolfner, M.F., 2003. An early role for the *Drosophila melanogaster* male seminal protein Acp36DE in female sperm storage. J. Exp. Biol. 206, 3521–3528.
Raubenheimer, D., 2011. Toward a quantitative nutritional ecology: the right-angled mixture triangle. Ecol. Monogr. 81, 407–427.
Raubenheimer, D., Simpson, S.J., Mayntz, D., 2009. Nutrition, ecology, and nutritional ecology: toward an integrated framework. Funct. Ecol. 23, 4–16.
Ravi Ram, K., Wolfner, M.F., 2007a. Sustained post-mating response in *Drosophila melanogaster* requires multiple seminal fluid proteins. PLoS Genet. 3, e238.
Ravi Ram, K., Wolfner, M.F., 2007b. Seminal influences: *Drosophila Acps* and the molecular interplay between males and females during reproduction. Integr. Comp. Biol. 47, 427–445.
Reinhardt, K., Naylor, R.A., Siva-Jothy, M.T., 2009. Ejaculate components delay reproductive senescence while elevating female reproductive rate. Proc. Natl. Acad. Sci. USA 106, 21743–21747.
Rice, W.R., 1998. Intergenomic conflict, interlocus antagonistic coevolution, and the evolution of reproductive isolation. In: Howard, D.J., Berlocher, S.H. (Eds.), Endless Forms: Species and Speciation. University Press, Oxford, pp. 161–270.
Rogers, D.W., Chase, R., 2001. Dart receipt promotes sperm storage in the garden snail *Helix aspersa*. Behav. Ecol. Sociobiol. 50, 122–127.
Rooney, J.A., Lewis, S.M., 1999. Differential allocation of male-derived nutrients in two lampyrid beetles with contrasting life-history characteristics. Behav. Ecol. 10, 97–104.
Rooney, J., Lewis, S.M., 2002. Fitness advantage from nuptial gifts in female fireflies. Ecol. Entomol. 27, 373–377.
Sakaluk, S.K., 1984. Male crickets feed females to ensure complete sperm transfer. Science 223, 609–610.
Sakaluk, S.K., 1986. Is courtship feeding by male insects parental investment? Ethology 73, 161–166.
Sakaluk, S.K., 2000. Sensory exploitation as an evolutionary origin to nuptial food gifts in insects. Proc. Biol. Sci. 267, 339–343.
Sauer, K.P., Lubjuhn, T., Sindern, J., Kullmann, H., Kurtz, J., Epplen, C., et al., 1998. Mating system and sexual selection in the scorpionfly *Panorpa vulgaris* (Mecoptera: Panorpidae). Naturwissenschaften 85, 219–228.
Schilthuizen, M., 2005. The darting game in snails and slugs. Trends Ecol. Evol. 20, 581–584.

Simmons, L.W., 1995. Male bushcrickets tailor spermatophores in relation to their remating intervals. Funct. Ecol. 9, 881–886.
Simmons, L.W., 2001. Sperm Competition and Its Evolutionary Consequences in the Insects. Princeton University Press, New Jersey.
Simmons, L.W., Achmann, R., 2000. Microsatellite analysis of sperm-use patterns in the bushcricket *Requena verticalis*. Evolution 54, 942–952.
Simmons, L.W., Bailey, W.J., 1990. Resource influenced sex roles of zaprochiline tettigoniids (Orthoptera:Tettigoniidae). Evolution 44, 1853–1868.
Simmons, L.W., Parker, G.A., 1989. Nuptial feeding in insects: mating effect versus paternal investment. Ethology 81, 332–343.
Simmons, L.W., Teale, R.J., Maier, M., Standish, R.J., Bailey, W.J., Withers, P.C., 1992. Some costs of reproduction for male bushcrickets, *Requena verticalis* (Orthoptera: Tettigoniidae): allocating resources to mate attraction and nuptial feeding. Behav. Ecol. Sociobiol. 31, 57–62.
Simmons, L.W., Craig, M., Llorens, T., Schinzig, M., Hosken, D., 1993. Bushcricket spermatophores vary in accord with sperm competition and parental investment theory. Proc. R. Soc. Lond. B Biol. Sci. 251, 183–186.
Simmons, L.W., Llorens, T., Schinzig, M., Hosken, D., Craig, M., 1994. Sperm competition selects for male mate choice and protandry in the bushcricket, *Requena verticalis* (Orthoptera: Tettigoniidae). Anim. Behav. 47, 117–122.
Simmons, L.W., Beveridge, M., Kennington, W.J., 2007. Polyandry in the wild: temporal changes in female mating frequency and sperm competition intensity in natural populations of the tettigoniid *Requena verticalis*. Mol. Ecol. 16, 4613–4623.
Sirot, L.K., Poulson, R.L., McKenna, M.C., Girnary, H., Wolfner, M.F., Harrington, L.C., 2008. Identity and transfer of male reproductive gland proteins of the dengue vector mosquito, *Aedes aegypti*: potential tools for control of female feeding and reproduction. Insect Biochem. Mol. Biol. 38, 176–189.
Sirot, L.K., Wolfner, M.F., Wigby, S., 2011. Protein-specific manipulation of ejaculate composition in response to female mating status in *Drosophila melanogaster*. Proc. Natl. Acad. Sci. USA 108, 9922–9926.
Smedley, S.R., Eisner, T., 1996. Sodium: a male moth's gift to its offspring. Proc. Natl. Acad. Sci. USA 93, 809–813.
Sonenshine, D.E., Bissinger, B.W., Egekwu, N., Donohue, K.V., Khalil, S.M., Roe, M., 2011. First transcriptome of the testis-vas deferens-male accessory gland and proteome of the spermatophore from *Dermacentor variabilis* (Acari: Ixodidae). PLoS One 6, e24711.
South, A., Lewis, S.M., 2011. The influence of male ejaculate quantity on female fitness: a meta-analysis. Biol. Rev. Camb. Philos. Soc. 86, 299–309.
South, A., Lewis, S.M., 2012a. Determinants of reproductive success across sequential episodes of sexual selection in a firefly. Proc. R. Soc. Lond. B. in press.
South, A., Lewis, S.M., 2012b. Effects of male ejaculate on female reproductive output and longevity in *Photinus* fireflies. Can. J. Zool. in press.
South, A., Sota, T., Abe, N., Yuma, M., Lewis, S.M., 2008. The production and transfer of spermatophores in three Asian species of *Luciola* fireflies. J. Insect Physiol. 54, 861–866.
South, A., Sirot, L.K., Lewis, S.M., 2011a. Identification of predicted seminal fluid proteins in *Tribolium castaneum*. Insect Mol. Biol. 20, 447–456.
South, A., Stanger-Hall, K., Jeng, M.-L., Lewis, S.M., 2011b. Correlated evolution of female neoteny and flightlessness with male spermatophore production in fireflies (Coleoptera: Lampyridae). Evolution 65, 1099–1113.

Stanger-Hall, K.F., Lloyd, J.E., Hillis, D.M., 2007. Phylogeny of North American fireflies (Coleoptera:Lampyridae): implications for the evolution of light signals. Mol. Phylogenet. Evol. 45, 33–49.

Stearns, S.C., 1992. The Evolution of Life Histories. Oxford University Press, Oxford.

Steele, R.J., 1986. Courtship feeding in *Drosophila subobscura*. I. The nutritional significance of courtship feeding. Anim. Behav. 34, 1087–1098.

Stutt, A.D., Siva-Jothy, M.T., 2001. Traumatic insemination and sexual conflict in the bed bug *Cimex lectularius*. Proc. Natl. Acad. Sci. USA 98, 5683–5687.

Swanson, W.J., Vacquier, V.D., 2002. Reproductive protein evolution. Annu. Rev. Ecol. Syst. 33, 161–179.

Swanson, W.J., Clark, A.G., Waldrip-Dail, H.M., Wolfner, M.F., Aquadro, C.F., 2001. Evolutionary EST analysis identifies rapidly evolving male reproductive proteins in *Drosophila*. Proc. Natl. Acad. Sci. USA 98, 7375–7379.

Tallamy, D.W., 1994. Nourishment and the evolution of paternal investment in subsocial arthropods. In: Hunt, J.H., Nalepa, C.A. (Eds.), Nourishment and Evolution in Insect Societies. Westview Press, Boulder, CO, pp. 21–56.

Thornhill, R., 1976. Sexual selection and paternal investment in insects. Am. Nat. 110, 153–163.

Thornhill, R., 1981. *Panorpa* (Mecoptera: Panorpidae) scorpionflies: systems for understanding resource-defense polygyny and alternative male reproductive efforts. Ann. Rev. Ecol. Syst. 12, 355–386.

Thornhill, R., Alcock, J., 1983. The Evolution of Insect Mating Systems. Harvard University Press, Cambridge, MA.

Trivers, R., 1972. Parental investment and sexual selection. In: Campbell, B. (Ed.), Sexual Selection and the Descent of Man. Aldine-Atherton, Chicago, pp. 136–179.

Vacquier, V.D., 1998. Evolution of gamete recognition proteins. Science 281, 1995–1998.

Vahed, K., 1998. The function of nuptial feeding in insects: a review of empirical studies. Biol. Rev. 73, 43–78.

Vahed, K., 2007. All that glisters is not gold: sensory bias, sexual conflict and nuptial feeding in insects and spiders. Ethology 113, 105–127.

van Der Reijden, E., Monchamp, J., Lewis, S.M., 1997. The formation, transfer, and fate of male spermatophores in *Photinus* fireflies (Coleoptera: Lampyridae). Can. J. Zool. 75, 1202–1205.

Wagstaff, B.J., Begun, D.J., 2005. Molecular population genetics of accessory gland protein genes and testis-expressed genes in *Drosophila mojavensis* and *D. arizonae*. Genetics 171, 1083–1101.

Wagstaff, B.J., Begun, D.J., 2007. Adaptive evolution of recently duplicated accessory gland protein genes in desert *Drosophila*. Genetics 177, 1023–1030.

Walsh, B., 2003. Population-genetic models of the fates of duplicate genes. Genetica 118, 279–294.

Walters, J.R., Harrison, R.G., 2010. Combined EST and proteomic analysis identifies rapidly evolving seminal fluid proteins in *Heliconius* butterflies. Mol. Biol. Evol. 27, 2000–2013.

Wedell, N., 1992. Protandry and mate assessment in the wartbiter *Decticus verrucivorus* (Orthoptera: Tettigoniidae). Behav. Ecol. Sociobiol. 31, 301–308.

Wedell, N., 1994. Variation in nuptial gift quality in bush crickets (Orthoptera: Tettigoniidae). Behav. Ecol. 5, 418–425.

Wedell, N., 2005. Sperm competition in butterflies and moths. In: Fellowes, M.D.E., Holloway, G.J., Rolff, J. (Eds.), Insect Evolutionary Ecology. CABI Publishing, London, pp. 49–81.

Wedell, N., Tregenza, T., Simmons, L.W., 2008. Nuptial gifts fail to resolve a sexual conflict in an insect. BMC Evol. Biol. 8, 204.

Welke, K.W., Schneider, J.M., 2012. Sexual cannibalism benefits offspring survival. Anim. Behav. 83, 201–207.

Wheeler, D., 1996. The role of nourishment in oogenesis. Annu. Rev. Entomol. 41, 407–431.

Wickler, W.Z., 1985. Stepfathers in insects and their pseudo- parental investment. Z. Tierpsychol. 69, 72–78.

Wigby, S., Chapman, T., 2005. Sex peptide causes mating costs in female *Drosophila melanogaster*. Curr. Biol. 15, 316–321.

Wolfner, M.F., 2007. "S.P.E.R.M". (seminal proteins (are) essential reproductive modulators): the view from *Drosophila*. Soc. Reprod. Fertil. Suppl. 65, 183–199.

Wolfner, M.F., 2009. Battle and ballet: molecular interactions between the sexes in *Drosophila*. J. Hered. 100, 399–410.

Xue, L., Noll, M., 2000. *Drosophila* female sexual behavior induced by sterile males showing copulation complementation. Proc. Natl. Acad. Sci. USA 97, 3272–3275.

Zeh, D.W., Smith, R.L., 1985. Paternal investment by terrestrial arthropods. Am. Zool. 25, 785–805.

# The Evolution of Inbred Social Systems in Spiders and Other Organisms: From Short-Term Gains to Long-Term Evolutionary Dead Ends?

LETICIA AVILÉS* and JESSICA PURCELL*,†

*DEPARTMENT OF ZOOLOGY AND BIODIVERSITY RESEARCH CENTRE, UNIVERSITY OF BRITISH COLUMBIA, VANCOUVER, BRITISH COLUMBIA, CANADA
†DEPARTMENT OF ECOLOGY AND EVOLUTION, BATIMENT BIOPHORE, UNIVERSITY OF LAUSANNE, LAUSANNE, SWITZERLAND

## I. INTRODUCTION

The question of why some species have evolved systems that lead to regular inbreeding is particularly intriguing given expected short- and long-term negative effects of this breeding system. In the short term, inbreeding is expected to result in the expression of recessive deleterious alleles with the consequent reduction in fitness (i.e., inbreeding depression) of the offspring of close relatives. Further, even after the most damaging recessive deleterious alleles have been removed from a population due to repeated cycles of inbreeding, inbred individuals are expected to suffer from reduced fitness due to loss of heterosis (Charlesworth and Charlesworth, 1987; Charlesworth et al., 1990; Lande and Schemske, 1985). The loss of genetic diversity due to inbreeding at the level of individual genomes extends to whole families or family lineages, which then may become more susceptible to disease and parasites, in parallel to what is observed in asexually reproducing organisms (Maynard Smith, 1978; Williams, 1975). As in asexual organisms, inbred lineages may also be subject to the detrimental effects of mutation accumulation (Felsenstein, 1974; Otto, 2009; West et al., 1999). For these reasons, inbred lineages may be more likely to be short lived in long-term evolutionary timescales. Despite these short- and long-term challenges, permanently inbred systems have arisen multiple times throughout the plant and animal kingdom (Stebbins, 1957, 1974; Takebayashi and Morrell, 2001). It is thus interesting to consider why behaviors that result in chronic inbreeding may have



0065-3454/12 $35.00
DOI: 10.1016/B978-0-12-394288-3.00003-4

become established in some species and not others and how processes acting at different timescales and levels may lead to the evolution of systems that in the long term may constitute evolutionary dead ends.

Species that are both inbred and social are particularly intriguing in this context, as most organisms that live in groups, whether temporarily or permanently, go to great lengths to avoid inbreeding, most often by the dispersal of one or both sexes (Perrin and Mazalov, 2000; reviewed by Pusey and Wolf, 1996). In mammals, for instance, males tend to be the dispersing sex; in birds, females disperse; and in many social insects, both sexes disperse (Clutton-Brock, 1989; Dobson, 1982; Greenwood, 1980; Lawson Handley and Perrin, 2007; Motro, 1991; Pamilo et al., 1997). Even in organisms whose colonies may undergo cycles of inbreeding, such as termites, a general dispersal phase and outbreeding are maintained (Korb, 2008; Shellman-Reeve, 1997; Thorne et al., 1999; Vargo and Husseneder, 2010). A few social organisms, however, such as the naked mole rats (Faulkes et al., 1997, 2003; Jarvis et al., 1994; Reeve et al., 1990), socially parasitic ants (Buschinger, 1989, 2009; Jansen et al., 2010), some psocids (Mockford, 1957; New, 1973), gall thrips (Chapman et al., 2000, 2008; McLeish et al., 2006), bark and ambrosia beetles (Kirkendall, 1983), spider mites (Saito, 1997, 2010), and the social spiders (Avilés, 1997; Lubin and Bilde, 2007; Riechert and Roeloffs, 1993) have evolved permanently inbred social systems. Spiders are particularly interesting because of the multiple times they have independently given rise to species that are both social and inbred (reviewed in Avilés, 1997; Lubin and Bilde, 2007; Riechert and Roeloffs, 1993). Thus, they provide an opportunity to study the forces and processes involved in the evolution of behaviors that may negatively impact some aspects of the fitness of individuals and also may be detrimental at longer evolutionary timescales.

Here, we use some recent studies on inbred spider sociality to address the following questions: (1) Why have the social spiders evolved such an unusual population structure for a social organism, and why have they done so repeatedly given the expected costs of inbreeding depression? (2) Given the number of independent origins of inbred spider sociality, why are there, nonetheless, so few social spider species overall? By reviewing some of the key literature on mole rats, parasitic ants, bark beetles, thrips, and others, we extend our discussion to other inbred social systems for which the questions of the association between sociality and inbreeding and the ultimate fate of inbred social lineages are relevant.

## II. The Inbred Social Spiders

Spider sociality, which is a relatively rare phenomenon, has been classified along two axes depending on whether colony members share a single communal structure or aggregate individual webs (nonterritorial vs. territorial species, respectively) and whether individuals remain together throughout their lives or disperse prior to mating (permanent social vs. periodic social) (for reviews, see Avilés, 1997; Buskirk, 1981; Lubin and Bilde, 2007; Uetz and Hieber, 1997). All inbred social species belong to the nonterritorial, permanent social category and will be referred to here simply as "social spiders." They tend to occur in clades where related species, which are thought to resemble the ancestral system from which the inbred social species originated, exhibit nonterritorial, periodic social systems, referred to here simply as "subsocial."

In addition to sharing a single communal nest and maintaining sociality throughout their lives, the inbred social spiders cooperate in prey capture and feeding and exhibit communal brood care. Most of the nonterritorial, permanent social species are highly inbred due to the fact that colony members of both sexes remain within the natal nest to mate generation after generation. Colonies thus grow by internal recruitment until they reach a size at which they give rise to daughter colonies by either fission (one species; Avilés, 2000) or the dispersal of adult inseminated females that establish new nests alone or in groups (the remaining species; Avilés, 1997; Lubin and Bilde, 2007). Throughout the processes of colony growth and proliferation, there appears to be little or no mixing among colony lineages (for an exception, see Avilés, 1994), as reflected in fixation indices ($F_{st}$) of the order of 0.8–0.9 in the more extreme cases (Agnarsson et al., 2010; Johannesen and Lubin, 2001; Lubin and Crozier, 1985; Smith and Engel, 1994; Smith and Hagen, 1996). Such strongly subdivided population structure, combined with a relatively high rate of colony turnover, has led to the evolution of the strongly biased primary sex ratios (of the order of 2–10 females per male, reviewed in Avilés, 1997; see also Avilés and Purcell, 2011; Lubin and Bilde, 2007) that characterize the inbred social spiders (Avilés, 1993a). Only one of the species classified as nonterritorial permanent social—*Tapinillus* sp.—appears to be outbred, as suggested by its 1:1 sex ratios (Avilés, 1994) and low $F_{st}$ (L. Avilés, unpublished allozyme data).

Phylogenetic evidence shows that the majority of the 20 or so inbred social spiders represent at least 14 independent origins of this social and breeding system (see Section VII for details), being derived from ancestors that were either solitary or subsocial (Agnarsson et al., 2006; Avilés, 1997; Johannesen et al., 2007; Lubin and Bilde, 2007) and thus expected to have been outbred. In subsocial species, the colonies consist of the offspring of a single mother who remain together for part of their life cycle but disperse prior to mating. These species are thus primarily outbred (e.g., Johannesen and Lubin, 1999, 2001; Smith, 1987). Two points of interest emerge from these patterns: (1) given the expectation of inbreeding depression, it is surprising how often the transition from an outbred to an inbred system has taken place; (2) given the number of origins, it appears surprising that there are relatively few inbred social spider species overall—compare, for instance, with ants: one origin, 14,095 species (Hölldobler and Wilson, 1990; see www.antweb.org for recent updates). So, why are there so many origins and, then, why so few species overall?

## III. Costs and Benefits in the Transition to Inbreeding

Waser et al. (1986) developed a model that predicts the conditions for the initial spread of an inbreeding allele in an initially outbred population. The model considers the number of offspring (or equivalent) contributed to the next generation by a male or a female who mates once with a relative in an outbred population. Fitness is estimated as the number of offspring produced by the inbred and any additional outbred matings, plus the offspring-equivalents produced by the related mate (number of offspring produced times the degree of relatedness between mates). Given this fitness measure, Waser et al. (1986) show that the critical value for the spread of inbreeding depends on (a) the costs of avoiding inbreeding through dispersal or other means ($c_m$ and $c_f$, for males and females, respectively); (b) the costs of inbreeding depression as a function of the degree of relatedness between mates, $r$, (d($r$)); and (c) the number of effective outbred matings an individual forgoes by mating with a relative ($\Delta N_m$ and $\Delta N_f$, for males and females, respectively). The costs of forgoing outbred matings, which may differ for males and females, depend on the mating system. For males, such costs can range from 0, when reproductive success is only limited by a male's ability to attract mates, to 1 when there is an ecological limit to the number of mates a male can attract or when males mate monogamously. For females, the costs of forgoing outbred matings can be assumed to be 1 since female reproductive success tends to be limited by resources, rather than mates. With the latter assumption, Waser et al. (1986) show that selection will favor female tolerance of inbreeding when

$$d(r) < 1 + [(c_f - 1) + r\Delta N_m(c_m - 1)]/(1 + r) \quad (1)$$

and male tolerance when

$$d(r) < 1 + [\Delta N_m(c_m - 1)] + r(c_f - 1)]/(1 + r) \quad (2)$$

In monogamous mating systems, $\Delta N_m = \Delta N_f = 1$, and assuming that the costs of inbreeding avoidance are the same for both sexes (i.e., $c_m = c_f$), the condition for the spread of inbreeding reduces to

$$d(r) < c \quad (3)$$

In polygynous mating systems, $\Delta N_f = 1$ and $\Delta N_m = 0$, since males are usually not ecologically limited by the number of mates they can acquire. In this case, still with $c_m = c_f$, Eqs. (1) and (2) reduce to

$$d(r) < \frac{r + c}{1 + r}, \quad \text{for females} \tag{4}$$

and

$$d(r) < \frac{1 + rc}{1 + r}, \quad \text{for males} \tag{5}$$

In the latter case, there is a much greater region of inbreeding tolerance for both males and females but also greater potential for sexual conflict (shown for $r = 0.5$ in Fig. 1).

Equations (1) and (2), and their simplified versions (3)–(5), provide a model for the estimation of the conditions that might have surrounded the origin of inbred spider sociality.

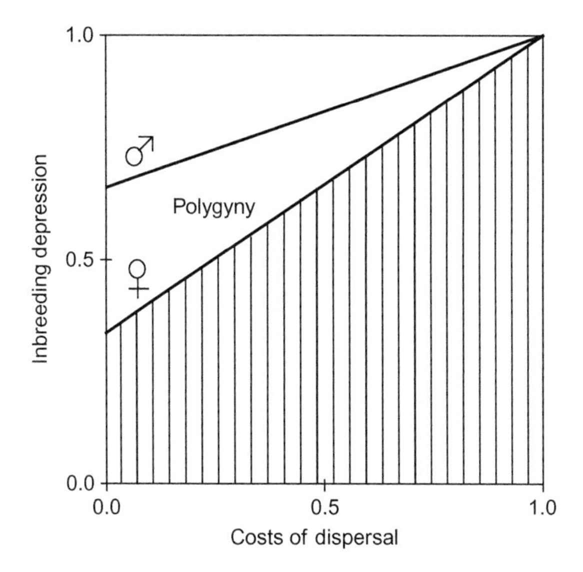


FIG. 1. Predicted region where inbreeding should be tolerated (hatched region) according to the Waser et al. (1986) model for a species with a polygynous mating system ($\Delta N_f = 1$ and $\Delta N_m = 0$, see text for details) and a relatedness of 0.5 (full sibs) among the related mates. For species with monogamous mating systems (and equal costs of inbreeding avoidance for males and females), the region of inbreeding tolerance is reduced to the area under the diagonal (not shown). Modified from Waser et al. (1986).

## IV. The Study Systems

The challenge of studying the origin of a trait, which is essentially a historical question, can be met by investigating extant species whose social and breeding systems resemble those expected in the ancestral species in which the trait—inbred sociality, in this case—might have originated. Two spider genera, in particular, are especially well suited for such a study, as they have given rise independently to several inbred social species within clades of mostly outbred subsocial ones. These are the genera *Anelosimus* in the New World and *Stegodyphus* in the Old World. In addition to phylogenetic evidence demonstrating the derived nature of the social species in these genera (Agnarsson, 2006; Agnarsson et al., 2007; Johannesen et al., 2007), molecular studies have confirmed the strongly subdivided population structure of the social species (Agnarsson et al., 2010; Johannesen et al., 2009a,b; Smith, 1986; Smith and Engel, 1994; Smith and Hagen, 1996) and the mostly outbred nature of the subsocial ones (e.g., Johannesen and Lubin, 1999, 2001; Smith, 1987). Additional independent evidence for the two breeding systems comes from the primary female-biased sex ratios of the social species and the 1:1 sex ratios of the subsocial ones (reviewed in Avilés, 1997; Lubin and Bilde, 2007). The task is then to estimate the parameters for Eqs. (1) and (2), or their simplified versions, in representative subsocial and social species.

## V. The Transition to Inbred Social Systems in Spiders

### A. Environmental Factors

Before assessing potential costs and benefits associated with the transition to inbreeding in spiders, it is useful to consider the environmental factors that may be responsible for the distribution of males and females in tightly clustered social groups, with relatively large distances between them, in the social species, and for the more or less scattered distribution of members of both sexes in the subsocial ones (Fig. 2). The geographical distribution of social and subsocial species in the genus *Anelosimus* provides some hints, as social species are restricted to low- to mid-elevation rainy environments in the tropics, whereas subsocial species occur at higher elevations and latitudes or in dry habitats (Avilés et al., 2007; Purcell and Avilés, 2007). That the social species are restricted to the highly productive low- to mid-elevation wet tropical habitats may be explained because only those areas contain a sufficient supply of larger insects to sustain large social colonies (Guevara and Avilés, 2007; Powers and Avilés, 2007). The reason is that, due to scaling

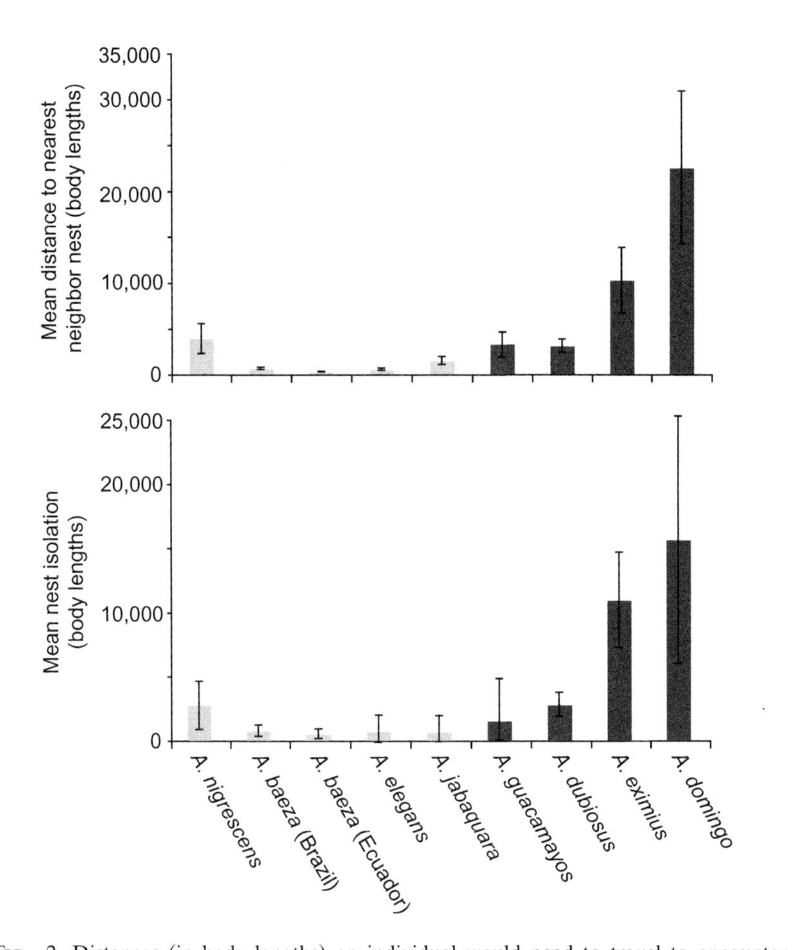


FIG. 2. Distances (in body lengths) an individual would need to travel to encounter a potentially unrelated mate for spiders with increasing levels of sociality and probable degrees of inbreeding (*A. nigrescens*: nearly solitary, probably outbred; *A. baeza* and *A. elegans*: subsocial, probably outbred; *A. jabaquara*: intermediate social–subsocial, probably mostly outbred; *A. guacamayos*: social, higher elevation, probably inbred; *A. dubiosus*: social, higher latitude, probably inbred; *A. eximius* and *A. domingo*: social, lowland tropical rainforest, highly inbred). (Top panel) Geometric mean distances between nests. (Bottom panel) Average of the shortest straight-line distance between nest clusters divided by the longest axis of the cluster, with nest clusters defined as groups of nests where no nest is separated from another by more than 10 m. Spiders in nests belonging to different clusters are more likely to be unrelated, as nests clusters, which are apparent in these environments, contain colonies often sharing a recent common ancestral colony (e.g., Agnarsson et al., 2010). For both measurements, the inbred social species (dark bars) tended to be more isolated from other nests or nests clusters than the outbred less social species (light bars). All distances were converted to units of body length and are shown plus and minus standard errors.

properties of tridimensional objects, web prey capture area per spider and thus number of insect prey *per capita* decline as colony size increases. In areas where a sufficient supply of large insects is present, as in the lowland rainforest, however, spiders are able to capture insects of increasing size as colony size increases (Yip et al., 2008). The net result is a prey intake *per capita* (Yip et al., 2008) and average individual fitness (Avilés and Tufiño, 1998; Bilde et al., 2007) that peak at intermediate colony sizes. Absence of subsocial species in the lowland rainforest, on the other hand, may be due to potentially high costs of dispersal and solitary living in this habitat, due to high predation rates (e.g., Henschel, 1998; Purcell and Avilés, 2008) and frequent destruction of webs by intense rains (e.g., Purcell and Avilés, 2008; Riechert et al., 1986). In temperate and high-elevation environments, in contrast, where prey are mostly small (Guevara and Avilés, 2007; Powers and Avilés, 2007), a declining biomass intake per spider as colony size increases should limit the size of the groups formed so that only groups that consist of the offspring of a single female can acquire sufficient resources for successful offspring development. In these environments, dispersal prior to reaching reproductive maturity—the defining feature of subsocial spiders—would provide an escape from competition for resources within the natal nests while being a viable option in environments where predation and disturbance by strong rains are lower than in the lowland rainforest (Purcell and Avilés, 2008). The outcome would be females clustered in social groups that are relatively isolated from one another in the highly productive, but riskier environment, and scattered as single individuals whose nests occur at relatively short distances from one another in environments where solitary or subsocial species occur (Fig. 2). Thus, a combination of large benefits of remaining in the natal group with high costs of dispersal in areas where the social species occur should explain philopatry and group living in the social species, at least in females.

What about males? Why do males remain past reproductive maturity in their natal nests when they no longer require resources to grow? Female spatial distribution should critically influence male dispersal decisions as males must not only escape local competition for resources but also find mates. Thus, while subsocial females become more or less scattered in the landscape following dispersal from the natal nest, the transition to permanent sociality would have resulted in the clustering of females in fewer and more isolated nests (Fig. 2; see also Fig. 4 of Jones et al., 2007 for a case in a subsocial species where density of nests decreases as the number of females they contain increases). According to the marginal value theorem (Charnov, 1976), males facing a constant risk of dispersal per unit distance would then have become increasingly philopatric as the distance between female clusters, and the size of those clusters (e.g., number of females in colonies), increased (e.g., Lubin et al., 2009). In fact, Corcobado-Marquez et al. (2012) have obtained evidence that not only the tendencies to disperse but also the ability to do so has decreased as level of sociality increases across species of the genus *Anelosimus*. Such reduction in dispersal abilities, which in males is accompanied by a reduction in relative leg length, should exacerbate dispersal costs in the environments where social species occur, thereby further reinforcing philopatry of males and females and thus inbreeding and sociality in these spider systems (Corcobado-Marquez et al., 2012).

This scenario explains not only why females would transition from a scattered subsocial state to a clustered social one but also why males disperse in the former but stay in their natal nest in the latter case. With both males and females remaining in their natal nests, inbreeding would be the natural outcome. The challenge is now to assess quantitatively the balance of costs and benefits potentially involved in this transition.

## B. Costs of Forfeiting Outbred Matings ($\Delta N_M$ and $\Delta N_F$)

In spiders, offspring care, when present, is performed exclusively by females (Foelix, 1996). Female reproductive success is therefore most likely limited by resources, while males are likely limited by their ability to acquire mates (e.g., Schneider and Lubin, 1998). Thus, the costs of forfeiting outbred matings in spiders can be approximated by $\Delta N_f = 1$, for females, and $\Delta N_m = 0$, for males (Table I), so that the simplified Eqs. (4) and (5) (Fig. 1) can be used to assess whether the conditions for a switch to inbreeding might have been met at the origin of inbred spider sociality. We note that the assumption of $\Delta N_f = 1$ for females is conservative relative to the hypothesis being tested. Were females to mate with multiple males as appears likely in both *Anelosimus* and *Stegodyphus* (e.g., Klein et al., 2005; Schneider and Lubin, 1996), then, given the possibility of postcopulatory female choice, $\Delta N_f$ would be less than 1. In such a case, the region of inbreeding tolerance for both females and males would be wider than shown in Fig. 1, with a smaller difference between the sexes. If males, on the other hand, were limited in the amount of sperm they have available to inseminate females, then $\Delta N_m$ would be greater than 0, in which case the region of tolerance for the sexes would be smaller than shown in Fig. 1, but still no smaller than the region under the diagonal, which corresponds to the limiting case of male and female monogamy (Eq. 3).

## C. Costs of Inbreeding Avoidance ($c_F$ and $c_M$)

At least two components of inbreeding avoidance need to be estimated to obtain values for parameter $c$ in Eqs. (4) and (5) for the transition to inbred spider sociality: the costs of dispersing from the natal nest for both males

TABLE I

ESTIMATES OF PARAMETERS FOR THE WASER ET AL. (1986) TRANSITION TO INBREEDING EQUATIONS FROM *Anelosimus* SPP. DATA

| Parameter: Costs of | Estimation | Study species/instar/locality | Females | Males |
|---|---|---|---|---|
| Forfeiting outbred matings | Inferred from mating and parental care system | Polygynous spider spp. with maternal care | $\Delta N_f = 1$ | $\Delta N_m = 0$ |
| Inbreeding avoidance: dispersal (CD) | $1 - \text{survival}_{\text{subsocial}}/\text{survival}_{\text{social}}$ | *A. arizona* subadults, Garden Canyon, AZ[a] and *A. eximius*, subadults, lowland tropical rainforest[b] | $1 - 0.14/0.43 = 0.68$ | $1 - 0.06/0.43 = 0.86$ |
| Inbreeding avoidance: forfeiting benefits of group living (CFGL) | $1 - \text{fitness}_{\text{solitary}}/\text{fitness}_{\text{social}}$ | *A. eximius* adults, lowland tropical rainforest[b] | $1 - 0.75/2.0 = 0.62$ | ? |
| Total costs inbreeding avoidance, $c_f$ or $c_m$ | $CD + (1 - CD) * CFGL$[c] | *A. arizona* and *A. eximius* subadults to adults | $c_f = 0.68 + (1 - 0.68) * 0.62 = 0.88$ | $0.88 > c_m > 1.0$ |
| Inbreeding depression, $d(r)$ | $1 - \text{fitness}_{\text{inbred}}/\text{fitness}_{\text{outbred}}$ | *A. arizona*, subadult to adult, Garden Canyon, AZ[d] | $d_{(0.2\text{–}0.5)} = 0.10\text{–}0.39$[d] | ? |

[a] Avilés and Gelsey (1998).
[b] Avilés and Tufiño (1998).
[c] Loss in fitness due to dispersal as subadults plus the fraction of the remaining fitness lost due to solitarily living as adults.
[d] Avilés and Bukowski (2006).

and females and, at least for females, the opportunity costs of living solitarily as adults. While dispersal is not the only mechanism of inbreeding avoidance (reviewed in Pusey and Wolf, 1996), whether selected for this purpose or not, it appears to be the main mechanism preventing close inbreeding in the subsocial spiders and the main mechanism that once suppressed would have given rise to the inbred social species (Avilés and Bukowski, 2006; Avilés and Gelsey, 1998; Bilde et al., 2005; Johannesen and Lubin, 2001; Powers and Avilés, 2003). Dispersal is also likely to be the mechanism incurring the highest fitness costs, thus representing a major component of the factors against which the costs of inbreeding depression must be pitted.

A challenge in measuring dispersal costs in the social–subsocial spider systems is that while the relevant costs are to be measured in the environments where social species occur, subsocial species tend to be absent from those habitats (e.g., Avilés et al., 2007). The alternative is to obtain estimates of dispersal costs in subsocial species in their native habitat, where dispersal naturally occurs, noting that these estimates are conservative relative to the hypothesis being tested.

Dispersal costs have thus been estimated for the subsocial *Anelosimus arizona* (referred to as *Anelosimus jucundus* or *A*. cf. *jucundus*, in Avilés and Gelsey, 1998; Avilés and Bukowski, 2006; Powers and Avilés, 2003). By periodically censusing mature and newly founded nests of this species in its native riparian habitat in Southern Arizona, Avilés and Gelsey (1998) estimated the number of spiders contained in predispersal nests and then recorded the fraction of those—13.5% of the females and 6.2% of the males—that successfully established individual nests and eventually reached maturity following dispersal. These estimates need to be calibrated against survival during the equivalent stages of a group-living species in its native habitat where dispersal at these stages does not happen. For this, we reanalyze data from Avilés and Tufiño (1998) who estimated the effect of colony size on fitness in natural and artificially established colonies of the social *Anelosimus eximius* in the lowland tropical rainforest. Here, $43 \pm 0.06\%$ of the offspring that reached late juvenile stages in colonies containing from 7 to 165 adult females plus their offspring ($N = 24$) survived to maturity, an estimate that we can assume to be the same for both sexes as young males and females occupy a common space and do not differ in size or behavior. With these values, we estimate the costs of dispersal as subadults as shown in Table I.

To these estimates, we need to add the opportunity costs of not living in groups as adults. Again using Avilés and Tufiño's (1998) data for *A. eximius*, we find that adult females living in groups of 23–165 individuals (females) have an average of $2.0 \pm 0.47$ offspring *per capita* surviving to maturity, compared to $0.75 \pm 0.75$, for females living solitarily or in pairs.

Given these two values and the same equation above, we find that the opportunity costs for adult females of living solitarily rather than in groups are 1–0.75/2.0 = 0.62 (Table I).

Overall costs that females avoid by remaining in their natal group can then be estimated as follows (Table I): 0.68 + (1 − 0.68) * 0.62 = 0.88, that is, the loss in fitness due to dispersal as subadults plus the fraction of the remaining fitness lost due to living solitarily as adults. We note that this is a conservative estimate given that dispersal costs, measured in a habitat where dispersal at subadult stages occurs naturally, are an underestimate of what those costs might be in the lowland rainforest. Costs for males are probably higher as they disperse greater distances than females in order to find mates (e.g., Klein et al., 2005). Depending on the degree of clustering and isolation of females of a given species (Fig. 2), such costs probably lie between the conservative 0.88 for females and 1.0 (Table I), in particular because the chances of a dispersing male encountering a mate should decline to the square power of the distance between nests and nest clusters.

There are not yet comparable data on dispersal costs for spiders in other genera, although what is known is generally consistent with the *Anelosimus* scenario. Rates of nest failure of solitary-living females of the social *Stegodyphus dumicola* and *Stegodyphus mimosarum*, for instance, are considerably higher than those of group-living females (Bilde et al., 2007; Henschel, 1993; Seibt and Wickler, 1988). Likewise, Roeloffs and Riechert (1988) found that solitary individuals of the African social spider *Agelena consociata* suffered high rates of predation and failed to establish successful nests. In at least two other permanent social spiders, the Papua-New Guinean *Achaearanea wau* (Lubin and Crozier, 1985; Lubin and Robinson, 1982) and the Amazonian *Aebutina binotata* (Avilés, 1993b, 2000), females were found to only establish colonies in groups, suggesting that solitary living for these species may be unsustainable in the habitats in which they occur.

We note that discrimination against kin as mates, an additional mechanism of inbreeding avoidance, appears absent in both *Stegodyphus* and *Anelosimus*. Thus, mating experiments in the subsocial *Stegodyphus lineatus* (Bilde et al., 2005) and *A. arizona* (T. Bukowski and L. Avilés, unpublished data) resulted in no difference in the tendency of males to court, female acceptance of the male, and mating success, when mating partners were either sibs or nonsibs (both species) or belonged to the same or a different habitat patch (*S. lineatus*).

### D. Costs of Inbreeding Depression [$d(r)$]

The extent of inbreeding depression that might have accompanied the transition from an outbred subsocial to an inbred social system can be indirectly assessed by artificially inbreeding extant outbred subsocial species, as done for species in the genera *Anelosimus* (Avilés and Bukowski, 2006) and *Stegodyphus* (Bilde et al., 2005). Avilés and Bukowski (2006) compared the size and survival of inbred (sib mated) and outbred offspring of the subsocial *A. arizona* during their social and solitary phases. Inbred and outbred offspring, respectively, were produced by experimentally mating spiders belonging to the same family or to a family collected from a locality several kilometers away ($N=34$ source families). The 43 inbred and 43 outbred sibships produced were allowed to develop in their communal nests under natural field conditions until just prior to dispersal. At this point, inbred and outbred offspring did not differ in their survival, rate of growth, or development time (Avilés and Bukowski, 2006). Following their subsequent solitary phase, which was completed in the laboratory, however, inbred females had a significantly smaller body size and later maturation date, with males having similar, albeit nonsignificant, trends. Estimates of inbreeding depression based on the loss in fecundity expected from the smaller size of inbred females ranged from 0.10 to 0.39 (Table I), depending on the equation used to predict fecundity (clutch size) from female size. A similar pattern appears likely for spiders in the genus *Stegodyphus* where only some traits with an indirect effect on fitness, such as offspring growth (38%) and adult body size (21%), exhibited inbreeding depression among the offspring of full sibs raised under greenhouse conditions (Bilde et al., 2005); no evidence of inbreeding depression was obtained for offspring survival or in crossings between individuals of the same local patch.

### E. Inbreeding Avoidance Versus Inbreeding Depression

Estimates from the genus *Anelosimus* (Table I) clearly show that the costs of avoiding inbreeding—0.88 for females and probably $>0.88$ for males (a conservative estimate)—appear to greatly exceed the potential costs of inbreeding depression—0.1–0.4 for females, at least after a single generation of inbreeding—thus falling well within the region in which a switch to inbreeding is predicted based on the Waser et al. (1986) model (Fig. 1). These findings suggest that the transition from an outbred to an inbred system in spiders, at least in the genus *Anelosimus*, may not have been as difficult as previously believed (Avilés, 1997; Riechert and Roeloffs, 1993).

Two additional factors have been proposed to further facilitate a transition to inbreeding in spiders: buffering of inbreeding depression by maternal and social effects (Avilés and Bukowski, 2006) and a mild history of inbreeding in the ancestral subsocial species (Bilde et al., 2005; Ruch et al., 2009). That maternal care and group living may have provided a buffer against inbreeding depression is suggested by the observation that no reduction in fitness was observed during the social phase of the life cycle of the artificially inbred *A. arizona* (see above; Avilés and Bukowski, 2006). Although alternative hypotheses may explain this pattern (e.g., inbreeding effects may be less evident at early stages of the life cycle, Husband and Schemske, 1996), it is certainly plausible that multiple individuals living communally may compensate for each other's deficiencies so that collectively they acquire sufficient care and resources to be sheltered to some extent from inbreeding depression. Additionally, a history of mild inbreeding in the ancestral species, which is suggested by the clumped distribution and relatively short dispersal distances in extant subsocial species (Avilés and Gelsey, 1998; Bilde et al., 2005; Bukowski and Avilés, 2002; Furey, 1998; Johannesen and Lubin, 1999, 2001; Lubin et al., 1998; Powers and Avilés, 2003; Ruch et al., 2009; Smith, 1987), may have purged the most damaging deleterious recessive alleles responsible for at least some aspects of inbreeding depression. Both the buffering and purging hypotheses, which are not mutually exclusive and may be relevant to both *Anelosimus* and *Stegodyphus* and other spider genera, require formal experimental testing.

In summary, the balance between potential costs of avoiding inbreeding, through dispersal and solitary living, and those of inbreeding depression is clearly favorable for a transition to inbreeding in the spider systems studied. These findings would thus explain the relatively numerous independent origins of inbred spider sociality. What about other inbred social systems?

## VI. Other Inbred Social Systems

Although avoiding inbreeding is the rule for most social organisms (Perrin and Mazalov, 2000; Pusey and Wolf, 1996), a handful of other social systems have turned to inbreeding. Here is a brief review of these systems.

### A. Mole Rats

Mole rats are of special interest as they have given rise to two eusocial species, one highly inbred—the naked mole rat—and one outbred—the Damaraland mole rat (Burland et al., 2002; Jarvis et al., 1994; O'Riain and Faulkes, 2008; Reeve et al., 1990). In both cases, it is suggested that eusociality may have evolved due to high costs of dispersal, constraints on independent breeding, and benefits of colonial living. Both species live in extremely arid environments with patchily distributed resources that can only be located during short and unpredictable periods of strong rainfalls when the hard rock soils in which they live soften enough for digging (Jarvis et al., 1994). Both species also form long-lived stable colonies with typically a single breeding female and nonbreeding colony members (Jarvis and

Bennett, 1993; Jarvis et al., 1994). The species, however, exhibit important differences that may influence their costs of avoiding inbreeding through rejection of related mates and/or dispersal to establish an independent nest. Although it is difficult to tease cause and effect apart, the smaller body size (30 vs. 131 g) and poikilothermic nature of the naked mole rats, along with their much larger colony sizes (up to 295 vs. up to 41 in the Damaraland mole rats) and primary mode of colony reproduction by budding (Jarvis et al., 1994; but see Braude, 2000), suggest that naked mole rats would have limited chances for successful independent breeding. Further, the fact that their colonies are widely distributed in space (Jarvis and Bennett, 1993) should make it difficult for dispersing males to locate a foreign nest. Thus, notwithstanding the findings that a rare disperser morph exists (Braude, 2000; O'Riain et al., 1996) and that, given a choice, outbreeding is preferred (Ciszek, 2000), costs of avoiding inbreeding are likely to be close to prohibitive in the naked mole rat or any ancestral species moving into its ecological niche. Again, although it is difficult to tease cause and effect apart, that the Damaraland mole rats have maintained the ancestral outbred system of species in their genus may be a result of their larger body size and their apparently greater chances of achieving reproduction upon dispersal. In fact, 8% of Damaraland mole rats achieve reproductive status upon dispersal compared to 0.1% in naked mole rats (Jarvis and Bennett, 1993), a difference we suspect it is related to the smaller size of their colonies, closer proximity in space, and greater frequency with which Damaraland mole-rat colonies are established. Additionally, the region of inbreeding tolerance (Fig. 1) may be more limited in the Damaraland mole rats given their monogamous mating system (i.e., $\Delta N_m = \Delta N_f \cong 1$), while multiple males (i.e., $\Delta N_f < 1$) may be present in naked mole-rat colonies. Finally, the much larger litter sizes of the naked mole rat (average 14 pups), which probably coevolved with its social and breeding system, might provide somewhat of a shelter against inbreeding depression (e.g., Stockley and Macdonald, 1998) when compared to a species with much smaller litter sizes, as are the Damaraland mole rats (average three pups). We are not aware of estimates of inbreeding depression in outbred African mole rats. In other rodents, as in prairie dogs, no evidence of inbreeding depression on five measures of reproductive success was found (Hoogland, 1992), while in deer mice a 0.44 depression in percent survival was detected (reviewed in Crnokrak and Roff, 1999). The latter estimate is comparable to our high estimate for the outbred subsocial spiders. Presumably, the costs of avoiding inbreeding would be lower than this figure for the Damaraland mole rats, but higher for the naked mole rats. Finally, despite their apparent long history of inbreeding, which is expected to have purged recessive deleterious alleles from naked mole-rat populations, inbreeding depression, in the form of greater susceptibility to disease due to low genetic diversity at the individual and colony level (i.e., loss of heterosis), has been observed (Ross-Gillespie et al., 2007).

### B. Socially Parasitic Ants

Among highly derived eusocial organisms, inbreeding has arisen in some socially parasitic ant genera (Buschinger, 1989, 2009; Jansen et al., 2010; Wilson, 1971). Species in the Myrmecine genus *Myrmoxenus* (formerly *Epymyrma*), for instance, have transitioned from slave making to a fully workerless parasitic lifestyle (Buschinger, 1989). Young queens of these parasitic species enter the colonies of their host species, eliminate the host's queen, and then rely on the host's workers to raise their brood. Sexuals in these and other species with similar life histories (e.g., *Myrmica* species, Jansen et al., 2010; *Plagiolepis xene*, Trontti et al., 2006; *Teleutomyrmex*, Wilson, 1971) have turned to intranidal mating and continuous inbreeding, with highly female-biased sex ratios arising secondarily. Buschinger (1989), following Wilson (1971), argues that intranidal mating eliminates the loss of reproductives during dispersal and ensures that the queen is fertilized in species that are necessarily rare given their socially parasitic lifestyle. In general, socially parasitic ants are expected to be monogynous and monoandrous (Brandt et al., 2005), although there is at least one species (*Plagiolepis xene*) wherein a queen will usually mate with two to four males (Thurin and Aron, 2011). There are little data on the number of matings that males may obtain in these species, but it should be small given the short life expectancy of male ants. To parameterize our simple model, then, we generally expect $\Delta N_{\mathrm{m}} = \Delta N_{\mathrm{f}} \cong 1$, but both males and females may be able to obtain a small number of additional matings depending on the species, so these values could also fall below 1 in some systems. Data are not currently available to reliably quantify the cost of inbreeding avoidance in socially parasitic ants species, but we do know that the proportion of host species nests that are usually parasitized is low (parasites are usually present in far less than 10% of host species colonies). Moreover, the inbred social parasites are generally highly specialized on one or two host species, and many species have morphologies, such as lost or reduced wing growth and small body sizes, that would reduce their ability to disperse effectively (Brandt et al., 2005). Estimates of inbreeding depression are also limited, but Schrempf et al. (2006) found that the naturally inbreeding ant *Cardiocondyla obscurior* exhibited a reduction in male sperm and in female longevity after prolonged sib mating. Based on the reduction in female life expectancy by one half after six generations of sib mating, it appears that inbreeding depression can be highly detrimental in ants, even in species with natural inbreeding, and must be counterbalanced by a high cost of inbreeding avoidance (Eq. 3).

### C. Termites

Inbreeding is known to occur in termites as a mechanism to maintain colony growth in species that are multiple-site or central-site foragers with relatively long-lived colonies (reviewed by Shellman-Reeve, 1997; Thorne et al., 1999; Vargo and Husseneder, 2010). In these species, colonies are typically initiated by nonrelated alate reproductives, which upon their death may be replaced by their offspring who will then mate incestuously within the colony. Inbreeding may continue for some time, but outbreeding is eventually restored at the population and species levels, as these colonies go on to produce alate reproductives that outbreed. Overall levels of inbreeding in the species studied, with a few exceptions, are thus low to moderate (Vargo and Husseneder, 2010). Estimating model parameters requires consideration of two stages in the colony life cycle. (a) At colony establishment, mating pairs, which mate monogamously (thus $\Delta N_m = \Delta N_m \cong 1$), are typically unrelated. Costs of inbreeding avoidance at colony establishment are probably low given the strong dispersal abilities of alate sexuals (reviewed in Vargo and Husseneder, 2010) and the fact that by mating in swarms (e.g., Husseneder et al., 2006) termites probably have greater assurance of finding a mate than individuals in the other systems considered here. The apparent lack of kin discrimination during pair formation also suggests that the chances of related individuals encountering one another are relatively low (e.g., Husseneder et al., 2006). In terms of inbreeding depression, DeHeer and Vargo (2006) observed that in the subterranean termite *Reticulitermes flavipes*, which has somewhat weaker dispersal abilities than other species, there is a drop in the number of sibling pairs formed, from about 26.2% during the nuptial flights to about 4.5% in established colonies. They suggest that this reduction may be the result of inbreeding depression (of the order of 0.17) in colony foundation success. (b) Once colonies are established, close inbreeding may occur following the death of one or both primary reproductives when replacement reproductives (neotenics), who are descendants of the original pair, mate with the surviving primary reproductive or with each other to maintain colony growth (reviewed by Shellman-Reeve, 1997; Thorne, 1997). Presumably in such cases, inbreeding avoidance would carry high costs, as it would imply the eventual death of the colony since the chances of outside reproductives entering the colonies appear to be extremely low (Vargo and Husseneder, 2010). Studies of the extent to which such cycles of inbreeding may lead to inbreeding depression are scarce. Calleri et al. (2006), however, found that after one generation of inbreeding, inbred grouped termites exposed to a fungal pathogen had significantly greater mortality and significantly higher cuticular microbial loads than outbred grouped individuals. Interestingly, two recent studies have discovered that some female replacement reproductives may be produced parthenogenetically by the queen. These parthenogenetic individuals, which are effective clones of the original queen, will produce outbred offspring when mating with the original king or a sexually produced descendant (Matsuura et al., 2009; Vargo et al., 2012). By combining asexual reproduction and inbreeding, some termite species may have found a way to maintain genetic diversity within the colonies without giving up on internal recruitment of new colony reproductives.

## D. Psocids

Among psocids (Insecta, Orden Psocoptera), the Archipsocidae is an especially intriguing and little known group of organisms where sociality and inbreeding appear to be associated in ways reminiscent of the social spider systems (e.g., Mockford, 1957; New, 1973). Colonies of these organisms live under dense webbing spun on the bark of trees, branches, and other structures, probably as a predator protection device. The colonies, which may grow to contain many hundreds of individuals, are initiated by a single, apparently inseminated, winged female whose offspring remain together throughout their lives. Mockford (1957) notes that if females mate before leaving the parental colony, then "nearly all matings in *Archipsocus* must result in close inbreeding." Consistent with this suggestion, sex ratios are highly female biased (Mockford, 1957). New (1973) also observed that colonies may fragment and that neighboring colonies may also coalesce. Not enough information is available on the biology of these organisms to gain insight into the factors responsible for their inbred systems. In a Florida species in the genus *Archipsocus*, Mockford (1957) notes that females copulate only once in their lifetimes, while males are capable of copulating many times. Thus, for this species at least, the region of inbreeding tolerance should be larger than the region under the diagonal (Fig. 1), as in the case of the social spiders. Also, as in the social spiders, costs of inbreeding avoidance should include costs of dispersal and, presuming that no nearby nests would be available to join, the loss of benefits of being in a group and the risks of not finding a mate. To our knowledge, no information on such costs or on those of inbreeding depression is available for these organisms.

### E. Bark and Ambrosia Beetles

Inbreeding and some form of sociality are also present in some bark and ambrosia beetles, spider mites, and thrips, although only at times in association with each other. With the exception of tree-killing bark beetles, which assemble in large aggregations to overpower the defenses of live trees, bark and ambrosia beetles are not particularly known for their social behavior (Kirkendall et al., 1997). Gregarious larval feeding, however, appears to be a prerequisite for predispersal inbreeding, which has evolved 8–10 times independently in nontree-killing species (Scolytidae) (Kirkendall, 1993). In inbreeding species, galleries are typically initiated by females. The resources they colonize tend to be rare, patchily distributed, and ephemeral in nature (e.g., fruits, seeds, and dead or dying twigs, branches, or trunks of woody plants) (Kirkendall, 1993; but see Jordal et al., 2000). Kirkendall (1993) suggests that premating dispersal should be favored in such cases as a means of hastening reproduction and as insurance against not finding a mate. Such matings are likely to be incestuous when the resources colonized are small enough to support only one or a few families and the architecture of the galleries or behavior of the larvae are conducive to siblings interacting with one another. Avoiding inbreeding in such cases should be costly because of a heightened risk of not finding a mate upon arriving to a new resource patch. To our knowledge, no estimates of such costs are available. Because monogamy is the most common mating system of the Scolytidae (Kirkendall, 1983, 1993), $\Delta N_{\mathrm{m}} = \Delta N_{\mathrm{f}} \cong 1$ is likely to have characterized the ancestral species from which various inbreeding clades might have originated. Two studies have estimated inbreeding depression. After several generations of artificially inbreeding a naturally outbred phloem feeding bark beetle, significant inbreeding depression, of the order of 0.39, was detected on the number of emerging adult offspring (calculated from Domingue and Teale, 2007; Table I), but not on other components of fitness. It is not difficult to envision that costs of avoiding inbreeding—which in these systems should involve primarily the risk of not finding a mate following rejection of a related mate—would be much greater than this value in species colonizing resources that are rare and unpredictable in time and space. Once inbreeding has been established, purging of deleterious recessive alleles is expected to reduce the impact of inbreeding depression, although long-term costs due to reduced genetic variability might remain. Interestingly, however, outbreeding depression, in the form of reduced hatching rates, but no inbreeding depression, was detected in a naturally inbred ambrosia beetle made to artificially outbreed (Peer and Taborsky, 2005). Outbreeding depression in highly inbred systems may reflect loss of local adaptations or of coadaptation within the genome when strongly differentiated inbred lineages are combined.

### F. Spider Mites

Inbreeding has also arisen multiple times in the Acari (Norton et al., 1993), but for the most part not in association with group living, except in spider mites (Prostigmata, Tetranychidae) and perhaps some taxa in the Mesostigmata (Mori et al., 1999; Saito, 1997, 2010). Some spider mite species, such as species in the Tetranychid genus *Stigmaeopsis* in Japan (Saito, 2010), live in extended family groups (two or three overlapping generations) housed by cooperatively built webs that are thought to have an antipredator function (Mori and Saito, 2005). Inbreeding seems to be common in these species, which, much like the social spiders, also exhibit highly female-biased sex ratios (Norton et al., 1993). Individuals in these species may remain in their natal nest throughout their life, or females may disperse to establish new nests. Male dispersal is thought to be rare and females likely disperse fairly short distances (Saito and Mori, 2005). Mitchell (1973) estimated that dispersal may be quite costly and showed that *Tetranychus urticae* females in uncrowded conditions will often forego dispersal. Females that do disperse will usually mate with a nest mate prior to leaving the natal nest (Mitchell 1973). Inbreeding depression, on the other hand, has been measured in the subsocial spider mite *Stigmaeopsis miscanthi*, which exhibits intermediate levels of inbreeding in nature. Saito et al. (2000) found no effect of inbreeding on the early survival of brood even after four generations of inbreeding, but they found that female fecundity decreased by at least 50% with increasing levels of inbreeding. They suggest that inbreeding depression due to recessive deleterious alleles may still be present even in species, such as are mites, with male haploidy. At the moment, there is no enough information to speculate on the balance between costs of inbreeding and of inbreeding avoidance or on the factors responsible for the inbred nature of these spider mite systems.

### G. Australian Gall-Forming Thrips

In the Australian gall-forming thrips, inbreeding appears to have preceded and perhaps to be associated with the origin of a soldier caste (Chapman et al., 2000, 2008; McLeish et al., 2006). McLeish et al. (2006) found relatively high levels of inbreeding in the solitary relatives of these eusocial thrips belonging to the genus *Kladothrips*. These authors suggest that inbreeding may be a pervasive feature of the Australian gall thrips, but with apparent reversals to outbreeding in some species, including, interestingly, a trend toward lower levels of inbreeding in the more highly derived eusocial species with nonreproductive soldiers (McLeish et al., 2006). Since moderate to high levels of inbreeding have been found in many of the solitary relatives of eusocial thrips, we consider the costs of inbreeding avoidance and inbreeding depression in both solitary and social *Kladothrips*. In this genus, dispersing females can either mate with a brother in their natal colony or seek an outbred mating during dispersal, with both strategies apparently present in most species (Chapman et al., 2008). Given that a single male may then be found with a female in a newly formed gall, it seems likely that both males and females would typically mate only once, so $\Delta N_m = \Delta N_m \cong 1$. Although the cost of dispersal has not been directly quantified to our knowledge, Acacia thrips experience a very short window for gall induction (Chapman et al., 2008). Thus dispersal and location of potential mates is expected to be risky. Inbreeding depression estimates are also not available in the thrips, but given the history of inbreeding in the group and their haplo-diploid genetic system, deleterious recessive alleles are likely to have been purged (McLeish et al., 2006). It would be interesting to investigate the relationship between dispersal risk, including the risk of not finding a mate, and the maintenance of inbreeding in these systems; some exceptional species, such as *Kladothrips waterhousei*, a social thrip that has lost inbreeding, may help to elucidate the balance between costs of inbreeding and of inbreeding avoidance.

## H. Trends

The patterns discussed above are consistent with the hypothesis that inbreeding should arise when the costs of inbreeding avoidance exceed those of inbreeding depression. Among the costs of inbreeding avoidance, the risk of not finding a mate appears particularly important. High risks of not finding a mate arise in some social systems as a result of the underlying distribution of resources to be colonized, which tend to be rare and unpredictable in space and/or time, as in bark and ambrosia beetles, socially parasitic ants, and naked mole rats. In other cases, as in the social spiders and possibly psocids and spider mites, it is the organisms themselves that create a distribution of social groups in space that is sparse enough to make the chances of finding an unrelated mate low (e.g., Fig. 2). In some cases, as in the gall-forming thrips and naked mole rats, it is probably the short window of time available for the location of resources that places a premium on colonization ability. The adversity of the local environment, which results in high costs of dispersal and of independent living, is also important, as is particularly clear in social spiders and the naked mole rats. In other systems, however, regular dispersal by females to colonize as new resource patch is maintained, as in bark and ambrosia beetles, parasitic ants, spider mites, psocids, and thrips. In order to better assess the balance between costs of inbreeding and of inbreeding avoidance in these systems, however, studies that quantify the costs of dispersal and risks of not finding a mate, as well as the potential costs of inbreeding depression, are needed.

The association of inbreeding with haplo-diploidy (males arise from unfertilized eggs or through paternal genome loss) in some of the systems, although by no means in all of them (Table II), is also noteworthy. This association poses a classic "chicken-and-egg" problem in terms of the direction of causality. On the one hand, male haploidy is expected to facilitate the transition to inbreeding by purging recessive deleterious alleles from the population (e.g., Mori et al., 2005), as may have been the case in ants and thrips, which belong to orders that are entirely haplo-diploid (Crespi, 1993). The direction of causality, however, may be the reverse in mites (Norton et al., 1993) and in bark and ambrosia beetles (Kirkendall, 1983, 1993) where diplo-diploidy is ancestral and inbreeding and male haploidy appear to have arisen more or less simultaneously multiple times. In these groups, it can be argued that the purging of recessive deleterious alleles resulting from inbreeding may have paved the way for haplo-diploidy, which in turn facilitates sex ratio control.

A final point to note is that once inbreeding has been established, the subsequent evolution of highly female-biased sex ratios in the majority of inbred social systems above discussed (naked mole rats and termites are the exception; Table II) should give an advantage to the inbred lineages in terms of rate of reproduction, much as occurs with asexual species. Thus, in the short term, it is not only the reproductive assurance and avoidance of dispersal costs that may be gained through a switch to inbreeding but also the increased rate of reproduction that results from the highly female-biased sex ratios that tend to evolve in association with this breeding system.

## VII. Long-Term Consequences of Inbreeding

While immediate demographic and ecological benefits of remaining in the natal nest for both males and females may have spurred the transition to inbred social systems in spiders and other social organisms, long-term negative effects of inbreeding may have eventually taken their toll. For the case of the spiders, at least, this suggests an answer to the question of why, despite the repeated origins of their inbred social system, there are so few inbred social species overall.

Although, barring new mutations, purging of recessive deleterious alleles is an expected outcome of chronic inbreeding (Charlesworth and Charlesworth, 1987; Charlesworth et al., 1990; Lande and Schemske, 1985),

TABLE II

CHARACTERISTICS OF SOME KEY INBRED SOCIAL SYSTEMS AND THEIR PHYLOGENETIC DISTRIBUTION, WHEN KNOWN

| Organism | Type of sociality | Type of inbreeding | Female biased-sex ratios | Genetic system | Geographic distribution | Habitat type | Number of inbred social species recorded | Minimum number of independent origins | Maximum average clade size |
|---|---|---|---|---|---|---|---|---|---|
| Social spiders | Multifemale groups | Continuous | Yes | DD | Tropics around the world | Low- to mid-elevation tropical rain forests (13 spp.); other (7 spp.) | 20[a] | 14 | 1.4 |
| Naked mole rats | Eusocial | Continuous | No | DD | Arid regions of East Africa | Areas of low unpredictable rainfall | 1[b] | 1 | 1 |
| Socially parasitic ants: | | | | | | | | | |
| *Myrmica* | Workless socially parasitic queens (inquilines) | Continuous | Yes | HD | *Myrmica*: Northern Hemisphere | Inside colonies of related ant species | *Myrmica* 5[c] | *Myrmica* 5 | *Myrmica* 1 |
| *Myrmoxenus*, formerly *Epimyrma* | | | | | *Myrmoxenus*: Mediterranean | | *Myrmoxenus* 6[d] | *Myrmoxenus* unresolved | *Myrmoxenus* unresolved |
| Termites | Eusocial | Periodic, due to replacement reproductives | No | DD | Mostly temperate and subtropical | Subterranean, soil feeding, drywood, dampwood, and grass eating | More common in lower termites[e] | Documented in at least six genera, but overall inbreeding low | ? |
| Spider mites (Acari: Prostigmata, Tetranychodidea) | Extended family groups | Continuous | Yes | HD | *Stigmaeopsis*: Japan, primary location where Acari sociality has been studied | *Stigmaeopsis celarius* and *S. longus*: Nest on Sasa bamboo; *S. miscanthi*: Miscanthus grass | 3[f] spp. so far shown to exhibit both social behavior and inbreeding | 1 in *Stigmaeopsis*, but inbreeding common in other nonsocial Acari taxa | 3 for *Stigmaeopsis* |
| Psocids: Archisocidae[g] | Multigeneration groups initiated by inseminated winged female | Continuous | Yes | ? | Tropical and subtropical | Arboreal, feed on vegetable matter over trunks of trees | ? | ? | ? |

*(Continued)*

TABLE II (*Continued*)

| Organism | Type of sociality | Type of inbreeding | Female biased-sex ratios | Genetic system | Geographic distribution | Habitat type | Number of inbred social species recorded | Minimum number of independent origins | Maximum average clade size |
|---|---|---|---|---|---|---|---|---|---|
| Australian gall thrips | Eusocial with soldier caste | Continuous | Yes | HD | Australia | Acacia trees | 5[h] | ? | ? |
| Ambrosia beetles: Xyleborini | Gregarious larval feeding; galleries colonized by one to few females; one sp. eusocial | Extreme sib mating | Yes | HD | Moist tropical forests around the world | Dead or dying twigs, branches, or trunks of woody plants, that is, patchy and ephemeral | >1300[i] | 1 | ~1300 |
| Other bark beetles: | | | | | | | | | |
| Hylesininae | | | Yes | HD | Hylesininae: mostly New World | Hylesininae: conifers or pith and seeds | 6[j] | 3 | 2 |
| Scolytinae, other than Xyleborini | | | Yes | | Other: various | Other: seeds or fruit-borers, pith, leaf stalks | 377[j] | 5 | 75.4 |

DD, diplo-diploid; HD, haplo-diploid.

[a]As per list given in Section VI, which is based on Lubin and Bilde (2007), but excluding *Tapinillus* sp., which is outbred, and *Anelosimus puravida*, whose inbred status is unconfirmed. Minimum number of origins based on Avilés (1997), Agnarsson et al. (2007), and Johannesen et al., (2007).

[b]Based on O'Riain and Faulkes (2008).

[c]Based on the Jansen et al. (2010) phylogeny, where 5 of 11+ presumably inbred, socially parasitic (inquiline) *Myrmica* species are included in the phylogeny: *M. hirsuta*, *karavajevi*, *laurae*, *rubra*, and *quebecensis*. All independent origins.

[d]Based on Buschinger (2009), phylogenetic relationships unresolved.

[e]Based on Vargo and Husseneder (2010), some degree of within colony inbreeding (or extended families), probably due to neotenic reproduction, documented in *Mastotermes* (Mastotermitidae), *Zootermopsis* (Termopsidae), *Incisitermes* (Kalotermitidae), *Reticulitermes* (Rhinotermitidae), *Coptotermes* (Rhinotermitidae), and *Nasutitermes* (Termitidae).

[f]Based on Sakagami et al. (2009).

[g]Based on Mockford (1957) and New (1973).

[h]Six of 22 documented species of Australian gall thrips are social, five of those inbred. Inbreeding, apparently basal to clade with social species, is also present in related taxa, but number of origins of inbreeding and average size of clades with inbred taxa yet undetermined. Based on Chapman et al. (2008).

[i]Based on Jordal et al. (2000).

[j]Based on Kirkendall (1993).

the resulting low genetic variability at the individual and population or colony levels is expected to make inbred lineages susceptible to disease and parasites and less able to respond to changing environments (Bijlsma et al., 1999; Hamilton, 1980; Shykoff and Schmid-Hempel, 1991). In the case of the spiders, the increased homozygosity of individuals and their social groups would have been compounded by the relatively large rate of turnover of their colony lineages, which should further reduce variability by either drift or selection at the metapopulation level (Avilés, 1997; but see Agnarsson et al., 2010). Finally, highly inbred populations may also be subject to genetic decay due to similar processes to those affecting asexual species, such as Müller's Ratchet (Heller and Maynard Smith, 1979; Lynch et al., 1995; Takebayashi and Morrell, 2001). By turning to inbreeding, social spiders may thus have evolved themselves into an evolutionary dead end (Agnarsson et al., 2006; Avilés, 1997).

Takebayashi and Morrell (2001) suggest that the phrase "evolutionary dead end" implies two major components: (1) the trait (e.g., inbreeding, asexual reproduction) cannot persist as a long-term strategy due to lineages that exhibit it having limited potential for adaptation and speciation, thus eventually going extinct, and (2) lineages exhibiting the trait cannot revert to the alternative state (e.g., from inbreeding to outcrossing). Both of these claims should result in what has been described as a "twiggy" or "spindly" distribution of the trait on a phylogenetic tree, that is, inbred (or asexual) species occurring more often in phylogenetically isolated terminal branches of the evolutionary tree than expected by chance (for asexual species, see Lynch et al., 1993; Maynard Smith, 1978; but Schwander and Crespi, 2009).

Given the currently confirmed list of inbred social spiders and their known phylogenetic relationships (Agnarson et al., 2006, 2007; Avilés, 1997; Johannesen et al., 2007; Lubin and Bilde, 2007), under the most conservative scenario, 9 of the 14 confirmed origins of inbred spider sociality correspond to single isolated species (*A. binotata*, *Mallos gregalis*, *S. dumicola*, *S. mimosarum*, *Stegodyphus sarasinorum*, *Achaearanea disparata*, *Anelosimus guacamayos*, *Anelosimus oritoyacu*, and *Theridion nigroannulatum*), 4 appear to correspond to species pairs (confirmed sister species pairs: *Achaearanea vervoorti* and *A. wau* and *Anelosimus lorenzo* and *Anelosimus rupununi*; congeneric species with unknown phylogenetic relationships *A. consociata* and *Agelena republicana* and *Diaea socialis* and *Diaea megagyna*), and only 1 could possibly represent a trio (*A. eximius*, *Anelosimus domingo*, and *Anelosimus dubiosus*) (list as in Lubin and Bilde, 2007, but excluding *Tapinillus* sp., which is outbred, and *Anelosimus puravida*, whose inbred status is unconfirmed). This gives the conservative estimate of an average clade size of the inbred social spiders of 1.4 species

(Table II), a surprisingly small number given the widespread phylogenetic distribution—in eight genera and six spider families (Avilés 1997; Lubin and Bilde, 2007)—of inbred spider sociality.

A quantitative test of the predictions of the dead-end hypothesis for all spiders is not feasible at the moment due to a lack of sufficiently detailed species level phylogenies of the relevant clades. Such a test, however, has been carried out for species in the family Theridiidae (which contains *Anelosimus*) and the genus *Stegodyphus* (Agnarsson et al., 2006; Johannesen et al., 2006), which, being the clades with the greatest concentration of inbred social species, yield conservative tests of the hypothesis.

Using the subset of the Theridiidae where extended maternal care is present, Agnarsson et al. (2006) showed that the observed ratio of 9 origins for 12 inbred social taxa is greater than expected under the null model of equal speciation and extinction for species of the two breeding systems. Likewise, the three inbred social *Stegodyphus* also represent isolated terminal branches, with no evidence of reversals, consistent with the dead-end hypothesis. Johannesen et al. (2006), however, note that the relatively high intraspecific sequence divergence of the three species (4–7% for the fastest evolving gene) suggests that the species may be a couple of million years old. Confirmation of the old age of these lineages, however, awaits eliminating the alternative hypothesis of a fast evolutionary rate at the molecular level for the whole genus, which cannot be discounted as only one outbred species from within a very limited geographical range (i.e., Israel) is available for comparison (3% sequence divergence). If the old age of the social *Stegodyphus* is confirmed, the question is then whether 2-million-years old is a surprisingly long time given the expectations of genetic degradation and/or susceptibility to disease and parasites of highly inbred taxa? Theoretical and empirical studies suggest it may not be. According to models for the evolution of sex that combine mutational and environmental factors, the age of persistence of asexual lineages (which should suffer a similar fate as highly inbred ones) is expected to vary according to parasite pressure (Howard and Lively, 1994; Meirmans and Neiman, 2006; Neiman et al., 2009). Thus, asexuals have been found to vary broadly in persistence times, with about 50% of asexual taxa lasting longer than 2 million years (e.g., Fig. 1 in Neiman et al., 2009). One possibility is thus that in the dry and seasonal habitats of the social *Stegodyphus*, parasite pressures may be low enough to allow for relatively long-lived inbred lineages, in contrast to habitats such as the lowland tropical rainforest where stronger parasite pressures, and thus shorter-lived inbred spider species, are expected. A test of this prediction for the social *Anelosimus* is in progress (I. Agnarsson, L. Avilés, and W. Maddison, unpublished).

## VIII. Is Inbreeding, in General, an Evolutionary Dead End?

The jury is still out in terms of inbreeding being, in general, an evolutionary dead end. This question was originally considered with respect to selfing in plants (Stebbins, 1957, 1974), where, albeit with some counterexamples (reviewed in Takebayashi and Morrell, 2001), there is support for the two components of the dead-end hypothesis—(1) selfing appears not to persist as a long-term strategy or to yield new lineages and (2) selfing lineages appear unlikely to revert to outcrossing. With the exception of the social spiders, however, the possible long-term consequences of inbreeding have not been formally explored in inbred social animals.

In the case of the mole rats, it is interesting that the highly inbred naked mole rat is the sole representative of its genus *Heterocephalus*, while its sister group, which includes the genus *Cryptomys* to which the Damaraland and other outbred social mole rats belong, is a species-rich clade of at least 14 species (Faulkes et al., 1997). This pattern is consistent with the inbred lineage diversifying much less than the outbred one, as expected under the dead-end hypothesis. However, the genetic distance between different populations of the naked mole rat (4.44% mean sequence divergence for cytochrome *b*, compared to 3.0 and 0.78 for two outbred species in the sister group) (Faulkes et al., 1997, 2003) indicates that the species is old (~3 million years old), as also suggested by the fact that its sister group is old enough to have diversified into many species (Faulkes et al., 1997). Faulkes et al. (1997) suggest that the lack of diversification of *Heterocephalus* may reflect its highly uniform and narrow ecological niche, which, if stable over millions of years, could also explain the old age of the lineage. Alternatively, it is not clear whether the isolated *Heterocephalus* lineage has been inbred throughout that time or whether it may have been able to revert to outbreeding under more clement environmental conditions, a possibility that has been suggested by some studies (Braude, 2000; Ciszek, 2000).

In the case of the socially parasitic ants, the five fully parasitic (and presumably inbred) inquiline *Myrmica* species included in the Jansen et al. (2010) phylogeny represented each an isolated and independent origin of this syndrome; a similar pattern may characterize species in the genus *Myrmoxenus* (formerly *Epimyrma*), although in this case phylogenetic relationships remain unresolved (Table II). Although the *Myrmica* patterns are consistent with the dead-end hypothesis, extreme host specialization of these taxa is likely to render them evolutionary dead ends for more reasons than their inbred breeding system. In termites, periodic inbreeding may be problematic in terms of the ability of individual colonies to resist disease and parasites (e.g., Shellman-Reeve, 1997), but because outbreeding at the population and species level is maintained there is no expectation of these organisms being evolutionary dead ends. Although considerable work has been done to understand the conditions surrounding inbreeding in some Acari taxa (e.g., Kaliszewski and Wrensch, 1993; Sabelis and Nagelkerke, 1993), further natural history and phylogenetic work are needed to assess the size and longevity of the clades resulting from the multiple derivations of strong inbreeding in these organisms.

The Australian gall-forming thrips may be an interesting counterexample to the dead-end hypothesis, as inbreeding, which appears relatively widespread in this clade, may have been reversed in some cases (Chapman et al., 2008; McLeish et al., 2006). However, the most glaring exception to the generalization that inbreeding leads to an evolutionary dead end is the ambrosia beetles. Remarkably, these haplo-diploid beetles form a clade of over 1300 species that are apparently all extreme (sib mating) inbreeders (Jordal et al., 2000). Interestingly, these beetles are able to use a great variety of plant taxa due to the fact that they cultivate fungi for food within their galleries. Jordal et al. (2000) argue that this extreme resource generalism, in conjunction with the colonization advantage conferred by haplodiploidy and inbreeding, may have promoted the rapid diversification of this clade. It thus appears that the ecological success and access to a new adaptive zone gained by these haplo-diploid beetles through their association with fungi may have been more than enough to compensate for any long-term detrimental consequences of their highly homozygous genomes.

With selection acting simultaneously at different levels of the biological hierarchy, it is clear that ecology and demography, on the one hand, and genetics, on the other hand, perform a balancing act in determining the fate of behavioral and other traits that may be favored at one level but disfavored at another. The diversity of social and breeding systems in the plant and animal kingdom represents the outcomes of different balances between these evolutionary forces. Organisms where both sociality and inbreeding have arisen independently multiple times, such as spiders, psocids, thrips, spider mites, and bark and ambrosia beetles, make ideal model systems in which the consequences of this balancing act can be explored.

## Acknowledgments

For comments on the chapter, we thank P.M. Waser, W.P. Maddison, M. Salomon, J. Guevara, K. Samuk, and G. Harwood. G. Harwood gathered some of the bibliographic materials on inbred social systems other than spiders. Funding for the research synthesized here was provided by the USA NSF (DEB-707474 and DEB-9815938 Grants) and NSERC Canada (Discovery Grants 22R81375 and 261354-2008 RGPIN) to L. A. J. P. was funded by an USA NSF Graduate Fellowship and a Swiss National Science Foundation grant (31003A-125306).

**References**

Agnarsson, I., 2006. A revision of the New World *eximius* lineage of *Anelosimus* (Araneae, Araneoidea, Theridiidae). Zool. J. Linn. Soc. Lond. 141, 447–626.

Agnarsson, I., Avilés, L., Coddington, J., Maddison, W.P., 2006. Sociality in Theridiid spiders: repeated origins of an evolutionary dead end. Evolution 60, 2342–2351.

Agnarsson, I., Maddison, W.P., Avilés, L., 2007. The phylogeny of the social *Anelosimus* spiders (Araneae: Theridiidae) inferred from six molecular loci and morphology. Mol. Phylogenet. Evol. 43, 833–851.

Agnarsson, I., Maddison, W.P., Avilés, L., 2010. Complete separation along matrilines in a social spider metapopulation inferred from hypervariable mitochondrial DNA region. Mol. Ecol. 19, 3052–3063.

Avilés, L., 1993a. Interdemic selection and the sex ratio: a social spider perspective. Am. Nat. 142, 320–345.

Avilés, L., 1993b. Newly-discovered sociality in the neotropical spider *Aebutina binotata* Simon (Araneae, Dictynidae). J. Arachnol. 21, 184–193.

Avilés, L., 1994. Social behavior in a web-building lynx spider, *Tapinillus* sp. (Araneae: Oxyopidae). Biol. J. Linn. Soc. Lond. 51, 163–176.

Avilés, L., 1997. Causes and consequences of cooperation and permanent-sociality in spiders. In: Choe, J.C., Crespi, B.J. (Eds.), The Evolution of Social Behavior in Insects and Arachnids. Cambridge University Press, Cambridge, pp. 476–498.

Avilés, L., 2000. Nomadic behavior and colony fission in a cooperative spider: life history evolution at the level of the colony? Biol. J. Linn. Soc. Lond. 70, 325–339.

Avilés, L., Bukowski, T.C., 2006. Group living and inbreeding depression in a subsocial spider. Proc. R. Soc. B Lond. 270, 157–163.

Avilés, L., Gelsey, G., 1998. Natal dispersal and demography of a subsocial *Anelosimus* species and its implications for the evolution of sociality in spiders. Can. J. Zool. 76, 2137–2147.

Avilés, L., Purcell, J., 2011. *Anelosimus oritoyacu*, a cloud forest social spider with only slightly female-biased primary sex ratios. J. Arachnol. 39, 178–182.

Avilés, L., Tufiño, P., 1998. Colony size and individual fitness in the social spider *Anelosimus eximius*. Am. Nat. 152, 403–418.

Avilés, L., Agnarsson, I., Salazar, P., Purcell, J., Iturralde, G., Yip, E., et al., 2007. Altitudinal patterns of spider sociality and the biology of a new mid-elevation social Anelosimus species in Ecuador. Am. Nat. 170, 783–792.

Bijlsma, R., Bundgaard, J., Van Putten, W.F., 1999. Environmental dependence of inbreeding depression and purging in Drosophila melanogaster. J. Evol. Biol. 12, 1125–1137.

Bilde, T., Lubin, Y., Smith, D., Schneider, J.M., Maklakov, A.A., 2005. The transition to social inbred mating systems in spiders: role of inbreeding tolerance in a subsocial predecessor. Evolution 59, 160–174.

Bilde, T., Coates, K.S., Birkhofer, K., Bird, T., Maklakov, A.A., Lubin, Y., et al., 2007. Survival benefits select for group living in a social spider despite reproductive costs. J. Evol. Biol. 20, 2412–2426.

Brandt, M., Foitzik, S., Fischer-Blass, B., Heinze, J., 2005. The coevolutionary dynamics of obligate ant social parasite systems—between prudence and antagonism. Biol. Rev. 80, 251–267.

Braude, S., 2000. Dispersal and new colony formation in wild naked mole-rats: evidence against inbreeding as the system of mating. Behav. Ecol. 11, 7–12.

Bukowski, T., Avilés, L., 2002. Asynchronous maturation of the sexes may limit close inbreeding in a subsocial spider. Can. J. Zool. 80, 193–198.

Burland, T.M., Bennett, N.C., Jarvis, J.U.M., Faulkes, C.G., 2002. Eusociality in African mole-rats: new insights from patterns of genetic relatedness in the Damaraland mole-rat (*Cryptomys damarensis*). Proc. Biol. Sci. 269, 1025–1030.
Buschinger, A., 1989. Evolution, speciation, and inbreeding in the parasitic ant genus *Epimyrma* (Hymenoptera, Formicidae). J. Evol. Biol. 2, 265–283.
Buschinger, A., 2009. Social parasitism among ants: a review (Hymenoptera: Formicidae). Myrmecol. News 12, 219–235.
Buskirk, R.E., 1981. Sociality in the Arachnida. Hermann, H.R. (Ed.), In: Social Insects, vol. 4. Academic Press, New York, pp. 282–367.
Calleri, D.V., Reid, E.M., Rosengaus, R.B., Vargo, E.L., Traniello, J.F.A., 2006. Inbreeding and disease resistance in a social insect: effects of heterozygosity on immnnocompetence in the termite Zootermopsis angusticollis. Proc. R. Soc. B 273, 2633–2640.
Chapman, T.W., Crespi, B.J., Kranz, B.D., Schwarz, M.P., 2000. High relatedness and inbreeding at the origin of eusociality in gall-inducing thrips. Proc. Natl. Acad. Sci. USA 97, 1648–1650.
Chapman, T.W., Crespi, B.J., Perry, S.P., 2008. The evolutionary ecology of eusociality in Australian gall thrips: a 'model clades' approach. In: Korb, J., Heinze, J. (Eds.), Ecology of Social Evoluion. Chapman and Hall, New York, pp. 57–83.
Charlesworth, D., Charlesworth, B., 1987. Inbreeding and its evolutionary consequences. Ann. Rev. Ecol. Syst. 18, 237–268.
Charlesworth, D., Morgan, M.T., Charlesworth, B., 1990. Inbreeding depression, genetic load, and the evolution of outcrossing rates in a multilocus system with no linkage. Evolution 44, 1469–1489.
Charnov, E.L., 1976. Optimal foraging, the marginal value theorem. Theor. Popul. Biol. 9, 129–136.
Ciszek, D., 2000. New colony formation in the "highly inbred" eusocial naked mole-rat: outbreeding is preferred. Behav. Ecol. 11, 1–6.
Clutton-Brock, T.H., 1989. Female transfer and inbreeding avoidance in social mammals. Nature 337, 70–72.
Corcobado-Marquez, G. Rodríguez-Gironés, M., Moya-Laraño, J. Avilés, L. (2012). Sociality level correlates with dispersal ability in spiders. Func. Ecol. in press.
Crespi, B.J., 1993. Sex allocation ratio selection in Thysanoptera. In: Wrensch, D.L., Ebbert, M.A. (Eds.), Evolution and Diversity of Sex Ratio in Insects and Mites. Chapman and Hall, New York, pp. 214–234.
Crnokrak, P., Roff, D.A., 1999. Inbreeding depression in the wild. Heredity 83, 260–270.
DeHeer, C.J., Vargo, E.L., 2006. An indirect test of inbreeding depression in the termites Reticulitermes flavipes and Reticulitermes virginicus. Behav. Ecol. Sociobiol. 59, 753–761.
Dobson, F.S., 1982. Competition for mates and predominant juvenile male dispersal in mammals. Anim. Behav. 30, 1183–1192.
Domingue, M.J., Teale, S.A., 2007. Inbreeding depression and its effect on intrinsic population dynamics in engraver beetles. Ecol. Entomol. 32, 201–210.
Faulkes, C.G., Bennett, N.C., Bruford, M.W., Obrien, H.P., Aguilar, G.H., Jarvis, J.U.M., 1997. Ecological constraints drive social evolution in the African mole-rats. Proc. Biol. Sci. 264, 1619–1627.
Faulkes, C.G., Abbott, D.H., O'Brien, H.P., Lau, L., Roy, M.R., Wayne, R.K., et al., 2003. Micro- and macrogeographical genetic structure of colonies of naked mole-rats *Heterocephalus glaber*. Mol. Ecol. 6, 615–628.
Felsenstein, J., 1974. The evolutionary advantages of recombination. Genetics 78, 737–756.
Foelix, R.F., 1996. Biology of Spiders. Oxford University Press, Oxford.

Furey, R.E., 1998. Two cooperatively social populations of the theridiid spider *Anelosimus studiosus* in a temperate region. Anim. Behav. 55, 727–735.

Greenwood, P.J., 1980. Mating systems, philopatry and dispersal in birds and mammals. Anim. Behav. 28, 1140–1162.

Guevara, J., Avilés, L., 2007. Multiple sampling techniques confirm differences in insect size between low and high elevations that may influence levels of spider sociality. Ecology 88, 2015–2033.

Hamilton, W.D., 1980. Sex versus non-sex versus parasite. Oikos 35, 282–290.

Heller, J., Maynard Smith, J., 1979. Does Muller's ratchet work with selfing? Genet. Res. 8, 269–294.

Henschel, J.R., 1993. Is solitary life an alternative for the social spider *Stegodyphus dumicola*? Namibia Sci. Soc. J. 43, 71–79.

Henschel, J.R., 1998. Predation on social and solitary individuals of the spider *Stegodyphus dumicola* (Araneae, Eresidae). J. Arachnol. 26, 61–69.

Hölldobler, B., Wilson, E.O., 1990. The Ants. Harvard University Press, Cambridge.

Hoogland, J.L., 1992. Levels of inbreeding among prairie dogs. Am. Nat. 139, 591–602.

Howard, R.S., Lively, C.M., 1994. Parasitism, mutation accumulation and the maintenance of sex. Nature 367, 554–557.

Husband, B.C., Schemske, D.W., 1996. Evolution of the magnitude and timing of inbreeding depression in plants. Evolution 50, 54–70.

Husseneder, C., Simms, D.M., Ring, D.R., 2006. Genetic diversity and genotypic differentiation between the sexes in swarm aggregations decrease inbreeding in the Formosan subterranean termite. Insect. Soc. 53, 212–219.

Jansen, G., Savolainen, R., Vepsäläinen, K., 2010. Phylogeny, divergence-time estimation, biogeography and social parasite-host relationships of the Holarctic ant genus *Myrmica* (Hymenoptera: Formicidae). Mol. Phylogenet. Evol. 56, 294–304.

Jarvis, J.U.M., Bennett, N.C., 1993. Eusociality has evolved independently in two genera of bathyergid mole-rats—but occurs in no other subterranean mammal. Behav. Ecol. Sociobiol. 33, 253–260.

Jarvis, J.U.M., O'Riain, M.J., Bennett, N.C., Sherman, P.W., 1994. Mammalian eusociality: a family affair. Trends Ecol. Evol. 9, 47–51.

Johannesen, J., Lubin, Y., 1999. Group founding and breeding structure in the subsocial spider *Stegodyphus lineatus* (Eresidae). Heredity 82, 677–686.

Johannesen, J., Lubin, Y., 2001. Evidence for kin-structured group founding and limited juvenile dispersal in the sub-social spider *Stegodyphus lineatus* (Araneae, Eresidae). J. Arachnol. 29, 413–422.

Johannesen, J., Johannesen, B., Griebeler, E.M., Baran, I., Tunc, M.R., Kiefer, A., et al., 2006. Distortion of symmetrical introgression in a hybrid zone: evidence for locus-specific selection and uni-directional range expansion. J. Evol. Biol. 19, 705–716.

Johannesen, J., Lubin, Y., Smith, D.R., Bilde, T., Schneider, J.M., 2007. The age and evolution of sociality in *Stegodyphus* spiders: a molecular phylogenetic perspective. Proc. Biol. Sci. 274, 231–237.

Johannesen, J., Moritz, R.F.A., Simunek, H., Seibt, U., Wickler, W., 2009a. Species cohesion despite extreme inbreeding in a social spider. J. Evol. Biol. 22, 1137–1142.

Johannesen, J., Wickler, W., Seibt, U., Moritz, R.F.A., 2009b. Population history in social spiders repeated: colony structure and lineage evolution in *Stegodyphus mimosarum* (Eresidae). Mol. Ecol. 18, 2812–2818.

Jones, T.C., Riechert, S.E., Dalrymple, S.E., Parker, P.G., 2007. Fostering model explains variation in levels of sociality in a spider system. Anim. Behav. 73, 195–204.

Jordal, B.H., Normark, B.B., Farrell, B.D., 2000. Evolutionary radiation of an inbreeding haplodiploid beetle lineage (Curculionidae, Scolytinae). Biol. J. Linn. Soc. Lond. 71, 483–499.

Kaliszewski, M., Wrensch, D.L., 1993. Evolution of sex determination and sex ratio within the mite cohort Tarsonemina (Acari: Heterostigmata). In: Wrensch, D.L., Ebbert, M.A. (Eds.), Evolution and Diversity of Sex Ratio in Insects and Mites. Chapman and Hall, New York, pp. 192–213.

Kirkendall, L.R., 1983. The evolution of mating systems in bark and ambrosia beetles (Coleoptera: Scolytidae and Platypodidae). Biol. J. Linn. Soc. Lond. 77, 293–352.

Kirkendall, L.R., 1993. Ecology and evolution of biased sex ratios in bark and ambrosia beetles. In: Wrensch, D.L., Ebbert, M.A. (Eds.), Evolution and Diversity of Sex Ratio in Insects and Mites. Chapman and Hall, New York, pp. 235–345.

Kirkendall, L.R., Kent, D.S., Raffa, K.A., 1997. Interactions among males, females and offspring in bark and ambrosia beetles: the significance of living in tunnels for the evolution of social behavior. In: Choe, J.C., Crespi, B.J. (Eds.), The Evolution of Social Behavior in Insects and Arachnids. Cambridge University Press, Cambridge, pp. 181–215.

Klein, B., Bukowski, T., Avilés, L., 2005. Male residency and mating patterns in a subsocial spider. J. Arachnol. 33, 703–710.

Korb, J., 2008. The ecology of social evolution in termites. In: Korb, J., Heinze, J. (Eds.), Ecology of Social Evolution. Springer-Verlag, Berlin, pp. 151–174.

Lande, R., Schemske, D.W., 1985. The evolution of self-fertilization and inbreeding depression in plants. I. Genetic models. Evolution 39, 24–40.

Lawson Handley, L.J., Perrin, N., 2007. Advances in our understanding of mammalian sex biased dispersal. Mol. Ecol. 16, 1559–1578.

Lubin, Y., Bilde, T., 2007. The evolution of sociality in spiders. Adv. Stud. Behav. 37, 83–145.

Lubin, Y., Crozier, R.H., 1985. Electrophoretic evidence for population differentiation in a social spider *Achaearanea wau* (Theridiidae). Insect. Soc. 32, 297–304.

Lubin, Y., Robinson, M.H., 1982. Dispersal by swarming in a social spider. Science 216, 319–321.

Lubin, Y., Hennicke, J., Schneider, J., 1998. Settling decisions of dispersing *Stegodyphus lineatus* (Eresidae) young. Israel J. Zool. 44, 217–226.

Lubin, Y., Birkhofer, K., Berger-Tal, R., Bilde, T., 2009. Limited male dispersal in a social spider with extreme inbreeding. Biol. J. Linn. Soc. Lond. 97, 227–234.

Lynch, M., Burger, R., Butcher, D., Gabriel, W., 1993. The mutational meltdown in asexual populations. J. Hered. 84, 339–344.

Lynch, M., Conery, J., Burger, R., 1995. Mutational meltdown in sexual populations. Evolution 49, 1067–1080.

Matsuura, K., Vargo, E.L., Kawatsu, K., Labadie, P.E., Nakano, H., Yashiro, T., et al., 2009. Queen succession through asexual reproduction in termites. Science 323, 1687.

Maynard Smith, J., 1978. The Evolution of Sex. Cambridge University Press, Cambridge.

McLeish, M.J., Chapman, T.W., Crespi, B.J., 2006. Inbreeding ancestors: the role of sibmating in the social evolution of gall thrips. J. Hered. 97, 31–38.

Meirmans, S., Neiman, M., 2006. Methodologies for testing a pluralist idea for the maintenance of sex. Biol. J. Linn. Soc. Lond. 89, 605–613.

Mitchell, R., 1973. Growth and population dynamics of a spider mite (*Tetranychus urticae* K., Acarina: Tetranychidae). Ecology 54, 1349–1355.

Mockford, E.L., 1957. Life history studies on some Florida insects of the genus Archipsocus (Psocoptera). Bull. Fl. St. Museum Biol. Sci. 1, 253–274.

Mori, K., Saito, Y., 2005. Variation in social behavior within a spider mite genus, *Stigmaeopsis* (Acari: Tetranychidae). Behav. Ecol. 16, 232–238.

Mori, H., Saito, Y., Tho, Y.P., 1999. Cooperative group predation in a sit-and-wait Cheyletid mite. Exp. Appl. Acarol. 23, 643–651.

Mori, K., Saito, Y., Sakagami, T., Sahara, K., 2005. Inbreeding depression of female fecundity by genetic factors retained in natural populations of a male-haploid social mite (Acari: Tetranychidae). Exp. Appl. Acarol. 36, 15–23.

Motro, U., 1991. Avoiding inbreeding and sibling competition: the evolution of sexual dimorphism for dispersal. Am. Nat. 137, 108–115.

Neiman, M., Meirmans, S., Meirmans, P., 2009. What can we learn from estimating asexual lineage age distributions? Ann. N. Y. Acad. Sci. 1168, 185–200.

New, T.R., 1973. The Archipsocidae of South America (Psocoptera). Trans. R. Ent. Soc. Lond. 125, 57–105.

Norton, R.A., Kethley, J.B., Johnston, D.E., O'Connor, B.M., 1993. Phylogenetic perspectives on genetic systems and reproductive modes of mites. In: Wrensch, D.L., Ebbert, M.A. (Eds.), Evolution and Diversity of Sex Ratio in Insects and Mites. Chapman and Hall, New York, pp. 8–99.

O'Riain, M.J., Faulkes, C.G., 2008. African mole-rats: eusociality, relatedness and ecological constraints. In: Korb, J., Heinze, J. (Eds.), Ecology of Social Evoluion. Chapman and Hall, New York, pp. 207–223.

O'Riain, M.J., Jarvis, J.U.M., Faulkes, C.G., 1996. A dispersive morph in the naked mole-rat. Nature 380, 619–621.

Otto, S.P., 2009. The evolutionary enigma of sex. Am. Nat. 174, S1–S14.

Pamilo, P., Gertsch, P., Thoren, P., Seppä, P., 1997. Molecular population genetics of social insects. Annu. Rev. Ecol. Syst. 28, 1–25.

Peer, K., Taborsky, M., 2005. Outbreeding depression, but no inbreeding depression in haplodiploid ambrosia beetles with regular sibling mating. Evolution 59, 317–323.

Perrin, N., Mazalov, V., 2000. Local competition, inbreeding, and the evolution of sex-biased dispersal. Am. Nat. 155, 116–127.

Powers, K.S., Avilés, L., 2003. Natal dispersal patterns of a subsocial spider *Anelosimus cf. jucundus* (Theridiidae). Ethology 109, 725–737.

Powers, K.S., Avilés, L., 2007. The role of prey size and abundance in the geographic distribution of spider sociality. J. Anim. Ecol. 76, 995–1003.

Purcell, J., Avilés, L., 2007. Smaller colonies and more solitary living mark higher elevation populations of a social spider. J. Anim. Ecol. 76, 590–597.

Purcell, J., Avilés, L., 2008. Gradients of precipitation and ant abundance may contribute to the altitudinal range limit of subsocial spiders: insights from a transplant experiment. Proc. Biol. Sci. 275, 2617–2625.

Pusey, A., Wolf, M., 1996. Inbreeding avoidance in animals. Trends Ecol. Evol. 11, 201–206.

Reeve, H.K., Westneat, D.F., Noon, W.A., Sherman, P.W., Aquadro, C.F., 1990. DNA "fingerprinting" reveals high levels of inbreeding in colonies of the eusocial naked mole-rat. Proc. Natl. Acad. Sci. USA 87, 2496–2500.

Riechert, S.E., Roeloffs, R.M., 1993. Evidence for and consequences of inbreeding in the cooperative spiders. In: Thornhill, N.W. (Ed.), The Natural History of Inbreeding and Outbreeding. University of Chicago Press, Chicago, pp. 283–303.

Riechert, S.E., Roeloffs, R., Echternacht, A.C., 1986. The ecology of the cooperative spider *Agelena consociata* in Equatorial Africa (Araneae, Agelenidae). J. Arachnol. 14, 175–191.

Roeloffs, R., Riechert, S.E., 1988. Dispersal and population genetic structure of the cooperative spider, *Agenlena consociata* in West African rainforest. Evolution 42, 173–183.

Ross-Gillespie, A., O'Riain, M.J., Keller, L.F., 2007. Viral epizootic reveals inbreeding depression in a habitually inbreeding mammal. Evolution 61, 2268–2273.

Ruch, J., Heinrich, L., Bilde, T., Schneider, J.M., 2009. The evolution of social inbreeding mating systems in spiders: limited male mating dispersal and lack of pre-copulatory inbreeding avoidance in a subsocial predecessor. Biol. J. Linn. Soc. Lond. 98, 851–859.

Sabelis, M.W., Nagelkerke, K., 1993. Sex allocation and pseudoarrhenotoky in Phytoseiid mites. In: Wrensch, D.L., Ebbert, M.A. (Eds.), Evolution and Diversity of Sex Ratio in Insects and Mites. Chapman and Hall, New York, pp. 512–541.

Saito, Y., 1997. Sociality and kin selection in Acari. In: Choe, J.C., Crespi, B.J. (Eds.), Evolution of Social Behavior in Insects and Arachnids. Cambridge University Press, Cambridge, pp. 443–457.

Saito, Y., 2010. Plant Mites and Sociality: Diversity and Evolution. Springer-Verlag, Tokyo.

Saito, Y., Mori, K., 2005. Where does male-to-male "aggression" compromise "cooperation"? Popul. Ecol. 46, 221–241.

Saito, Y., Sahara, K., Mori, K., 2000. Inbreeding depression by recessive deleterious genes affecting female fecundity in a haplo-diploid mite. J. Evol. Biol. 13, 668–678.

Sakagami, T., Saito, Y., Kongchuensin, M., Sahara, K., 2009. Molecular phylogeny of *Stigmaeopsis*, with special reference to speciation through host plant shift. Ann. Entomol. Soc. Am. 102, 360–366.

Schneider, J.M., Lubin, Y., 1996. Infanticidal male eresid spiders. Nature 381, 655–656.

Schneider, J.M., Lubin, Y., 1998. Intersexual conflict in spiders. Oikos 83, 496–506.

Schrempf, A., Aron, S., Heinze, J., 2006. Sex determination and inbreeding depression in an ant with regular sib-mating. Heredity 97, 75–80.

Schwander, T., Crespi, B.J., 2009. Twigs on the tree of life? Neutral and selective models for interacting macroevolutionary patterns with microevolutionary processes in the analysis of asexuality. Mol. Ecol. 18, 28–42.

Seibt, U., Wickler, W., 1988. Why do "family spiders", Stegodyphus (Eresidae), live in colonies? J. Arachnol. 16, 193–198.

Shellman-Reeve, J.S., 1997. The spectrum of eusociality in termites. In: Choe, J.C., Crespi, B.J. (Eds.), The Evolution of Social Behavior in Insects and Arachnids. Cambridge University Press, Cambridge, pp. 52–93.

Shykoff, J.A., Schmid-Hempel, P., 1991. Parasites and the advantage of genetic variability within social insect colonies. Proc. R. Soc. Lond. B 243, 55–58.

Smith, D.R., 1986. Population genetics of *Anelosimus eximius* (Araneae, Theridiidae). J. Arachnol. 14, 201–217.

Smith, D.R., 1987. Genetic variation in solitary and cooperative spiders of the genus *Anelosimus* (Araneae: Theridiidae). In: Eder, J., Rembold, H. (Eds.), Chemistry and Biology of Social Insects. Verlag J. Peperny, Munich, pp. 347–348.

Smith, D.R., Engel, M.S., 1994. Population structure in an Indian cooperative spider, *Stegodyphus sarasinorum* Karsch (Eresidae). J. Arachnol. 22, 108–113.

Smith, D.R., Hagen, R.H., 1996. Population structure and interdemic selection in the cooperative spider *Anelosimus eximius*. J. Evol. Biol. 9, 589–608.

Stebbins, G.L., 1957. Self fertilization and population variability in higher plants. Am. Nat. 91, 337–354.

Stebbins, G.L., 1974. Flowering Plants: Evolution Above the Species Level. Harvard University Press, Cambridge.

Stockley, P., Macdonald, D.W., 1998. Why do female common shrews produce so many offspring? Oikos 83, 560–566.

Takebayashi, N., Morrell, P.L., 2001. Is self-fertilization an evolutionary dead end? Revisiting an old hypothesis with genetic theories and a macroevolutionary approach. Am. J. Bot. 88, 1143–1150.

Thorne, B.L., 1997. Evolution of eusociality in termites. Annu. Rev. Ecol. Syst. 28, 27–54.

Thorne, B.L., Adams, E.S., Traniello, J.F.A., Bulmer, M., 1999. Reproductive dynamics and colony structure of subterranean termites of the genus Reticulitermes (Isoptera Rhinotermitidae): a review of the evidence from behavioral, ecological, and genetic studies. Ethol. Ecol. Evol. 11, 149–169.

Thurin, N., Aron, S., 2011. No reversion to single mating in a socially parasitic ant. J. Evol. Biol. 24, 1128–1134.

Trontti, K., Aron, S., Sundstrom, L., 2006. The genetic population structure of the ant Plagiolepis xene—implications for genetic vulnerability of obligate social parasites. Cons. Genet. 7, 241–250.

Uetz, G.W., Hieber, C.S., 1997. Colonial web-building spiders: balancing the costs and benefits of group living. In: Choe, J.C., Crespi, B.J. (Eds.), Evolution of Social Behavior in Insects and Arachnids. Cambridge University Press, Cambridge, pp. 458–475.

Vargo, E.L., Husseneder, C., 2010. Genetic structure of termite colonies and populations. In: Bignell, E.L. et al., (Ed.), Biology of Termites: A Modern Synthesis. Springer, pp. 321–347.

Vargo, E.L., Labadie, P.E., Matsuura, K., 2012. Asexual queen succession in the subterranean termite *Reticulitermes virginicus*. Proc. R. Soc. B 279, 813–819.

Waser, P.M., Austad, S.N., Keane, B., 1986. When should animals tolerate inbreeding? Am. Nat. 128, 529–537.

West, S.A., Lively, C.M., Read, A.F., 1999. A pluralist approach to sex and recombination. J. Evol. Biol. 12, 1003–1012.

Williams, G.C., 1975. Sex and Evolution. Princeton University Press, Princeton.

Wilson, E.O., 1971. The Insect Societies. Harvard University Press, Cambridge.

Yip, E., Powers, K.S., Avilés, L., 2008. Cooperative capture of large prey solves scaling challenge faced by large spider societies. Proc. Natl. Acad. Sci. USA 105, 11818–11822.

# The Behavior of Wild White-Faced Capuchins: Demography, Life History, Social Relationships, and Communication

SUSAN PERRY

DEPARTMENT OF ANTHROPOLOGY, BEHAVIOR, EVOLUTION AND CULTURE PROGRAM, UNIVERSITY OF CALIFORNIA, LOS ANGELES, CALIFORNIA, USA

## I. INTRODUCTION

The primary impetus for studying nonhuman primate social systems in anthropology has been to provide insights into human behavioral evolution. During the 1960s and 1970s, behavioral primatology focused on catarrhines, and particularly on great apes, because of their phylogenetic proximity to humans, and thus their similarity to hominins resulting from homology. However, to understand the selective pressures that shaped key behavioral, morphological, and life history traits, it is critical to take a broader comparative perspective, seeking evolutionarily independent appearances of the traits of interest and investigating the cross-species correlates of these traits. Although the New World primates were among the first primate species to be studied by early primatologists in the 1930s (Carpenter, 1934), it is only quite recently that their utility as strategic models has been discovered. The New World primates (platyrrhines) exhibit an astonishing array of social systems and behavioral strategies that both include and extend beyond the range of social systems present in the most commonly studied species (chimpanzees with their fission–fusion systems and female dispersal, and savanna baboons and macaques with their female-bonded social systems and male dispersal) that have figured most prominently as models for human evolution (Strier, 1994). Because the presence of these social systems in various platyrrhine taxa represents independent evolutionary events from their counterparts in the Old World, the use of platyrrhine data to inform models of social evolution allows researchers to



0065-3454/12 $35.00
DOI: 10.1016/B978-0-12-394288-3.00004-6

address questions about the selective pressures favoring particular social strategies that cannot be adequately addressed using ape models and homological reasoning alone.

Capuchin monkeys, in particular, have recently captured the interest of biological anthropologists because they are useful models for certain aspects of social evolution, due to their many evolutionary convergences with humans, chimpanzees, and Old World monkeys. Their utility as models has made capuchins popular study subjects for biologists, psychologists, and anthropologists interested in explaining the evolution of life history strategies, cooperation, cultural capacities, communication (particularly gestural communication and nonconceptive sex), hunting of vertebrates, complex foraging techniques, and intelligence.

Capuchins (*Cebus*) and the closely related squirrel monkeys (*Saimiri*) have the highest encephalization quotients of any nonhuman primates (Fragaszy et al., 2004; Stephan et al., 1970). Like humans, capuchins have a long life span (up to 55 years for white-faced capuchins in captivity [Hakeem et al., 1996]) and long juvenile periods (Fragaszy et al., 2004). Capuchins and some of the squirrel monkeys are the only female-bonded primates among the New World Monkeys, thus converging with most of the Old World Monkeys. Capuchin monkeys stand out among the New World monkeys as having frequent and highly coordinated coalitions and long-term alliances. The nontufted capuchin species have an elaborate evolved repertoire of stereotyped signals for communicating about coalition formation, and aggressive coalitions are a daily occurrence. They cooperate in infant care, defense against predators, and defense against other members of their species. White-faced capuchins have an extraordinarily high rate of coalitionary lethal aggression, roughly comparable to that of humans and eastern chimpanzees (Gros-Louis et al., 2003; Perry and Manson, 2008). Capuchins are a particularly interesting taxon in which to study cooperation and conflict because their intraspecific conflicts are so frequent and intense and because cooperation is so critical for defense of food and access to mates, as well as for defense and care of offspring. This chapter will explore the role that these extreme levels of cooperation and conflict have played in shaping or being shaped by (a) the demographic structure of white-faced capuchin populations, (b) life history strategies of males and females, (c) social interaction patterns within and between groups, and (d) the communicative repertoires of these monkeys.

Most of what we know about the life histories of white-faced capuchin monkeys comes from two long-term research projects, both located in tropical dry forest sites in Guanacaste, Costa Rica: the Lomas Barbudal Monkey Project (referred to as "Lomas"), established in 1990 by Susan Perry (Perry and Manson, 2008; Perry et al., 2012), and a project in Santa

TABLE I
DATA AVAILABLE FROM PRIMARY STUDY GROUPS AT LOMAS BARBUDAL

| Group | Years of behavioral sampling | Genetic data | Years of demographic data | Comments |
|---|---|---|---|---|
| Abby (AA) | 1990–2011 | Yes | 1990–2011 | Habituated in 1990 |
| Flakes (FL) | 2003–2011 | Yes | 2003–2011 | Fissioned from AA in 2003 |
| Rambo (RR) | 1996–2011 | Yes | 1996–2011 | Habituated in 1996 |
| Splinter (SP) | 2010–2011 | Yes | 2000–2011 | Fissioned from RR in 1999 |
| Musketeers (MK) | 2004–2011 | Yes | 2004–2011 | Fissioned from RR in 2004 |
| Cupie (CU) | 2007–2011 | Yes | 2007–2011 | Fissioned from MK in 2007 |
| Newman (NM) | 2008–2011 | Yes | 2002–2011 | Habituated in 2008 |
| Pelon (FF) | 2002–2011 | Yes | 2001–2011 | Habituated in 2002 |
| Rafiki (RF) | 2007–2011 | Yes | 2007–2011 | Fissioned from FF in 2007 |
| Lost Boys (LB) | 2010–2011 | Yes | 2010–2011 | Fissioned from RR in 2009 |
| Solo's (SO) | | Partial | 2003–2011 | Males migrated from AA in 2003 |
| Badoodie (BD) | | Partial | Sporadic | Males migrated from RR in 2007 |
| Canal (CN) | | | Sporadic | |
| El Salto (ES) | | | Sporadic | |
| Chingo (CH) | | | Sporadic | |
| San Ramon (SR) | | | Sporadic | |

Rosa National Park, started in 1983 by Linda Fedigan (Fedigan and Jack, 2012). Shorter studies of social behavior have been conducted at Barro Colorado Island in Panama (Crofoot et al., 2008; Mitchell, 1989; Oppenheimer, 1968), La Trujillo in Honduras (Buckley, 1983), and three other Costa Rican sites: Palo Verde (Panger, 1997; Perry et al., 2003a), La Selva (Boinski and Campbell, 1995), and Curu Wildlife Refuge (Baker, 1999; Perry et al., 2003a). Unless otherwise stated, the data in this chapter come from the Lomas Barbudal Monkey Project. Over the course of the past 21 years, this research project has collected data on 445 individual monkeys from 12 social groups, amassing a total of over 79,000 h of behavioral data. Table I summarizes the periods of data collection for each study group and the types of data collected.

## II. Population Structure

### A. Kinship and the Structure of Capuchin Social Groups

White-faced capuchins live in social groups averaging 15.2 monkeys per group at Santa Rosa (Fedigan and Jack, 2012) and averaging 18.8 monkeys per group, with a range of 5–40 monkeys per group, at Lomas Barbudal

(Perry et al., 2012; S. Perry, unpublished data). Groups are typically composed of 2–11 related adult females, one or more adult males, and juveniles and infants. Most groups contain multiple males (ranging from 1 to 13 per group), and at Lomas Barbudal, adult male to female sex ratios can vary from 0.22 to 1.44. The mean sex ratio at Santa Rosa is 0.85 (Fedigan and Jack, 2011), with the maximum for that site being 2.25 (Fedigan et al., 1996). Average sex ratios for other capuchin sites range from 0.54 to 0.88 (Fragaszy et al., 2004). All-male groups of up to 8 males have been observed at Lomas (S. Perry, unpublished data) but not in Santa Rosa (K. Jack, personal communication), and although these groups can persist for more than a year, all-male groups are more unstable in their composition than are the groups consisting of both sexes.

The first large-scale genetic analysis of a white-faced capuchin population was conducted at Lomas by Muniz (2008) and included 167 genotyped individuals, in which 73% of the individuals from six social groups were genotyped over a period of 389 group-months of observation. Muniz found that female–female dyads were more closely related than male–male dyads on average, though there are rare exceptions to this general rule for some groups during some time periods, as a consequence of comigration by closely related males. Over time, there was less within-group variation (i.e., more stability) in the degree of genetic relatedness for female–female dyads than for male–male dyads. There was also somewhat less between-group variation in relatedness for female–female dyads relative to male–male dyads. Not surprisingly, dyads consisting of individuals who were members of the same group were more closely related than dyads consisting of members of different groups.

White-faced capuchins are characterized by high survivorship (see Section II.B), long alpha male tenures (see Section III.A.4), and a high degree of male reproductive skew (see Section III.A.3). Consequently, individuals typically live together with many close relatives, such as full- and half siblings (Godoy, 2010; Muniz, 2008; Perry et al., 2008). Table II shows the mean number of maternal and paternal siblings and peers (i.e., group members who differ in age by less than a year) typically coresident with 6-year-old capuchins at Lomas Barbudal in a sample of 29 focal females and 19 focal males from 9 social groups (I. Godoy and S. Perry, unpublished data). The number of maternal sisters tends to increase as females age (Perry et al., 2008), though the number of paternal sisters available changes less predictably with age. Eighty-three percent of females and 74% of males still coresided with their mothers at age 6 years, whereas only 17% of females and 21% of males still coresided with their fathers. At age 6 years, 79% of capuchins of either sex had at least one paternal half sibling, 90% had at least one maternal full- or half sibling, and 92% had at least one

TABLE II
AVERAGE NUMBER OF KIN AND PEERS CORESIDENT FOR 6-YEAR-OLD FEMALE ($N=29$) AND MALE ($N=19$) CAPUCHINS FROM NINE SOCIAL GROUPS AT LOMAS BARBUDAL

| | Female | Male |
|---|---|---|
| Maternal sisters (full or half) | 1.3 | 1.1 |
| Paternal half sisters | 2.0 | 1.7 |
| Maternal brothers (full or half) | 1.4 | 1.3 |
| Paternal half brothers | 1.6 | 1.8 |
| Female peers | 1.6 | 1.4 |
| Male peers | 1.0 | 1.2 |

peer still coresident with them. It is difficult to find comparable data for other primate species, especially with regards to paternal kinship, but it does appear that the availability of close female kin is somewhat higher in Lomas capuchins than in Amboseli baboons (Silk et al., 2006a), for which 6-year-old females have on average one maternal sister and three peers.

White-faced capuchin groups occasionally split, apparently as a consequence of group coordination difficulties when group size gets large, and kinship affects the dynamics of the fission process. At Lomas, females typically remain with their most closely related female kin during group fissions (Muniz, 2008; S. Perry, L. Muniz and I. Godoy, unpublished data). In an analysis of three group fissions, average female–female dyadic relatedness increased in three fission products, remained the same in two, and decreased in one (Muniz, 2008). In five fissions that we have observed at Lomas, the fission occurred along matrilines, and the only anomalous instance in which daughters did not accompany their mothers was a case in which a female abandoned all of her sons and daughters to accompany her own mother and sisters to a new subgroup. Males are less predictable in their choices: in 16 out of 34 cases, natal males have eventually (within 5 months of the fission) become stable members of the subgroup in which they have weaker kin ties with females. During the fission process, it appears that males experience more indecision than females do regarding which subgroup to join, moving frequently back and forth between the two groups during the initial stages of the fission process. We have seen several cases at Lomas in which young males initially joined their mother's subgroup but soon thereafter moved back into the subgroup having fewer closely related females.

Although we do not yet have complete kinship data for the Lomas population, we do know that monkeys tend to have large numbers of kin in adjacent groups. Typically, when a large group fissions, the two fission products remain neighbors with an overlap zone between their ranges, and because the largest groups are typically those with the longest term alpha males, this means that adjacent groups are heavily populated by daughters of the same alpha male. We have seen at least 15 cases in which the son of one group's alpha male became the alpha male of the adjacent group. Thus, males whose fathers were long-term alphas may have a higher probability of migrating to (and breeding in) a group containing paternal half sisters than males whose fathers were short-term alphas and hence produced fewer offspring.

## B. Causes of Mortality

White-faced capuchins can live up to 55 years in captivity (Hakeem et al., 1996), though we estimate that only about 15 of the 427 monkeys who have been members of the monitored groups at Lomas have thus far survived past age 30, and 10 of those died before reaching age 40. We estimate that only 5 of the 236 of the individuals currently living in the study groups at Lomas are over 30 years old and that the oldest is 37 years old; thus, only 2% of individuals are older than 30 years. All five of these elderly monkeys are females, and two of them seem to be postreproductive, not having produced infants in the past 7 years. However, these estimates of mortality beyond age 21 are not entirely accurate because we do not have precise birthdates prior to 1990. Ages of animals for whom we do not have birthdates are estimated using a combination of the following techniques: (a) comparing photos to those of known aged individuals and (b) making educated guesses of minimum age based on demographic assumptions regarding the ages of her progeny, assuming that females first give birth when they are 6 years and have 2-year interbirth intervals (IBIs) (e.g., a female who has four offspring in the group according to our genetic results, at the time the female's group is first studied, must be at least 12 years old). At Lomas, the annual mortality rates drop to below 5% per year by the second year of life and to less than 1% by age 9 years (Perry et al., 2012).

As in most studies of wild mammals, the cause of death is rarely observed directly. However, the Lomas Barbudal Monkey Project has accumulated a sample of 48 deaths for which the cause of mortality is known (28 cases in which the cause of death is known reliably, and 20 cases in which there is fairly solid circumstantial evidence for cause of death). The primary causes of death vary somewhat according to age–sex class, but it is clear that, in general, the two primary sources of mortality are humans and other capuchin monkeys. All adult female deaths (two confirmed and four suspected cases) were due to recreational hunting by humans. For adult males, the primary cause of death was fights with other male capuchin monkeys (seven confirmed and three suspected cases), but there were two deaths caused indirectly by human development (a case of electrocution on uninsulated electrical cables and a car accident during an intergroup encounter on a highway). Although one juvenile was eaten by a boa constrictor, the other four juvenile deaths were human related: one was hit by a car and the rest (two confirmed, one suspected) were victims of poachers.

At Lomas, approximately 26–30% of 262 infants died during their first year of life (Perry et al., 2012). Of these, we had data on the cause of death for 25 infants. Three infant deaths were due to congenital problems, and two were due to diseases. There were also two suspected cases of poaching and one suspected predation by a snake. However, infanticide by adult male capuchins appears to be the primary cause of infant death (see Sections III.B.4 and III.A.4), accounting for 17 cases, or 43–68% of infant deaths (depending on the quality of evidence demanded to attribute infanticide as the cause). We have directly observed six cases of infanticide and additionally observed five cases in which the alpha male stalked the infant before its death and/or the mother alarm called at the alpha male soon before or after the infant was killed (Manson et al., 2004; Perry and Manson, 2008; Perry et al., 2012). In addition, there have been six cases of infant death during periods in which the alpha male has killed other infants and/or the infant was seen wounded shortly before it died, during a period of recent alpha turnover.

## III. Life History Strategies

### A. Male Life History Strategies

The main challenges a male must overcome in order to reproduce successfully are (a) locating himself in a group that has as many non-kin females as possible who will breed with him, (b) staying alive during the dangerous migration process, (c) rising to alpha status so that he can breed, (d) defending his access to reproductive females, and (e) defending his offspring from infanticidal males and predators. With the possible exception of (a), having solid relationships with allies is helpful, if not critical, for accomplishing all of these goals. The primary cause of mortality for adult males is conflicts with other adult males, and infanticide is the primary cause of death for a male's offspring (see Sections II.B and III.A.4). Given the prominence of coalitionary aggression in this species, a male is far less likely to succeed reproductively if he competes on his own than if he recruits allies (S. Perry, unpublished data; see Section III.A.1). Therefore, males not only need fighting skills to compete but must also have the intelligence to manage their social relationships with their male allies so that they receive continued support. It seems likely that these social challenges have been a major selective force in creating the extraordinarily large relative brain sizes in this species (Perry and Manson, 2008). White-faced capuchin males employ a variety of strategies to increase their fitness, depending on their demographic circumstances.

*1. Dispersal: When and Where to Go, and with Whom?*

Males first migrate at a median age of 7 years (mean 7.3 years) at Lomas ($N=56$, range: 20 months to 12 years) (Table III). At Santa Rosa, the range of natal dispersal ages is approximately the same (19 months to 11 years); however, the mean age of first migration is much younger (4.5 years, $N=30$) if disappearances are assumed to all be dispersal events, and 5.5 years ($N=14$) if only confirmed dispersals are included in the sample (Jack et al., 2011). Both field sites have some males who remain in their natal groups at 12 or more years of age (Jack et al., 2011; S. Perry, unpublished data). It does not appear to be the case that young males are evicted from the group either at Lomas or at Santa Rosa (Jack and Fedigan, 2004a). However, takeover events, that is, changes in alpha male caused by an influx of migrants, are often associated with emigrations of natal males at both field sites. At Santa Rosa, natal males are 18.7 times as likely to disperse in the aftermath of a takeover event, holding other variables constant (natal male's age, number of maternal brothers, number of adult males, and group size) (Jack et al., 2011). A strictly comparable analysis has not yet been carried out for Lomas, but slightly less than half of all natal migrations occur in the wake of a takeover or the death of the migrants' father.

TABLE III
LIFE HISTORY PARAMETERS OF LOMAS BARBUDAL CAPUCHINS

| | Median | SD | Min. | Max. | $N$ |
|---|---|---|---|---|---|
| Age at first reproduction (female) | 6.05 years | 0.62 years | 5.50 years | 7.55 years | 30 |
| Age at first reproduction (male) | | | 7.72 years | | 5 |
| Dispersal age (males) | 7 years | 2.25 years | 20 months | 12 years | 56 |
| IBI when first infant lives | 767 days | 145 days | 381 days | 982 days | 41 |
| No. of offspring produced by females >20 years old | 8 | 2.12 | 5 | 13 | 18 |
| Tenure length (alpha male, uncensored intervals) | 14 days | 2.86 years | <1 day | 18 years | 86 |
| Maximum lifespan (female) | | | | 37 years | 41 |
| Maximum lifespan (male) | | | | 35 years | 57 |
| Infant mortality rate in first year of life | | | 26% | 30% | 262 |

As males reach dispersal age, they begin visiting neighboring groups, lurking on the periphery for a period of a few hours to a few days, and then return to the natal group (S. Perry, personal observation). Eventually, they make a more permanent transfer. At Santa Rosa, it has been found that males disperse to groups in which they will have a more favorable sex ratio of females to males, and in which they have higher dominance rank than they had in their previous groups (Jack and Fedigan, 2004b).

Males disperse multiple times during the course of their lifetime, often comigrating with other males from their group. At Lomas, 80% of males migrate with at least one other male, which is similar to the 74% comigration rate seen in males at Santa Rosa (Jack and Fedigan, 2004b). Males are slightly more prone to comigrate during their natal migration (100% of 50 natal migrations at Lomas [S. Perry, unpublished data] and 82% of 11 natal migrations at Santa Rosa [Jack and Fedigan, 2004a]). Males sometimes comigrate with the same males multiple times, both at Santa Rosa (Jack and Fedigan, 2004b) and at Lomas. By age 15, 10 out of 14 males followed from birth in the Lomas population were still coresiding with at least one male relative, despite having made multiple group transfers. Comigrants are often kin, but they do not always have to be kin. At Lomas Barbudal, we have observed at least 14 cases of comigration of brothers (Table IV). Using our kinship data for groups in which we have complete genealogical relationships for multiple generations (aside from the relatedness of the oldest females, for whom we assume zero relatedness, thereby probably reducing the relatedness of some dyads via the maternal line beyond its true value), it seems that, on average, the mean dyadic relatedness of a comigrant cluster is approximately at the level of half sibling ($r=0.24$) for those comigrant clusters that enter bisexual groups and $r=0.28$ for comigrants in all-male groups, or $r=0.25$ if you combine the two samples (Table IV). Most of these estimates are from natal migrations rather than secondary migrations, for which relatedness may be expected to be a bit lower; however, for our small sample of three secondary migrations for which genetic data are available, the $r$ value is the same ($r=0.24$, $N=3$) if only those dyads with known parentage are used. We can very tentatively assess the probable relatedness of comigrant clusters for whom we do not know natal groups, using estimates based on assumptions of the likelihood of encountering these combinations of alleles based on their frequency in the population (Queller and Goodnight, 1989): in three comigrant dyads arriving from outside our study groups, we obtained $r$ values of 0.67, 0.58, and 0.38 (mean 0.54), which tentatively suggest that these are at least half-sibling, or more likely full-sibling, dyads. In all three of these cases, the comigrants made multiple migrations together. Muniz (2008) conducted a preliminary analysis of the impact of kinship on comigration decisions using resampling techniques and found that in this

TABLE IV
KIN RELATIONSHIPS OF COMIGRANTS

| Number of males | Mean coefficient of relatedness | Number of primary kin dyads of each type[a] |
|---|---|---|
| *Comigrants into other groups* | | |
| 3 | 0.42 | 1 FS, 2 PHS |
| 2 | 0.16 | |
| 2 | 0.125 | |
| 5 | 0.16 | 6 PHS |
| 2 | 0.25 | 1 PHS |
| 5[b] | 0.24 | 2 FS |
| 3 | 0.29 | 3PHS |
| 1 | 0.125 | |
| 1 | 0.34 | 1 PHS |
| 5[b] | 0.20 | 1 FS, 2 PHS |
| 1 | 0.031 | 1 PHS |
| *All-male groups* | | |
| 8[b] | 0.18 | 1 FS, 2 MHS, 4 PHS |
| 4 | 0.21 | 2 MHS, 1 PHS |
| 2 | 0.28 | 1 PHS |
| 3 | 0.08 | 1 PHS |
| 2 | 0.50 | 1 FS |
| 3 | 0.42 | 1 PHS, 2 PO |

[a]FS, full sibling; MHS, maternal half sibling; PHS, paternal half sibling; PO, parent–offspring (i.e., father–son).

[b]All males in this comigration alliance are sons, grandsons, or great-grandsons of the same alpha male.

small sample of 13 migration events, the average dyadic relatedness of comigrant dyads was significantly higher than the average relatedness of randomly selected male–male dyads from the same group of origin.

Although no detailed quantitative analysis has yet been performed regarding the effects of comigration on male reproductive success, it is the case that all of the 10 well-documented successful alpha takeovers by nonresident males at Lomas were accomplished by comigrant teams rather than by single migrants (S. Perry, unpublished data). Thus, it seems likely that allies are a critical part of male reproductive strategies.

### 2. *All-Male Groups*

Of 24 males who have been the subjects of study at Lomas from the time of their birth through their 12th year of life, 14 males (58%) have spent some time in all-male groups. Although stable all-male groups of long duration have not been observed at Santa Rosa (K. Jack, personal communication), they are relatively common at Lomas. It is not yet clear whether this is a true difference between the sites, or whether there has just been less emphasis on documenting the activities of extragroup males at Santa Rosa. All-male groups are extremely difficult to find and monitor, at least when the males are habituated and therefore refrain from alarm calling at humans, because they are small groups composed of individuals who rarely vocalize, and they have larger home ranges than do multi-male, multi-female groups. All-male groups at Lomas vary in size from 2 to 8 males, and the ages of these males typically vary from 5 to 16 years of age. Most of these all-male groups seem to be in search of a group to migrate to, but they do go through periods when they do not seem to be making serious attempts to join other groups and, in fact, avoid other groups. It is difficult to know the precise duration of these groups, because typically there are periods both before the first sighting and after the last sighting of the group during which the members of that group are not seen as members of another group. However, one large all-male group at Lomas ("Lost Boys," or "LB," which numbered eight males at its peak size) was known to exist for 20 months before the group disappeared from the study area in May 2011. Other all-male groups that were not as continuously monitored have existed for multiple months as well.

Males in all-male groups have a different life style than males who are resident in groups with females. The daily routine involves far less conflict. Males who normally pick fights regularly in the presence of females are far calmer when they are resident in an all-male group. The activity budget of males in an all-male group involves more foraging and resting and less social interaction than is the case for males in a bisexual group. There seems to be little if any striving for dominance, and males groom and huddle together during rest periods. These long periods of calm are punctuated by brief periods of intense chaos when the all-male group encounters a group with females in it. Sometimes, the all-male group silently flees the bisexual group, but more often the males of the all-male group try to interact with members of the other group. Males who are resident in groups with females are almost invariably hostile to all-male groups, and thus encounters between bisexual groups and all-male groups often result in wounding of males. In 20 months of visiting other groups, the members of LB only succeeded in invading two groups, and their success was short lived on both occasions (lasting just 1 day in one instance, and less than a month in the other).

It is still not known what the typical ranging patterns are for an all-male group, because only one such group (LB at Lomas) has been monitored continuously for long periods of time. LB group had a home range that encompassed or at least overlapped with the home ranges of at least 17 other groups, whereas a typical multi-male, multi-female group has just 3–6 neighboring groups with whom its range overlaps. After males from LB suffered severe wounds from encounters with other groups, they could commonly be found in one of two core areas, where they seemed to have lower rates of intergroup encounters.

### 3. *Male Reproductive Skew*

Reproductive skew is high in the Lomas population, with 31% of adult males siring all of the offspring and alpha males siring 74% of offspring ($N=188$ offspring genotyped from nine groups) (Perry et al., 2011). An alpha male's ability to monopolize reproduction shifts over the course of his tenure as alpha male. Early in his tenure, an alpha male sires almost all offspring produced. However, capuchins appear to have a strong aversion to father–daughter inbreeding (Godoy, 2010; Muniz et al., 2006; Perry et al., 2012): although alpha males sire 92% of offspring produced with unrelated females, they sire only 6% of offspring produced with their female descendants. Therefore, as an alpha male's tenure extends to the point when his female descendants form a high proportion of the reproductive aged females, reproductive skew diminishes (Muniz et al., 2010). Adult female descendants of the alpha male breed primarily with immigrant males, but they also breed with paternal half siblings and other more distantly related natal males (Godoy, 2010). Male reproductive skew is most significantly affected by the kin relationship between alpha males and adult females, but skew is also somewhat reduced when there are larger numbers of adult males and adult females in the group (Muniz et al., 2010). Less is currently known about the impact of dominance rank on male reproductive success in the Santa Rosa population, because paternity data are currently available for only 17 infants from two social groups; however, alpha males sired 14 of these 17 infants (Jack and Fedigan, 2006), which is consistent with the Lomas results.

### 4. *Variation in Alpha Male Tenure Length and the Consequences of Alpha Male Replacements*

Bisexual groups go through phases of stability, when the same male is in the alpha male position for up to 18 years (i.e., three generations), and phases of extreme chaos, when the group can have a different alpha male every few days for a couple of years before one really seizes firm control (Muniz et al., 2010; Perry and Manson, 2008). The median alpha male tenure length is 2 weeks, with a few long tenures bringing the average up to 0.98 years ($N=86$ completed tenures) (Perry et al., 2011; S. Perry, I. Godoy and W. Lammers, in preparation). Both at Santa Rosa and at

Lomas, it is obvious that demographic turnover in the form of alpha male replacements is a key factor affecting virtually every other aspect of animals' social strategies, because of the high rates of infanticide that occur in the wake of such turnovers. Infanticide is a behavior that happens rarely and quickly, and often at night, which means that both luck and extreme dedication on the part of researchers are necessary to document the cause of death for even a small proportion of infant disappearances. Nonetheless, several such events have been observed, and the general demographic circumstances surrounding infant disappearances have been well documented at both sites. At Santa Rosa, using a sample of 87 infants for whom birth and death dates were reliably known, infants were far more prone to disappear in the wake of alpha male replacements than during other times (Fedigan, 2003). During years following takeovers, 82% of infants died, whereas only 12% of infants died during peaceful years (Fedigan, 2003). In a larger demographic sample from Santa Rosa ($N=95$ infants), 59% of all infant deaths occurred in the context of alpha male replacement (Fedigan et al., 2008). At Lomas, using a sample of 210 infants, 49% died in the wake of takeovers and only 18% died in peaceful periods (Perry et al., 2012). Although replacement of the alpha male had a statistically significant impact on infant mortality rates, migration-related changes in male group membership that did not involve alpha male rank changes had no significant impact on infant mortality rates (Perry et al., 2012).

The available evidence is consistent with the idea that infanticide is a sexually selected male reproductive strategy. At both field sites, females who have lost an infant, either due to infanticide or due to some other cause, conceive significantly sooner than females whose infants survive (Fedigan, 2003; Fedigan et al., 2008; Perry and Manson, 2008; Perry et al., 2012). At Lomas, the only males observed to kill infants are alpha males. Alpha males sometimes kill infants even when they have been members of the group long before attaining alpha status, which implies that long-term familiarity between an alpha male and the females is not sufficient to protect the females' infants from infanticide. Relations between subordinate adult males and infants are typically amiable. Thus far, there have been no cases in which alpha males killed their own infants, but we only have a sample of five cases for which we (a) know both the identity of the killer and (b) have established via genetic analysis the identity of the victim's father (Perry and Manson, 2008; S. Perry and L. Muniz, unpublished data). Females do generally seem receptive to mating with the killer of their infants (Perry and Manson, 2008). Although alpha males can have extraordinarily long tenures in this species, the median alpha male tenure is quite short (with 78% of alpha tenures lasting less than a year), and committing infanticide significantly shortens the victim's mother's time to conception (Perry et al., 2012). Thus, selective pressure is expected to cause males to commit infanticide in order to avoid losing what may be their only chance at reproduction.

The amount of time that an alpha male has maintained the alpha position also has profound effects on the breeding prospects for subordinate males, and hence their probable willingness to help the alpha male defend the group from external males. Because alpha males do almost all of the breeding during the first 6 years of their tenures, subordinates will have better breeding opportunities if they are in a group with an alpha male who has already been in place for many years and has mature daughters or granddaughters in the group. The presence of subordinate males who can help with group defense is useful to the alpha male, particularly when there are enough females in his group to provide a strong temptation for foreign males to invade. At Santa Rosa, the threat of group takeover is lower in groups that have more adult males and/or a high ratio of males to females (Jack and Fedigan, 2004b, 2011).

### 5. *Conclusions*

There is high variance in male reproductive success, with most males leaving few or no offspring and a very small number of males having spectacular reproductive success. Alpha males sire most of the offspring, particularly early in their tenures as alpha. Those infants sired by non-alpha males are generally the offspring of females who are directly descended from the alpha male. Alliances with other males seem critical to attaining and retaining access to reproductive females: males who comigrate are more likely to attain alpha status, and allies provide aid in intergroup encounters or encounters with potential male invaders. Conflicts with other males are the primary source of adult male mortality, and infanticide is a commonly practiced alpha male reproductive strategy. Males frequently comigrate with males who are often closely related to them.

## B. Female Life History Strategies

### 1. *Dispersal*

Females are the more permanent group members in this species, typically remaining with the same female kin their entire lives, unless their group fissions. None of the 90 adult females monitored in the Lomas population have ever been seen to disperse. (Note that there is an error in an earlier paper from Lomas stating that there was one female immigration [Manson et al., 1999]: subsequent genetic testing proved this suspected immigrant female, JJ, to be identical to an infant born into that group.) However, 5 of the 36 adult females that are part of the Santa Rosa study were observed to disperse, and in 4 of these cases, it was known that the females were dispersing during an alpha male takeover of their group (Jack and Fedigan, 2009). Three of these females dispersed in the company of juveniles and/or their defeated alpha male, and two dispersed alone.

### 2. *Timing of Reproduction*

Females reproduce for the first time at a mean age of 6.2 years (SD = 0.58), with the earliest known first parturition occurring at age 5.5 years at Lomas Barbudal (Perry et al., 2012); the age of first reproduction is quite similar at Santa Rosa (6.5 years: Fedigan and Jack, 2012). Females typically give birth to singletons; although twinning has been observed on one occasion at Lomas (Perry and Manson, 2008), it caused obvious physiological stress for the mother, and the twins did not survive their first year of life. Once they have begun reproducing, females give birth approximately once every 2 years until they are in their 30s, at which point reproductive rate slows or reproduction ceases altogether. Of nine Lomas females estimated to be above 30 years of age, three are still reproducing at a normal rate, two had postreproductive phases of 4 and 5 years at the time of their deaths, one was still apparently nulliparous, and the remaining three have not reproduced for 3, 7.3, and 7.4 years, respectively. Thus far, the maximum number of offspring produced by a female at Lomas is 13, of which 8 (6 females and 2 males) survived to reproductive age. Although white-faced capuchins are capable of living up to 55 years in captivity (Hakeem et al., 1996), the maximum estimated age of a wild capuchin at Lomas is 37 years. In a sample of eight Lomas females who have lived at least 20 years and are dead or postreproductive, the mean number of offspring produced is $8.0 \pm 2.7$ and the mean number of offspring surviving to age 5 years is $5.9 \pm 1.4$.

### 3. *Infant Development During IBIs*

Mothers carry their infants about 75% of the time during the first 3 months of life (see Fig. 1 for an example of how the rate of maternal carrying declined over the course of the first 2 years of life for a male and a female infant). The rate of maternal carrying declines rapidly during months 4–6 when infants start to become more independent and alloparents become more active in carrying infants. By the time they are 6 months old, the infants are moving independently almost all the time and are spending about 10% of their time alone, that is, with no other monkey within 10 body lengths of them. The amount of time they spend alone increases steadily for infants of both sexes up until the end of the second year of life (Fig. 1), which is when the infant's next sibling typically is born. Infants nurse until

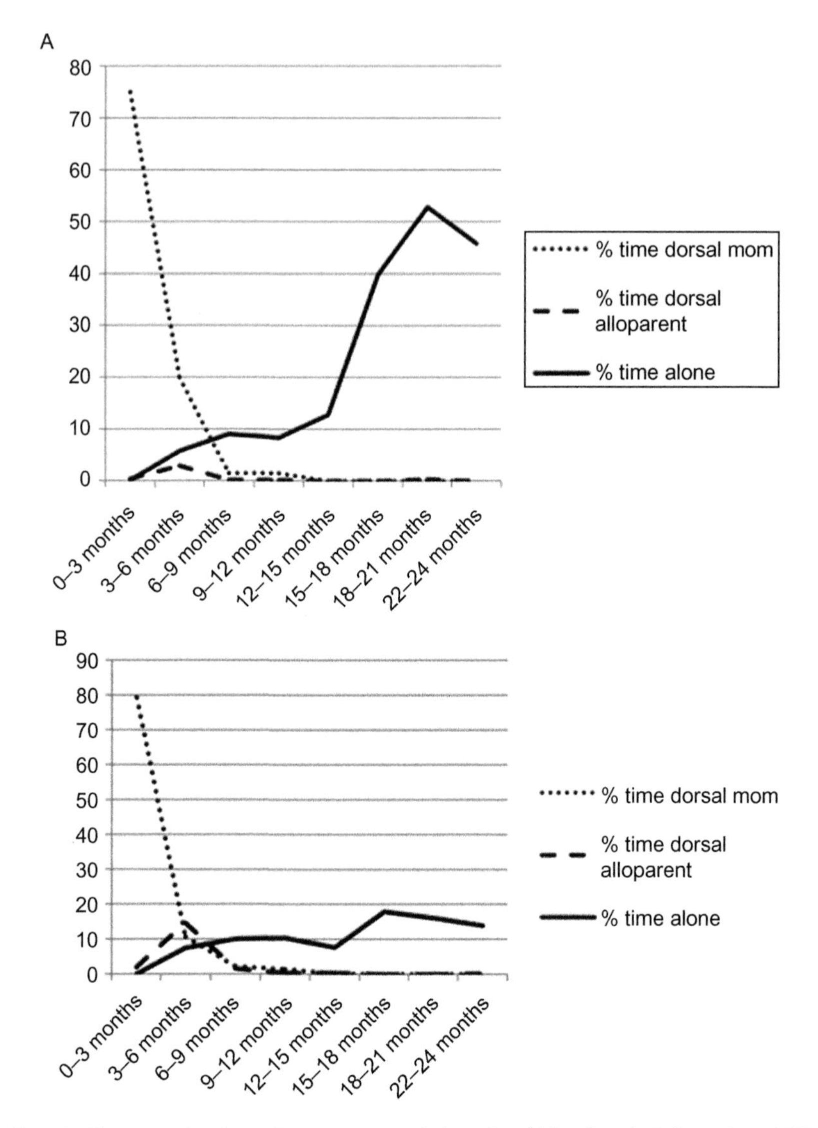


FIG. 1. These graphs show the percentage of time that (A) a female infant of a middle-ranking mother and (B) a male infant of a middle-ranking mother were carried by their mothers (dotted lines) and alloparents (dashed lines) over the course of the first 2 years of life. Solid lines demonstrate the percentage of time they spent alone, that is, outside 10 body lengths of other monkeys.

the next infant is born, with nursing rates declining steadily throughout the 2-year period (see nursing profile for a female infant from a middle-ranking mother in Fig. 2). One female at Lomas managed to conceive her next offspring when the first infant was only 7.4 months old and still nursing regularly. The mean IBI is 763 days at Lomas (S. Perry, unpublished data), and ~821 days at Santa Rosa (Fedigan and Jack, 2012), if we include only those intervals for which the first infant lives.

### 4. *Factors Affecting Length of IBIs*

Although female dominance rank appears not to exert a significant influence on infant survival or on the length of IBIs in this species, at least at Santa Rosa, females who have a large number of female matrilineal kin (mother, grandmother, maternal siblings, and offspring) have significantly shorter IBIs (Fedigan et al., 2008). Given the importance of female kin for enhancing fitness, it is perhaps not surprising that female dispersal is so rare. The chief determinant of IBI length is survival of the previous infant, which is influenced by stability of the alpha male position (Fedigan et al., 2008; Perry et al., 2012). Rainfall is also a significant factor, with heavier rain producing shorter IBIs during the following year (Fedigan et al., 2008). Using a larger data set (park-wide censuses over a 25-year period, where female reproductive success is estimated by calculating the difference between observed and expected ratios of immatures to adult females), the Santa Rosa team determined that female reproductive success is higher in

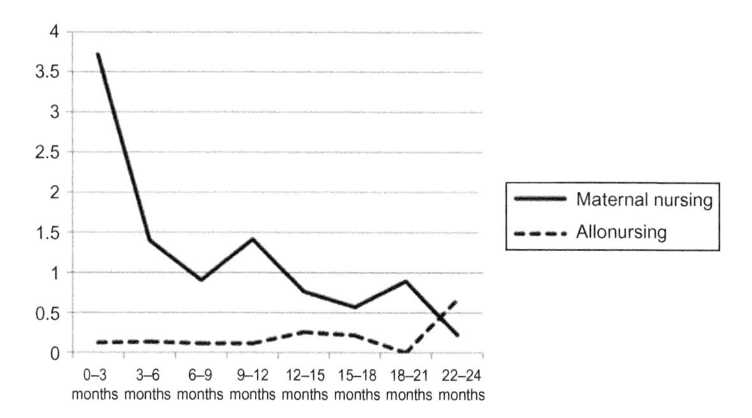


FIG. 2. This figure shows the number of seconds/hour that an infant female of a middle-ranking mother spent nursing from her mother (solid lines) and from allomothers (dashed lines) over the course of the first 2 years of life.

groups that have larger numbers of adult males (Fedigan and Jack, 2011). This presumably means that the advantage of having more males to defend the group from takeover is greater than the disadvantage of having more feeding competition in larger groups. Given the prevalence of alloparenting in this species (see Section III.B.4), it would be worthwhile to investigate the effects that different levels of alloparenting have on mothers' IBIs; however, such analyses have not yet been completed at either site.

### C. Life History, Demography, and Selection Pressures on Social Strategies

It should be clear from the evidence presented above that the timing and frequency of alpha male takeovers have a profound impact on many aspects of male and female reproductive strategies and on the demographic composition of social groups. The available data indicate that infanticide is the most frequent cause of infant mortality and that the stability of the alpha male position is the most important determinant of female reproductive success. Because male rank and kinship with the group's females are the two most important factors affecting male reproductive success, the length of alpha male tenures is a key component of male strategies as well: if a male cannot succeed in becoming alpha, his next best bet is to be a subordinate male in a group that has a long-term alpha with adult female descendants whom he can breed with. Being a subordinate brother of an alpha male is likely to be a particularly good strategy, due to the inclusive fitness benefits accruing to subordinates during the early years when they are not yet breeding themselves. Because the current evidence indicates that inbreeding aversion is present between fathers and their daughters or granddaughters, but inbreeding avoidance seems weaker or perhaps even nonexistent between females and other types of male relatives such as paternal brothers, cousins, and uncles (Godoy, 2010; Perry et al., 2012), subordinate brothers of alphas are likely to have breeding opportunities once the daughters of their alpha brothers have reached sexual maturity. The issue of whether males form more stable and cooperative alliances with their male kin than with non-kin males, as might be predicted from kin selection theory, is an empirical question yet to be resolved in capuchins. Capuchins' social structure is fairly similar to that of lions, in which males primarily form comigration alliances with male kin (Packer and Pusey, 1982). However, it seems that within a male lion alliance, kinship does not exert a consistent effect on cooperative behavior (Packer and Pusey, 1982); Grinnell et al. (1995) suggest that male cooperation in defense of the pride against intruder males is not conditional on kinship but is based on mutualism. However, in red howler monkeys, Pope (1990) demonstrated that alliances consisting of related or potentially related males lasted longer and exhibited less rank instability than did alliances composed of unrelated males. This topic is currently under investigation at Lomas.

Although infanticide is the most striking consequence of alpha male replacements, the consequences of alpha male replacements extend beyond the immediate issue of infant survival. The type of rearing environment available in a stable group is quite different from the type of rearing environment experienced by a monkey in a group that suffers frequent alpha male turnover. In a group with a stable alpha male, there are far more immature animals who survive, which means that there will be far more full siblings and paternal half siblings available as social partners. Although this has not yet been quantified at either site, it seems likely that having a large number of surviving siblings and cohort mates will mean that an immature monkey will have far more play experience during early development, and this is likely to impact social and fighting skills, and hence breeding success as adults. Although we do not yet understand the causal mechanism, it is intriguing to note that preliminary results from Lomas indicate that 63% of 27 males who lived in demographically stable groups as juveniles became alpha males after migrating, whereas only 11% of 18 males whose groups exhibited alpha male turnover in the first 5 years of their lives became alpha males (Fisher's exact test, $P=0.002$). It is not yet clear whether males natal to stable groups do better because they have (a) a superior level of social experience (e.g., play opportunities), (b) a wider range of compatible social partners with whom they can engage in cooperative activities, or (c) less exposure to debilitating levels of stress during early development. Males from the Lomas population who spent their first 5 years of life in a stable group with a long-term alpha male migrated significantly later than males whose groups exhibited alpha male turnover (stable groups: $8.27 \pm 0.43$ years, $N=26$; unstable groups: $6.41 \pm 0.34$ years, $N=28$, $P=0.008$). This result mirrors findings from Santa Rosa (Jack et al., 2011) that males disperse later when their putative fathers remain in the group longer (though this factor explained only 15% of the variance in age of dispersal and was nonsignificant when one outlier was removed). Ongoing research by the Lomas research group will explore the many ways in which demographic conditions, personality, and various types of early experience affect various components of individuals' life history strategies. The experience of living in a group with frequent alpha turnover and male migration is quite different from being in a more stable group and might be expected to require different social strategies. Whether or not we should expect environmentally contingent developmental strategies according to the stability of the early rearing environment depends on past selection history, that is, the degree to which both stable (i.e., long-term alpha) and unstable (i.e., short-term alpha) conditions were common features of the environment in which the species evolved.

Only six alpha males at Lomas (out of 42 males, in 11 groups) have had long enough tenures to be in a position to breed with their daughters, and we lack paternity data from one of those groups. On the surface, it would seem that these long-tenured males are just outliers, who probably have little impact on the evolutionary history of the species. However, it is important to note that (a) these males have sired approximately one-third of all the offspring produced in the past two decades, from the nine breeding groups we have genotyped, and (b) these long-term alphas have provided stable rearing environments for many immature monkeys, by protecting them from infanticide and enabling them to have large numbers of young kin to play with and (eventually, for males) comigrate with. In other words, it is important to think not only of the consequences of these long-term alpha males' behavior for their own reproductive success but also about the impact of their strategies on the lives of other members of the population: short-term alphas have less time to impact social dynamics of other group members than do long-term alphas. Even if very few males are extraordinarily successful, as long as there are a few of them in each population, they will have a large impact on the selective environment for other individuals in the population. One way of thinking about their probable impact is to document, for each year of life for each monkey in the population, whether it was coresident with a short- or a long-term alpha male (i.e., to count the number of "monkey-years" in each condition). This helps us estimate the probability that an individual will experience each of these conditions over the course of its lifespan. In a sample of 2792 "monkey-years" from monkeys of all ages in 10 social groups at Lomas, we find that 26% of monkeys' time is spent residing in a group that has already had a stable alpha male in place for more than 6 years. Looking just at completed tenures, we find that 63% of monkey-years are spent residing in groups that have alpha males who end up with tenures exceeding 6 years, and 37% of monkey-years are spent in groups that have short-term alpha males. If these conditions are typical of the species generally, then long alpha tenures are likely to have a huge impact on the evolution of the behavioral biology of the species, even if very few individual males attain these long alpha male tenures, because of the impact long tenures have on the social strategies of other members of their groups.

It is always difficult to know how much to generalize from one study population to the rest of the species and how much to infer about the evolutionary history of the species from the demographics of single studies. Although results from Lomas and Santa Rosa are generally convergent, there are some interesting differences between the two sites with regard to the age of male dispersal, the prevalence of internal (vs. external) takeovers of the alpha position, and the prevalence of long alpha male tenures. Thus far, our knowledge of the overall kinship structure comes primarily from Lomas, because genetic data from a large sample of the Santa Rosa monkeys are not yet available, and the results from Lomas are likely to differ from the Santa Rosa, due to the larger number of long-term alphas at Lomas. The picture that emerges from the Lomas population regarding the kinship structure is fascinating because it is so rare among mammals for alpha males (i.e., the primary breeding males) to be in coresidence with their daughters and granddaughters once they are reproductively mature (Clutton-Brock, 1989). High reproductive skew (Muniz et al., 2006, 2010), combined with extraordinarily long alpha male tenures of up to 18 years (Perry et al., 2012), produces a complex kinship structure yielding far more paternal half- and full-sibling dyads than are typical among primate societies, and an unusually high degree of relatedness within groups (Perry et al., 2008). These demographic features, if typical throughout evolutionary time, might be expected to shape the psychology of the species. For example, one might expect to see more fine-tuned kin recognition mechanisms for recognizing paternal kinship in a species that had frequent opportunities to choose between paternal kin and paternal non-kin as alliance partners or breeding partners. There is much work yet to be done in this area, but thus far, the evidence from Lomas for paternal kin recognition is mixed: daughters and fathers discriminate against one another in a mating context (Muniz et al., 2006), but females do not discriminate in favor of paternal half siblings when selecting grooming partners (Perry et al., 2008).

## IV. Social Relationships

### A. Male–Male Relationships

#### 1. *Costs and Benefits of Male–Male Relationships*

Relative to females, males spend very little time in close proximity, grooming one another and engaging in relaxed affiliative interactions (Perry, 1996a). However, they do rest in contact, groom, play, and exhibit quirky intimate rituals (Perry, 1998b; Perry et al., 2003a,b) with other males at low rates. The alpha male occupies a quite distinct role from the other males. He is more central than the subordinate males with respect to association patterns and engages in higher rates of urine rubbing and display behaviors (Campos et al., 2007; Perry, 1998a,b). Subordinate males are quick to provide aid to the alpha male, even though they are often the targets of mild aggression from the alpha. The alpha male is typically inconsistent with respect to which male he supports in within-group conflicts between males (Perry, 1998b). Coalitions of subordinate males against the alpha male were never seen in the early years of the Lomas study (Perry, 1998b) and have been seen only very rarely in later years of the project.

Males assist one another in ~63–67% of all requests for coalitionary aid against coresident males and 51–65% of requests for aid against heterospecifics (typically predators) (Perry, 1998b). (See Section V.A for a description of signals used to request coalitionary aid.) Males are far more dependable as allies when the opponents are extragroup males. In all but 1 of 44 cases from Lomas (AA group) in which at least one male participated in intergroup aggression, he received assistance from at least one other male from his group (Perry, 1996b), and in all cases in which one male solicited aid from another against a foreign male, he received it (Perry, 1998b). Although precisely comparable statistics are not available from other groups, this rate of cooperation in intergroup encounters seems fairly typical of the Lomas population in general. Even in cases when two coresident males have been regularly fighting to the point of inflicting severe wounds on one another, they will cooperate in intergroup encounters against foreign males, stacking in the characteristic "overlord" behavior, which puts the bottom male in a vulnerable position relative to the male on top (see Fig. 3). It seems clear that external threats such as members of

Fig. 3. Two sisters form an "overlord" against a third female (photo: Susan Perry). (For color version of this figure, the reader is referred to the online version of this chapter.)

other groups or other species are perceived as a potential focal point of solidarity: redirected aggression against non-monkeys is extremely common in tense social situations such as feeding competition, and two competitors will often cease hostilities when one solicits the other's aid against a third party.

In order to survive and attain high reproductive success, males must effectively manage their social relationships and form solid alliances. Males' relationships with other males are characterized by less frequent interaction and greater tension than females' relationships with other females. Male–male relationships involve quite high stakes and are often characterized by intimate and bizarre rituals that seem to function as Zahavian bond tests (Perry and Manson, 2008; Perry et al., 2003a; see Section V.B for further discussion). Because of the high degree of reproductive skew and the high mortality risk involved in male–male competition, males need one another's aid in order to successfully compete for reproductive opportunities. Allies are essential in order to reduce mortality risk during the dispersal process, and also to attain breeding positions, particularly when males are attempting to invade a group from the outside to attain the alpha position (see Section III.A.1). Once a male has succeeded in obtaining an alpha position, he still needs allies to keep reproductive rivals from invading and/or usurping his position. New alpha males typically commit infanticide, so loss of the alpha rank means not only a reduction in future breeding opportunities but also reduced survival of existing offspring. By having more subordinate males in the group, the alpha male has a larger number of allies in intergroup encounters, thereby reducing the risk of takeover from the outside; however, he increases the chance of being usurped in rank by one of his coresident males and also has some (minimal) risk of subordinate males inseminating females.

*2. Adult Males' Relationships with Immature Males*

Both adult and juvenile males show a strong interest in infant males that begins as early as the infant's first month of life, and males are enthusiastic alloparents of male infants (Perry and Manson, 2008). At Lomas, in a developmental study of six male and seven female infants over the first year of life, males showed a threefold preference for grooming male rather than female infants as early as 3 months of age. Similarly, in a study of the impact of infant presence on male–female relationships at Santa Rosa ($N=4$ mothers of male infants and 4 mothers of female infants), it was found that males increased their affiliation with mothers of male infants while decreasing their affiliation with mothers of female infants (Sheller et al., 2009). Subordinate adult males are typically highly tolerant of infants and juveniles, actively playing with juveniles and giving infants rides.

Although infanticide is the most common source of infant mortality, the danger of infanticide is limited to quite specific circumstances: at Lomas, in all cases for which we know both the identity of the killer and the genetic relationship between killer and victim, the killer was the alpha male and the infant was not the offspring of the killer. Alpha male fathers and subordinate non-fathers are highly tolerant of young males and take an active role in developing affiliative relationships with them. Subordinate males are very rarely the fathers of the young males with whom they are nurturing these relationships (S. Perry, L. Muniz, and I. Godoy, unpublished data); so in most cases, these relationships cannot be viewed as a form of paternal care. More likely, they are establishing relationships with young males who can later become useful allies during their attempts to acquire and defend breeding positions. Indeed, young males often choose to comigrate with their older male playmates rather than stay with their mothers when groups are disrupted by fission or changes in alpha male. Although very young males are probably not particularly effective supporters in coalitionary aggression, they do eventually grow up and become helpful in intergroup encounters.

## B. Female–Female Relationships

### 1. *Dominance Interactions*

Females exhibit relatively stable linear dominance hierarchies, as defined by spontaneous submissive behaviors in dyadic contexts (Bergstrom and Fedigan, 2010; Perry, 1996a). Female rank reversals are rare, and an adult female can expect to be challenged approximately once every 2.3 years (Santa Rosa: Bergstrom and Fedigan, 2010). Although females tend to support their closer kin against less closely related individuals (Perry et al., 2008), white-faced capuchins do not follow the strict "youngest ascendancy" rule of rank acquisition seen in most cercopithecines, in which a female's youngest daughter assumes the rank immediately below that of her mother, with her siblings occupying the ranks right below her according to their ages (Bergstrom and Fedigan, 2010; Perry et al., 2008). Females tended to deviate from predictions of the "youngest ascendancy" model by about 1.5 ranks over a 10-year period in AA group at Lomas during which the number of adult females increased from 5 to 10 (Perry et al., 2008). In a sample of 22 sister pairs from 5 social groups at Lomas, there were 18 cases in which the eldest sister was dominant at least initially (reversing ranks permanently in 6 cases later in life) and 4 cases in which the youngest sister was dominant when she first climbed the hierarchy but reversed ranks later in life (Perry et al., 2008; S. Perry, unpublished data).

In this same data set, most changes in dominance rank were due to the physical maturation of the alpha female's daughters and granddaughters, who assumed ranks higher than some of the subordinate females in the group. However, two adult females dropped below the ranks of their own daughters. A study of female dominance ranks at Santa Rosa (including 33 females from 3 groups, 9 of which were included in the rank acquisition analysis) demonstrated that there was a high correlation between early adult rank and later adult rank, and that older sisters are dominant to younger sisters ($N=8$), with the exception of one dyad in 1 year (Bergstrom and Fedigan, 2010). Because mothers were present in only 33% of these cases, they were not available to help their youngest daughters defeat their older daughters in the dominance hierarchy. Thus, lack of support is a possible reason for why young daughters failed to ascend above their older sisters (Bergstrom and Fedigan, 2010). In 21 of 22 cases at Lomas, however, the mother was still alive when the younger sister acquired her initial adult rank, and yet 82% of oldest sisters retained their dominance over younger adult sisters (S. Perry, unpublished data).

### 2. *Grooming and Association Patterns*

At both Lomas Barbudal and Santa Rosa, female–female dyads groom at significantly higher rates than do male–female dyads, and male–female dyads groom significantly more than do male–male dyads (Manson et al., 1999). As is the case for many species of cercopithecines, females with newborn infants receive more grooming from females than do females without newborn infants (Manson et al., 1999).

Females structure their social networks according to dominance rank and kinship. A 10-year study of a single social group at Lomas demonstrated that the relative effects of these two variables vary according to the number of adult females in the group (Perry et al., 2008). When group size is small, so that all females are closely related and there is very little variation in the coefficient of relatedness among dyads, kinship does not significantly influence association and grooming patterns. However, rank distance does: more closely ranked individuals spend more time together and groom more (Perry et al., 2008). When group size becomes larger, there is greater variation in coefficients of relatedness among female dyads. Under these conditions, dominance rank is no longer a significant variable, but kinship is: more closely related dyads spend more time together and groom more (Perry et al., 2008). A more extensive analysis of proximity patterns at Lomas Barbudal (Kajokaite, 2011), incorporating data from five social groups over a 5-year period, demonstrated that both kinship (coefficient of relatedness) and rank difference strongly influenced proximity scores, with kinship exerting a far stronger effect than rank. As predicted, females associated preferentially with more closely ranked individuals and closer relatives. In this study, there was no effect of age (i.e., individuals more closely aligned in age did not preferentially associate). In this same study (Kajokaite, 2011, using a partner stability index similar to that described in Silk et al., 2006b), it was found that 12 of the 18 females for which there were at least 4 years of data available were consistent in their top three partner choices, and all but one had at least one long-term bond, that is, at least one female who was consistently among her top three associates at a degree higher than chance over a 4- to 5-year period. Long-term bonds were formed by all full-sibling ($N=8$) and maternal half-sibling dyads ($N=2$) and by 38% of mother–daughter ($N=16$), 27% of aunt–niece ($N=11$), and 40% of paternal half-sibling ($N=5$) dyads, whereas only 25% of the more distantly related dyads ($N=16$) formed long-term bonds (Kajokaite, 2011).

Grooming tends to be directed up the hierarchy at Lomas Barbudal (Manson et al., 1999; Perry, 1996a), but this pattern is less consistent among the groups studied at Santa Rosa. The one group–period at Santa Rosa that exhibited a trend toward grooming down the hierarchy was in the middle of a dominance struggle involving the two highest ranked females; all of the dyads that exhibited down-hierarchy grooming included one of these females that was vying for alpha status, so it may be possible that the alpha and beta females were competing for social support by grooming more than usual (Manson et al., 1999). Indeed, grooming rates tend to be correlated with coalitionary support among females of this species (Perry, 1996a).

### 3. *Coalitions*

Most of the detailed analyses available regarding female coalition formation came from a 2-year study of a single social group, AA group from Lomas (Perry, 1995, 1996a, 2003; Perry et al., 2004, 2008). Females regularly formed coalitions with one another against a variety of targets, males being the most common target (Perry, 1996a). When females formed coalitions against other females, they never formed "revolutionary" coalitions of two subordinates against a dominant. Ninety-six percent of female–female coalitions against females were against a target who was subordinate to both coalition partners, and the remaining 4% were against targets intermediate in rank to the two partners (Perry, 1996a). A quarter of these coalitions against females occurred in the context of feeding competition, a third of these coalitions happened during completely unprovoked fights, and the remainder were in a variety of social situations. The unprovoked fights were typically between a young female who was ascending the dominance hierarchy and the female who was adjacently ranked in the hierarchy; thus, unprovoked fights may serve to establish or maintain dominance rank (Perry, 1996a).

Higher-ranking females are more likely to form coalitions against either sex than are lower-ranking females (Perry, 1996a). Dyads that groom more are also more likely to form coalitions against either sex; this result remains significant after controlling for dominance rank for female–female coalitions against females, but not against males (Perry, 1996a). In a study of the effects of matrilineal kinship on coalition formation among triads of females in AA group over a 10-year period (Perry et al., 2008), it was found that females were completely consistent in their choice of whom to support within triads, and that they supported the more closely related female in 17 of 18 triads.

### 4. *Alloparental Care*

Females provide care for one another's infants in the form of allonursing and allocarrying (Manson, 1999; Perry, 1996a; Perry and Manson, 2008). When mothers are still living and actively engaged in mothering, this care occurs at low levels (e.g., allonursing only 1–5 times per 100 h), but when mothers are lost for prolonged periods or die, alloparental care increases dramatically and adoptions may even occur (Perry, 1996a; Perry and Manson, 2008). For example, an infant whose mother was killed by a poacher when he was aged 8 months survived till age 3 years (when he was also killed by a poacher), because he received extensive nursing and carrying from a wide variety of alloparents (Perry and Manson, 2008). Also, a 6-month infant whose mother died in November 2011 has thus far survived over 2 months with alloparental nursing and carrying from adult males and females as well as juveniles and seems to be in excellent physical condition (S. Perry, unpublished data). In another case, an adolescent first-time mother abandoned her infant, leaving it with an older niece who was a more experienced mother and had an infant born the same week. The allomother successfully reared both infants to age 5 months, when her own infant vanished, and continued rearing the adopted infant to age 14 months, when it too vanished for unknown reasons (both infants were in excellent health when they disappeared). Although these adoption cases are rather extreme, it is likely that allonursing is crucial for infant survival in many cases, because infants are still dependent on milk long past the age at which are frequently carried by the mother. Because group members are highly dispersed while foraging, with infants traveling together as a pack in the center of the group and females foraging on the periphery, it is common for low-ranking mothers to become separated from the group for periods of many hours or even days, on occasion. Thus, infants often need to request milk from non-mothers, when they have difficulty in locating the mother.

The Santa Rosa researchers have discovered that females who have more primary female kin in the group have shorter IBIs, and they speculate that this relationship may be due to a greater availability of allomothers (Fedigan et al., 2008). Certainly, this topic should be explored in more detail, as the precise relationship between kinship and alloparental care frequencies has not yet been examined in this species.

In a study of allonursing in a single group (AA, from Lomas) over a 2-year period (Perry, 1996a), all females nursed infants that were not their own, and higher-ranking females allonursed less often than lower-ranking females. However, allonursing was not strictly parasitic, as infants did not nurse preferentially from females who were lower ranking than their mothers. There was no evidence of reciprocity between allomothers, which is perhaps not surprising, since females rarely witness other females nursing their infants and therefore cannot accurately track the distribution of social favors.

### 5. *Conclusions*

The general picture of female–female relationships that emerges is that females structure their relationships on the basis of kinship and rank, but not age. Female's relationships with one another are remarkably stable over time, and they work to maintain stability in their social groups, forming coalitions to reinforce existing dominance relationships both within the female hierarchy and among males, by supporting the alpha male. Whenever possible, females prefer to interact with matrilineal kin, though they seem not to discriminate in favor of patrilineal kin (Perry et al., 2008). Females help one another in successfully rearing infants, both by alloparenting and by forming coalitions against males in defense of their offspring.

## C. Male–Female Relationships

Adult males weigh about 45% more than adult females do (Ford, 1994; Mitani et al., 1996), and males' maximum canine tooth height is 60% higher than females' (Masterson, 2003). Whereas males often inflict deep wounds on their opponents with their canine teeth, wounds inflicted by females' canines are much more rarely severe (S. Perry, unpublished data), and all observed incidents of lethal aggression at Lomas have been caused by adult males' canines rather than females' (Gros-Louis et al., 2003; Perry and Manson, 2008). Thus, it is not surprising that females consistently defer to males in dyadic contests (Perry, 1997). Male aggression against infants poses a major challenge to adult females, as infanticide during alpha male replacements is the leading cause of infant deaths (Fedigan, 2003; Fedigan et al., 2008; Perry et al., 2012), and hence the biggest obstacle to females in increasing their lifetime reproductive success is male reproductive strategies.

There is very little that a female can do about infanticidal males, given the superior fighting ability of males. Females can (and often do) discourage new males from joining the group by behaving aggressively toward them, and by refraining from grooming them. Mothers and siblings of young infants are wary of extragroup males, newly immigrant males, and particularly new alpha males, and are usually careful to keep their infants at a safe distance, at least during the early stages of their relationships with these males. Females avidly form coalitions with both males and females, to counter attacks by potential infanticidal males (Perry and Manson, 2008). Such coalitions can be successful over short time periods, and hence probably do save some infants from infanticide during temporary overthrows of alpha males who then make a comeback; but if a new male succeeds in overthrowing the alpha male and then remains in the group as alpha for a period of several months, it is impossible for a mother with a young infant to be sufficiently vigilant to keep her infant away from the alpha and near her female allies 24 h a day, every day. Eventually, a determined alpha male succeeds in killing the infant. Although all of the infant's kin respond with aggression toward the infanticidal male immediately following the killing, they soon resume normal interactions with him, and females whose infants have been killed do subsequently breed with the killer if he is still resident in the group when they resume cycling (Perry and Manson, 2008). As predicted by Sarah Hrdy's sexually selected infanticide model (Hrdy, 1974), it seems that the best option for a female who has lost an infant to infanticide is to breed with the killer, because he has just shown himself to be the winner of male–male competition and is therefore likely to be the best protector of his own babies. Capuchin males who kill other males' infants do not show aggressive tendencies toward their own infants; in fact, they often prove to be quite tender fathers and grandfathers (S. Perry, personal observation).

For these reasons, the best strategy for a female capuchin is to promote group stability as best she can, by actively supporting the alpha male's efforts to remain alpha and keep rival males from entering the group. Female support appears to be helpful and perhaps even critical in order for alpha males to retain their positions (Gros-Louis et al., 2003; Perry, 1998a). The alpha male is the preferred coalition partner of females and also the favored grooming partner (Perry, 1997). Not only does the alpha male help defend females and their infants from extragroup males, but he also assists them in coalitionary aggression against the subordinate males in the group (i.e., the alpha male's own allies in defense of mating access), by helping the females evict subordinate males from feeding sites. Although precise quantitative measures of these interaction patterns are published only for AA group at Lomas, the general pattern in which the alpha male is the preferred social partner of females holds true across all of the study groups at Lomas Barbudal. One of the striking features of male–female relationships is the rapidity with which females change partner preferences when an alpha male loses his rank to another male. Females are intensely loyal to their group's alpha male up until the time when he loses his rank, but that loyalty does not persist once he becomes a subordinate (Perry, 1998a; Perry and Manson, 2008). Not only do females switch their preference from the old alpha to the new alpha with regard to proximity, grooming, and coalition formation, but they also begin to breed with the new alpha rather than the former one, even if the old alpha male remains in the group (Muniz et al., 2010).

Although females are relatively powerless when it comes to long-term problems such as preventing male immigration or protecting their babies against attack from a determined alpha male, they are more effective at defeating males in short-term conflicts such as feeding competition (Perry, 1997). Females reliably form coalitions with one another against males, successfully evicting them from feeding trees. Although most such coalitions are against subordinate males, coalitions of females sometimes even succeed against alpha males. Because females, compared to males, have more numerous and reliable allies due to their extensive kin networks, they generally wield more power in the group than do subordinate males, and sometimes even more than an alpha male. Although capuchins generally support dominants over subordinates, both males and females support females in contests against males 85% of the time (Perry, 2003).

Although males inflict costs on females in the form of feeding competition (Rose, 1994) and infanticide, they also confer some benefits on females by serving as alloparents and as "hired guns" in defense of the group against predators and other capuchin males. Males assist in alloparenting, by offering frequent rides to infants, playing with them, and protecting them from predators. Males are somewhat better than females in detecting predators (Rose, 1994) and also take an active role in mobbing them (Perry and Manson, 2008; Rose, 1994). Adult males, and alpha males in particular, exhibit higher levels of vigilance than females do (Rose and Fedigan, 1995). However, the most important role that males play is probably defense against extragroup males. There is some suggestive evidence that male vigilance is primarily directed toward males from other groups, as it is heightened in home-range overlap zones, where intergroup encounters are more common (Rose and Fedigan, 1995). All resident males in a group cooperate in keeping new males from migrating in and taking over the alpha position (Perry, 1996b; Perry and Manson, 2008).

It seems that having more subordinate males in the group is advantageous to females, because groups with a higher percentage of adult males are characterized by higher female reproductive success (as measured by the observed minus expected ratio of immatures to adult females); this is likely due to the greater capacity of alpha males to ward off infanticidal intruders when they have the help of subordinate males in group defense (Fedigan and Jack, 2011). If subordinate males are useful to females, then it might be expected that they will provide some incentives for them to remain in the group. Even though females associate with and groom the alpha male far more than they do subordinate males, they form affiliative relationships with subordinate males as well (Carnegie et al., 2006; Perry, 1997; Perry and Manson, 2008). They also mate with subordinate males, though these matings are almost always nonconceptive, occurring during pregnancy and other nonovulatory periods (Manson et al., 1997; Perry, 1997) unless the females are direct descendants of the alpha male (Godoy, 2010; Muniz et al., 2006, 2010). Grooming is a behavior that is almost unidirectional in male–female dyads, with females grooming males at a rate of about 1–2 orders of magnitude greater than the rate at which males groom females. Males ignore 88% of females' requests to be groomed, and when they do groom females, they rarely do so for more than a second or two.

In summary, the alpha male is the most important ally of females, because of his role in defending their infants (i.e., his offspring and grandoffspring) from infanticide. Subordinate males are also important, because of their aid in intergroup encounters, but a subordinate male who assumes the alpha position is also likely to kill infants, even if he knew them and interacted amiably with them when he was a subordinate. This probably explains why females form strong bonds with the alpha male and assist him in maintaining his position, thereby ensuring enhanced infant survival.

## D. Relationships Between Members of Different Groups

There is considerable home-range overlap between neighboring capuchin groups, and groups encounter one another at a rate varying from once a week (Perry, 1996b) to as many as nine times per week for a group that has many neighboring groups (Meunier et al., 2012). Encounter rates may be slightly higher in the dry season than in the rainy season at Lomas (Perry, 1996b), perhaps because monkeys concentrate their activities around permanent water sources more often in the dry season.

Relationships between groups are typically hostile. In a typical intergroup encounter, the females of the two groups grab their infants and flee, along with the juveniles, while most if not all of the adult males run toward one another. The males display at one another for periods up to 2 h, and they occasionally engage in contact aggression (Perry, 1996b). Although contact aggression is rare, it does sometimes escalate to lethal aggression: we have seen males killed during intergroup encounters on at least three occasions at Lomas, and there are many other cases in which wounded males disappear after intergroup encounters and may well have died rather than emigrating far away (Gros-Louis et al., 2003; Perry and Manson, 2008; S. Perry, unpublished data). Rigid intergroup dominance hierarchies do not exist in this species, and several factors interact to predict the outcome of an intergroup encounter. In the Barro Colorado Island population in Panama, group size (both number of adult males and number of adults) was a significant predictor of probability of displacing another group during an intergroup encounter (Crofoot et al., 2008). However, the location of the encounter was also an important factor: smaller groups could defeat larger groups if the encounter occurred closer to the center of the smaller group's home range (Crofoot et al., 2008). Playback experiments at Lomas (Meunier et al., 2012) have revealed that adult males are more likely to participate aggressively in a simulated intergroup encounter if (a) there is another adult male from their group in view of them at the time they hear the vocal challenge from a neighboring group and/or (b) another group member is actively threatening or moving toward the direction of the playback speaker that is emitting the vocal threats from the other group. At Lomas, the relative number of males in the two neighboring groups does not seem to affect the willingness of adult males to challenge the neighbors (Meunier et al., 2012).

Adult females and immature individuals participate very rarely in intergroup encounters. At Lomas, the most noteworthy cases of participation, in which females and juveniles have not only participated but also physically attacked their opponents, are generally intergroup encounters between recent fission products. These incidents are puzzling, because the female and juvenile participants are imposing costs on their matrilineal kin. Relations between fission products are unusually hostile for the first few months after the fission and then subside to normal levels (i.e., virtual noninteraction between females and immatures of neighboring groups). Although no females participate regularly, old postreproductive females seem more prone to participate (at least vocally) than the younger females who have infants to defend.

Adult males, especially subordinate males, sometimes exhibit nonhostile interactions with males of other groups. Sometimes they will pause their aggressive display behaviors to exchange odd vocalizations that are not typically heard in an intragroup context. Frequently, they exhibit elements of the courtship repertoire, "dancing" with the extragroup males, making "duck faces," and, on rare occasion, even sexually mounting males from the opponent group. Such behaviors are most commonly seen when some males sneak back to the site of the encounter after the other group males have returned to the females. Although it is not entirely clear how these behaviors function, it seems likely that they serve some sort of assessment function. Possibly, males use intergroup encounters as a way of assessing their options to form alliances with new males and migrate to different groups.

In general, males are hostile to males from other groups whom they encounter. Oddly, however, we have found that alpha males from small groups sometimes visit one another, even staying overnight and leaving their females unattended. Two unrelated alpha males from adjacent groups at Lomas formed a close friendship, characterized by frequent visits, play, and cosleeping, and a conspicuous absence of aggression. They were the only adult males in their groups, at least in the early years of their relationship, and it could be that they missed having male company. After one of these alpha males was eventually overthrown by extragroup males and evicted from his group, he joined his friend's group as a subordinate male.

Intergroup encounters, as well as encounters with all-male groups, are probably the most important context in which males need dependable allies. It is in these contexts that males have a chance to assess the competitive strength of other alliances and to investigate the number of reproductive females in neighboring groups, to determine the probable costs and benefits of attempting to disperse to a new group. Allies are critical both to invade a new group and to defend breeding access in a group of residence.

## V. Social Learning, Traditions, and Communication

### A. Species-Typical Signals for Communicating About Coalitionary Aggression

The gestural repertoire of white-faced capuchins includes a rich array of signals that seem to function to recruit allies or demonstrate alliances between two or more individuals in the context of aggressive conflicts (Gros-Louis et al., 2008; Oppenheimer, 1973; Perry and Manson, 2008). Capuchins demonstrate solidarity during coalitionary aggression by aligning their bodies in particular ways while simultaneously directing open-mouth threats toward a common opponent. There are three main ways in which they do this: (a) by stacking on top of one another in the "overlord" position (Oppenheimer and Oppenheimer, 1973), similar to the infant mount position, with the heads stacked as in a totem pole and the top monkey's arms clasping the chest of the lower monkey (see Fig. 3); (b) in an embrace, with one monkey's arm and perhaps other parts of the body draped over the partner (see Fig. 4); and (c) the "cheek-to-cheek" posture, in which the monkeys stand side by side with their faces and perhaps other parts of the body pressed against one another (see Fig. 5). Recruitment of partners can be accomplished by head flagging, in which the recruiter jerks the head back and forth rapidly between the solicitee and the target while threatening and often bouncing at the target of aggression. Mild vocal threats and/or screams (Gros-Louis et al., 2008) may also play a role in recruitment, though they are also used in dyadic aggression. Overlords can be initiated either by backing into the solicited individual or by pouncing on his or her back.

All of these signals seem to be universal among white-faced capuchin populations and seem to extend to other species of nontufted capuchins as well (*Cebus olivaceus*: Oppenheimer and Oppenheimer, 1973; *Cebus albifrons*: Defler, 1979; Thomas Defler, personal communication; *C. albifrons aequatorialis*: F. Campos, personal communication). Such behaviors are far more rare in the wild tufted capuchin populations (*Sapajus* or *Cebus nigritus*, formerly known as *Cebus apella* (Ferreira et al., 2006; Jessica Lynch Alfaro, personal communication; see Lynch Alfaro et al., 2012 for a clarification of

FIG. 4. An adult female embraces an adult male while both threaten the same target (photo: Susan Perry). (For color version of this figure, the reader is referred to the online version of this chapter.)

FIG. 5. The alpha male and alpha female do a cheek-to-cheek posture (photo: Susan Perry). (For color version of this figure, the reader is referred to the online version of this chapter.)

the phylogeny)), as might be expected from the lower frequency of coalition formation in *Sapajus* (see Section VI). Judging from personal communication with *Sapajus* researchers (Jessica Lynch Alfaro [Caratinga, Brazil], Renata Ferreira and Patricia Izar [semifree ranging capuchins in Parque do Tietê, Brazil]), the range of behaviors used to signal about coalition formation is narrower in the tufted capuchins than in the nontufted capuchins. These researchers report that the "overlord" position seems to be missing from the behavioral repertoire for most wild populations of *Sapajus* (*Cebus*) *nigritus*. Embraces, head flags, and cheek-to-cheek postures are sometimes seen at low levels, though perhaps in a less stereotyped manner than they are seen in *Cebus capucinus*. For example, in most photos of coalitionary behaviors taken in *Sapajus*, embraces and cheek-to-cheek postures involve less body contact than is seen in *C. capucinus*; also, head flags in *S. nigritus* are described as alternation of glances between opponent and supporter. In wild populations of *Sapajus* [*Cebus*] *libidinosus* such as Fazenda Boa Vista, the cheek-to-cheek posture is seen fairly often, and sometimes, the monkeys "embrace" by draping a tail (but not an arm) over the coalition partner. Head flags (as described for *S. nigritus*) are also observed, but overlords are not seen (Michele Verderane and Patricia Izar, personal communication). Observations of a captive group of *C.* [*Sapajus*] *apella* present a slightly different picture, in which coalitionary recruitment signals are still rare compared to the frequency with which they are observed in *C. capucinus*

(Colleen Gault, personal communication), but all of the same signals are likely to be present in the repertoire at low levels, with overlords being one of the rarest forms of recruitment (T. McKenney, personal communication).

The presence of so many coalitionary recruitment signals in the evolved communicative repertoire of *C. capucinus* is suggestive of a long evolutionary history of coalitionary aggression, indicating that intensive cooperation for the purpose of competing has been an important selective pressure for capuchins, and for the nontufted capuchins in particular.

## B. Group-Specific Gestural Communication Used in Relationship Negotiation

One of the more striking features of *C. capucinus* behavioral biology that has emerged from the decades of collaborative investigation between researchers at various white-faced capuchin field sites is the finding that these monkeys are prone to developing social conventions of a risky and intimate nature (Perry et al., 2003a,b). Several innovations have been described that spread through social networks across as many as three links in a social transmission chain, becoming frequently expressed elements of a group's behavioral repertoire before they finally vanish from the group's repertoire after a period of up to 11 years. These behaviors are sometimes unique to a single social group but in some cases are invented in multiple social groups (Perry, 2011; Perry et al., 2003a). By definition, traditional communicative rituals are never present in all white-capuchin social groups (Perry et al., 2003a).

All of the group-specific dyadic social rituals that have been described for *C. capucinus* involve some element of risk or discomfort. Hand sniffing (Fig. 6) is the social convention that has been invented independently in the largest number of groups. In this behavior, one monkey inserts the partner's fingers in its nostrils and maintains that pose for long periods of time (often exceeding 10 min). Although the basic structure of this behavior is common to all dyads in a social network that performs this behavior, there is considerable variation between dyads and/or between groups in the way this is performed. Sometimes it is mutual, with both individuals inserting their fingers in the other's nostrils. Sometimes, the hand is just cupped over the muzzle, with no fingers deeply inserted in orifices. Other dyads insert one finger in the nose and another finger in the mouth or eye socket (Fig. 7), inserting the finger as deeply as the first knuckle. Because the fingernails tend to be long, dirty, and sharp, most variants of hand sniffing must cause pain, discomfort, and/or the risk of infection to the delicate membranes they contact. A similar kind of social convention is the sucking of body parts (ears, fingers, or tails), which can be performed mutually (for fingers and

FIG. 6. Two females hand sniff, with their fingers in one another's nostrils (photo: Susan Perry). (For color version of this figure, the reader is referred to the online version of this chapter.)

FIG. 7. An adult female inserts the alpha male's finger deep into her eye socket, while inserting her finger in his nostril (photo: Wiebke Lammers). (For color version of this figure, the reader is referred to the online version of this chapter.)

tails, but not ears) for periods of up to an hour. The other observed social conventions fall into a class of behaviors termed "games," which have similar underlying structure: one individual grasps some object (the partner's finger or tail, a tuft of hair plucked from the partner's body, or an inedible object such as bark or an unripe fruit) firmly between its teeth. Then the partner struggles to open the mouth and seize the object, using mouth, feet, and hands to pry open the jaws. When the object or body part has been retrieved, they repeat the ritual, generally switching roles, and continuing the game for several minutes (Perry, 2011; Perry and Manson, 2008; Perry et al., 2003a). These rituals that involve placing a body part between the jaws of a partner are potentially risky, given the demonstrated capacity of capuchin males to inflict injuries on one another; that is, they are rituals that an individual should be willing to perform only if he trusts the partner not to bite too hard.

The most plausible explanation for these risky intimate social conventions, given the currently available evidence, is that they are Zahavian bond tests. Zahavi (1977) proposed that some signals are used to elicit affective displays from social partners that will inform the individual who imposes the "test" about the status of the relationship. Such signals often involve elements of discomfort or risk and are the sorts of behaviors that can be perceived as pleasurable or aversive, depending on the status of the relationship (an example Zahavi uses in his 1977 paper is tongue kissing in humans). For example, if the recipient of the Zahavian bond-testing signal responds enthusiastically, then the signaler knows the quality of the relationship is high, but if the recipient squirms away or responds aggressively, then this indicates that the recipient's perception of the status of the relationship is not so high that the signaler should view the recipient as a reliable social partner for future cooperative endeavors.

Although this species already has various stereotyped bond-testing signals (e.g., the use of courtship dances and nonconceptive sex in relationship negotiation contexts [Manson et al., 1997; Perry and Manson, 2008]), it may be the case that these traditional dyadic rituals, which are co-constructed by the members of the dyad over a period of several months, provide a richer source of information about commitment to that particular dyadic relationship than the stereotyped signals do (Perry, 2011; Perry et al., 2003a).

The distribution of social conventions across space and time is intriguing. Eleven of 16 groups studied at four different sites in northwestern Costa Rica have exhibited between one and five different types of social conventions during the period of time when they were observed (Perry, 2011; Perry et al., 2003a). Typically, a group that has been studied for several years will have periods during which no social conventions are exhibited, in addition to the periods in which such behaviors are common. This temporal distribution suggests that there may be particular circumstances in the political and demographic histories of groups that make bond testing more or less necessary.

Putting *C. capucinus* into comparative perspective, it is noteworthy that there is no other species for which so many traditions have been reported in the domain of social communicative rituals, despite the fact that many other species have rich repertoires of traditions in other behavioral domains (Laland and Galef, 2009; Perry and Manson, 2003). It is particularly striking that these behaviors have never been described in a population of wild tufted capuchins (P. Izar, personal communication; Perry, 2011), despite the fact that researchers of this species have actively been seeking to document such behaviors in their cataloging of traditions for the genus. It is not entirely clear whether the paucity of traditional dyadic social rituals is due to methodological differences between studies (e.g., a tendency for researchers to overlook such behaviors) or whether *C. capucinus* really is an outlier in this respect. It seems possible that the extreme prevalence of coalitions, and of lethal coalitionary aggression in particular, makes alliances a more critical aspect of the social strategy for white-faced capuchins than is the case for most other mammalian species. A greater need for alliances might select for a richer variety of techniques for testing the quality of social bonds so that individuals could have a better idea of whom they can depend on in life-threatening situations. Certainly, coalition formation is more prevalent in *C. capucinus* than in most (if not all) tufted capuchin populations (Perry, 2011).

## VI. Group Life Cycles and Comparisons with Tufted Capuchins (*Sapajus*)

One of the most important discoveries to come out of the Lomas and Santa Rosa white-faced capuchin projects, as well as the long-term study of the Iguazu *C.* [*Sapajus*] *nigritus* population (Janson et al., 2012), is that groups go through fairly predictable cycles of change triggered by male takeovers and infanticide sprees, in which short periods of chaos are followed by long periods characterized by low mortality rates and stability in the identity of the alpha male (Fedigan et al., 2012; Perry and Manson, 2008). The distribution of reproductive benefits and the patterning of cooperative activities differ considerably between these stable and unstable periods, and most individuals spend a portion of their lives experiencing each state. For these reasons, cross-sectional studies that do not take individual life histories and group histories into account are bound to produce incomplete explanations of social structures and reproductive strategies and are also likely to overemphasize between-group or between-site differences.

Takeovers almost always result in infanticide of the younger infants and often trigger male dispersal events (particularly emigration of natal males) and group fission (Fedigan, 2003; Jack et al., 2010; Janson et al., 2012;

Perry and Manson, 2008), and, at Santa Rosa, have even resulted in female emigration (Jack and Fedigan, 2009). The new groups formed in the wake of takeover events tend to be much smaller than their parent groups, and their members (at least the females) tend to be more closely related on average (Janson et al., 2012; Muniz, 2008). At Lomas (S. Perry, unpublished data) and Iguazu (Janson et al., 2012), the alpha males of the newly formed groups tend to be younger than the previous alpha males. During the first year of a stable alpha male tenure, most females reproduce during the same year as a consequence of having lost the previous infant to infanticide (Janson et al., 2012; Lomas: S. Perry, unpublished data). Under the protection of a stable alpha male, the group grows in size, and the increasing availability of mature females attracts and/or retains male immigrants. A large group of females presents a particularly attractive target for males who are seeking breeding opportunities, and thus, at this point in the group cycle, another takeover becomes probable. In *C. capucinus*, the probability of a takeover occurring seems to depend not only on the fighting ability of the alpha male but also on his success in recruiting male allies for group defense. However, subordinate males are not active in group defense in *Sapajus*, and conflicts over the alpha position are typically dyadic (Janson, 1986a,b; Janson et al., 2012).

One of the most striking factors regarding *C. capucinus* that demands explanation is its high level of coalitionary aggression. Although *Sapajus*, the most closely related genus, also exhibits coalitionary behavior, it does so to a far lesser extent than white-faced capuchins do. Relative to *C. capucinus* or *C. albifrons*, *Sapajus* also exhibits a less elaborate array of gestures for coalitionary recruitment (see Section V.A), and an absence of parallel dispersal (i.e., comigration of two or more individuals from one group to another) (Janson et al., 2012). *Sapajus* also lacks the traditional bond-testing rituals such as hand sniffing and games that are so often observed in *C. capucinus*. Clearly, a life history perspective is needed in order to make sense of these differences, and the only *Sapajus* field site that has long-term data is Iguazu. Unfortunately, most of the fine-grained data on coalition formation for *Sapajus* are from other sites rather than Iguazu, but there are good data on the frequency of parallel dispersal and some data on coalition formation from Iguazu. *S. nigritus* males disperse singly: aside from one possible codispersal at Iguazu (which the researchers believed to be an instance in which local tourists captured two males on the same day, rather than a true dispersal), joint dispersal has not been seen (Janson et al., 2012).

The differences between *S. nigritus* and *C. capucinus* with regard to propensity to form coalitions are more puzzling when we consider the many convergences in their social lives: both species form multi-male, multi-female groups in which males disperse, whereas females are philopatric and mothers and daughters have similar ranks as adults (Janson et al., 2012). Births at

Iguazu are mildly seasonal, with IBIs tending to be approximately 2 years unless the previous infant dies. Infanticide by new alpha males is the primary cause of infant mortality in both species (Janson et al., 2012). Although comparable data are not available for causes of male mortality, male tufted capuchins do wound and kill one another during rank reversals, and dispersal appears to be a dangerous time for tufted capuchins (Janson et al., 2012).

Infanticide has been argued to be a factor selecting for male–female coalitions in primates generally (van Schaik, 1996), and thus coalitions between the alpha male and adult females should be expected in both species. Precisely analogous analyses of patterns of coalition formation have not been conducted in the two species. In general, coalitions in *S. nigritus* tend to be loser-support coalitions (Ferreira et al., 2006), in contrast with coalitions in *C. capucinus*, which are more often winner-support coalitions (Perry, 2003). So, although a *S. nigritus* alpha male is likely to provide much support for his offspring and females, it seems unlikely that females are regular supporters of the alpha male. Coalitionary support of the breeding male by females does occur on occasion in *S. nigritus*, however: After an alpha male at Iguazu was defeated, injured, and evicted by a challenger, all but one of the group's females supported their former alpha male against the new alpha male when the former alpha returned to the group the month after his eviction (Janson et al., 2012).

Janson (1986a,b) attributed the higher level of male–male cooperation among *C. albifrons* males, relative to *Sapajus* males, to more equitable distributions of both food resources and mating opportunities. One process that could reduce male reproductive skew in a female-philopatric species is long alpha male tenure. When alpha males coreside with their adult daughters, inbreeding avoidance will generate mating opportunities for subordinate males. Thus, subordinate males eventually gain a benefit after enduring multiyear periods of very low reproductive success in a reproductive queue waiting for an alpha male's daughters to mature, while alpha males gain from the coalitionary support provided by these subordinates in confrontations with extragroup males. An alpha male's adult and subadult resident sons also gain inclusive fitness benefits from enhancing their father's continued reproductive success by aiding in protecting his vulnerable offspring from infanticidal males. Once an alpha male's daughters reach sexual maturity, subordinate males, now with mating opportunities, have a larger reproductive stake in the group and are therefore even more highly motivated to defend it from potentially infanticidal outsiders. As an alpha male's tenure lengthens even further, reducing the number of unrelated females available to him as mates, he still gains from the presence of subordinate allies who protect his descendants even though the breeding opportunities are going primarily to the subordinate males.

If reduced reproductive skew due to long alpha male tenure length were the mechanism by which the nontufted capuchins achieved male–male cooperation in the joint defense of mates and offspring, then you would expect to see much shorter alpha tenures in the tufted capuchins, who engage in less male–male cooperation. However, at this point, we probably do not have sufficient data from multiple sites to know whether alpha tenures are consistently longer in *C. capucinus* than in *S. nigritus*, and there are slight methodological differences in the ways tenure lengths are scored (with some sites discarding the really short tenures) that make direct between-site comparisons difficult. Lomas has the longest recorded maximum alpha tenures, at about 18 years, but the Iguazu capuchins actually have a longer mean tenure length of about 5 years and quite frequently remain resident long enough to coreside with their adult daughters, with the maximum alpha tenure thus far being at least 11.6 years (Janson et al., 2012). It appears that there are typically more short tenures during chaos phases before a new alpha male seizes firm control over a group at Lomas than at Iguazu. In order to test the hypothesis that father–daughter inbreeding avoidance during long alpha male tenures is the mechanism by which males reduce competition and enhance cooperation in group defense against foreign males, it would also be necessary to conclusively demonstrate father–daughter inbreeding avoidance in *Sapajus*. Behavioral observations of overt female courtship in *Sapajus* at Iguazu suggest that daughters do avoid breeding with their fathers in this species (Di Bitetti and Janson, 2001). However, it should be noted that initial observations of matings in *C. capucinus* (e.g., Perry, 1997, which was based on behavioral observations rather than genetic paternity assessment) indicated a far more equitable distribution of paternity than actually exists in this species (Muniz et al., 2010); so caution is still warranted in making assumptions about the *Sapajus* mating system until genetic paternity data are available for both species.

It is currently difficult to evaluate the trade-offs of employing different subordinate male strategies (e.g., migrating singly vs. jointly, attempting external vs. internal takeovers with and without the assistance of allies, or peaceful inheritance of a breeding position after a period of waiting in a reproductive queue). The relative merits of different options would depend on the individual qualities of individual males relative to their current male competitors and the sets of options available to them. Presumably, in many cases, the best option for a male who is not in peak condition is to be a subordinate helper in a group where he has some hope of breeding opportunities (e.g., with the alpha's daughters), or gaining inclusive fitness benefits by helping rear the offspring of an alpha male brother, rather than risking the dangerous option of challenging a more powerful male for the alpha position or enduring another perilous migration phase. A more sophisticated analysis of these trade-offs, employing a variety of modeling techniques that can be tested with long-term data, is currently underway.

Hopefully, further data collection on both genera of capuchins will reveal whether the differences between populations are as great as they seem based on currently available data sets, or whether they are due to temporal sampling bias. Capuchins have such variable behavioral strategies, and their lives are so long, that it would take several decades to accurately sample the full range of behavioral strategies within a particular population. In the meantime, various modeling strategies, including agent-based modeling, may prove useful to determine the conditions that favor the evolution of cooperation in capuchins.

**Acknowledgments**

The following field assistants contributed a year or more of data to the Lomas Barbudal Monkey Project data set: B. Barrett, L. Beaudrot, M. Bergstrom, R. Berl, A. Bjorkman, L. Blankenship, J. Broesch, J. Butler, F. Campos, C. Carlson, M. Corrales, N. Donati, C. Dillis, G. Dower, R. Dower, K. Feilen, K. Fisher, A. Fuentes J., M. Fuentes A., C. Gault, H. Gilkenson, I. Godoy, I. Gottlieb, J. Gros-Louis, L. Hack, S. Herbert, S. Hyde, W. Lammers, L. Johnson, S. Lee, S. Leinwand, T. Lord, M. Kay, E. Kennedy, D. Kerhoas-Essens, E. Johnson, S. Kessler, J. Manson, W. Meno, C. Mitchell, Y. Namba, A. Neyer, C. O'Connell, J.C. Ordoñez J., N. Parker, B. Pav, K. Potter, K. Ratliff, H. Ruffler, S. Sanford, M. Saul, I. Schamberg, C. Schmitt, J. Verge, A. Walker-Bolton, E. Wikberg, and E. Williams. I am particularly grateful to H. Gilkenson, W. Lammers, C. Dillis, M. Corrales, and R. Popa for managing the field site. E. Wikberg and W. Lammers contributed a year or more of effort to organizing the data set. The genetic analysis was conducted by L. Muniz and I. Godoy in Linda Vigilant's lab. This project is based on work supported by the funding provided to SEP by the Max Planck Institute for Evolutionary Anthropology, the National Science Foundation (Grants No. SBR-9870429, 0613226, and 6848360, a graduate fellowship, and an NSF-NATO postdoctoral fellowship), five grants from the L.S.B. Leakey Foundation, three grants from the National Geographic Society, The Wenner-Gren Foundation, Sigma Xi, an I.W. Killam postdoctoral fellowship, and several faculty development or student grants and fellowships from University of California, Los Angeles and The University of Michigan. Any opinions, findings, and conclusions or recommendations expressed in this chapter are those of the author and do not necessarily reflect the views of the National Science Foundation or other funding agencies. I thank the Costa Rican park service (MINAET and SINAC, currently), Hacienda Pelon de la Bajura, Hacienda Brin d'Amor, and the residents of San Ramon de Bagaces for permission to work on their land. This research was performed in compliance with the laws of Costa Rica, and the protocol was approved by the University of Michigan IACUC (protocol #3081) and the UCLA animal care committee (ARC #1996-122 and 2005-084 plus various renewals). S. Caro, I. Godoy, C. Gault, K. Jack, K. Kajokaite, W. Lammers, J. Manson, and J. Mitani provided helpful discussion and comments on the chapter. My thinking about capuchin cooperation has been further stimulated by discussions with F. Campos, T. Defler, L. Fedigan, R. Ferreira, P. Izar, C. Janson, K. Kajokaite, J. Lynch Alfaro, M. Verderane, F. de Waal, and T. McKenney.

**References**


Baker, M., 1999. Fur Rubbing as Evidence for Medicinal Plant Use by Capuchin Monkeys (*Cebus capucinus*): Ecological, Social, and Cognitive Aspects of the Behavior. Ph.D. Thesis, University of California-Riverside.

Bergstrom, M.L., Fedigan, L.M., 2010. Dominance among female white-faced capuchin monkeys (*Cebus capucinus*): hierarchical linearity, nepotism, strength and stability. Behaviour 147, 899–931.

Boinski, S., Campbell, A.F., 1995. Use of trill vocalizations to coordinate troop movement among white-faced capuchins: a second field test. Behaviour 132, 875–901.

Buckley, J.S., 1983. The Feeding Behavior, Social Behavior, and Ecology of the White-Faced Monkey, *Cebus capucinus*, at Trujillo, Northern Honduras, Central America. Ph.D. Thesis, University of Texas, Austin.

Campos, F.C., Manson, J.H., Perry, S.E., 2007. Urine washing and sniffing in wild white-faced capuchins (*Cebus capucinus*): testing functional hypotheses. Int. J. Primatol. 28, 55–72.

Carnegie, S., Fedigan, L.M., Ziegler, T., 2006. Post-conceptive mating in white-faced capuchins: hormonal and sociosexual patterns in cycling, non-cycling and pregnant females. In: Estrada, A., Garber, P., Pavelka, M., Luecke, L. (Eds.), New Perspectives in the Study of Mesoamerican Primates: Distribution, Ecology, Behavior and Conservation. Springer Press, New York, pp. 387–409.

Carpenter, C.R., 1934. A field study of the behavior and social relations of howling monkeys (*Alouatta palliata*). Comp. Psychol. Monogr. 10, 1–168.

Clutton-Brock, T.H., 1989. Female transfer and inbreeding avoidance in social mammals. Nature 337, 70–72.

Crofoot, M.C., Gilby, I.C., Wikelski, M.C., Kays, R.W., 2008. Interaction location outweighs the competitive advantage of numerical superiority in *Cebus capucinus* intergroup contests. Proc. Natl. Acad. Sci. USA 105, 577–581.

Defler, T., 1979. On the ecology and behavior of *Cebus albifrons* in eastern Colombia: II. Behavior. Primates 20, 491–502.

Di Bitetti, M.S., Janson, C.H., 2001. Reproductive socioecology of tufted capuchins (*Cebus apella nigritus*) in northeastern Argentina. Int. J. Primatol. 22, 127–142.

Fedigan, L.M., 2003. The impact of male takeovers on infant deaths, births and conceptions in *Cebus capucinus* at Santa Rosa, Costa Rica. Int. J. Primatol. 24, 723–741.

Fedigan, L.M., Jack, K.M., 2011. Two girls for every boy: the effects of group size and composition on the reproductive success of male and female white-faced capuchins. Am. J. Phys. Anthropol. 144, 317–326.

Fedigan, L.M., Jack, K.M., 2012. Tracking monkeys in Santa Rosa: long-term lessons from a regenerating tropical dry forest. In: Kappeler, P., Watts, D. (Eds.), Long-Term Field Studies of Primates. Springer, Heidelberg, pp. 165–184.

Fedigan, L.M., Rose, L.M., Morera Avila, R., 1996. Tracking capuchin monkey populations in a regenerating Costa Rican dry forest. In: Norconk, M.A., Rosenberger, A.L., Garber, P.A. (Eds.), Adaptive Radiations of Neotropical Primates. Plenum Press, New York, pp. 289–307.

Fedigan, L.M., Carnegie, S.D., Jack, K.M., 2008. Predictors of reproductive success in female white-faced capuchins (*Cebus capucinus*). Am. J. Phys. Anthropol. 137, 82–90.

Ferreira, R.G., Izar, P., Lee, P.C., 2006. Exchange, affiliation, and protective interventions in semifree-ranging brown capuchin monkeys (*Cebus apella*). Am. J. Primatol. 68, 765–776.

Ford, S., 1994. Evolution of sexual dimorphism in body weight in platyrrhines. Am. J. Primatol. 34, 221–244.

Fragaszy, D.M., Visalberghi, E., Fedigan, L.M., 2004. The Complete Capuchin: The Biology of the Genus *Cebus*. Cambridge University Press, Cambridge, UK.

Godoy, I., 2010. Testing Westermarck's Hypothesis in a Wild Primate Population: Proximity During Early Development as a Mechanism of Inbreeding Avoidance in White-Faced Capuchin Monkeys (*Cebus capucinus*). Master's Thesis, University of California, Los Angeles.

Grinnell, J., Packer, C., Pusey, A.E., 1995. Cooperation in male lions: kinship, reciprocity, or mutualism? Anim. Behav. 49, 95–105.

Gros-Louis, J., Perry, S., Manson, J.H., 2003. Violent coalitionary attacks and intraspecific killing in wild capuchin monkeys (*Cebus capucinus*). Primates 44, 341–346.

Gros-Louis, J., Perry, S., Fichtel, C., Wikberg, E., Gilkenson, H., Wofsy, S., et al., 2008. Vocal repertoire of *Cebus capucinus*: acoustic structure, context and usage. Int. J. Primatol. 29, 641–670.

Hakeem, A., Sandoval, R.G., Jones, M., Allman, J., 1996. Brain and life span in primates. In: Birren, J.E., Schaie, K.W. (Eds.), Handbook of the Psychology of Aging. fourth ed. Academic Press, San Diego, pp. 78–104.

Hrdy, S., 1974. Male-male competition and infanticide among the langurs (*Presbytis entellus*) of Abu, Rajasthan. Folia Primatol. 22, 19–58.

Jack, K.M., Fedigan, L.M., 2004a. Male dispersal patterns in white-faced capuchin, *Cebus capucinus*. Part 1: patterns and causes of natal emigration. Anim. Behav. 67, 761–769.

Jack, K.M., Fedigan, L.M., 2004b. Male dispersal patterns in white-faced capuchin, *Cebus capucinus*. Part 2: patterns and causes of secondary dispersal. Anim. Behav. 67, 771–782.

Jack, K.M., Fedigan, L.M., 2006. Why be alpha male? Dominance and reproductive success in wild white-faced capuchins (*Cebus capucinus*). In: Estrada, A., Garber, P.A., Pavelka, M.S. M., Luecke, L. (Eds.), New Perspectives in the Study of Mesoamerican Primates: Distribution, Ecology, Behavior, and Conservation. Springer, New York, pp. 367–386.

Jack, K.M., Fedigan, L.M., 2009. Female dispersal in a female-philopatric species, *Cebus capucinus*. Behaviour 146, 471–498.

Jack, K.M., Sheller, C., Fedigan, L.M., 2011. Social factors influencing natal dispersal in male white-faced capuchins (*Cebus capucinus*). Am. J. Primatol. 73, 1–7.

Janson, C., 1986a. Capuchin counterpoint. Nat. Hist. 95, 44–53.

Janson, C., 1986b. The mating system as a determinant of social evolution in capuchin monkeys (*Cebus*). In: Else, J.G., Lee, P.C. (Eds.), Primate Ecology and Conservation. Cambridge University Press, Cambridge, UK, pp. 169–179.

Janson, C., Baldovino, M.C., Di Bitetti, M., 2012. The group life cycle and demography of Brown Capuchin Monkeys (*Cebus [apella] nigritus*) in Iguazú National Park, Argentina. In: Kappeler, P., Watts, D. (Eds.), Long-Term Field Studies of Primates. Springer, Heidelberg, pp. 185–214.

Kajokaite, K., 2011. Variation in Strength and Stability of Female-Female Long-Term Relationships in Wild Capuchin Monkeys (*Cebus capucinus*). Undergraduate Honors Thesis, University of California, Los Angeles Anthropology Department.

Laland, K.N., Galef, B.G. (Eds.), 2009. The Question of Animal Culture. Harvard University Press, Cambridge, MA.

Lynch Alfaro, J.W., Boubli, J.P., Olson, L.E., Di Fiore, A., Wilson, B., Gutiérrez-Espeleta, G. A., et al., 2012. Explosive Pleistocene range expansion leads to widespread Amazonian sympatry between robust and gracile capuchin monkeys. J. Biogeogr. 39, 272–288.

Manson, J.H., 1999. Infant handling in wild *Cebus capucinus*: testing bonds between females? Anim. Behav. 57, 911–921.

Manson, J.H., Perry, S., Parish, A.R., 1997. Nonconceptive sexual behavior in bonobos and capuchins. Int. J. Primatol. 18, 767–786.

Manson, J.H., Rose, L., Perry, S., Gros-Louis, J., 1999. Dynamics of female-female social relationships in wild *Cebus capucinus*: data from two Costa Rican sites. Int. J. Primatol. 20, 679–706.

Manson, J.H., Gros-Louis, J., Perry, S., 2004. Three apparent cases of infanticide by males in wild white-faced capuchins (*Cebus capucinus*). Folia Primatol. 75, 104–106.

Masterson, T.J., 2003. Canine dimorphism and interspecific canine form in *Cebus*. Int. J. Primatol. 24, 159–178.

Meunier, H., Molina-Vila, P., Perry, S., 2012. Participation in group defence: proximate factors affecting male behaviour in wild white-faced capuchins. Anim. Behav. 83, 621–628. doi:10.1016/j.anbehav.2011.12.001.

Mitani, J.C., Gros-Louis, J., Richards, A.F., 1996. Sexual dimorphism, the operational sex ratio, and the intensity of male competition in polygynous primates. Am. Nat. 147, 966–980.

Mitchell, B., 1989. Resources, Group Behavior, and Infant Development in White-Faced Capuchin Monkeys, *Cebus capucinus*. Ph.D. Thesis, University of California, Berkeley.

Muniz, L., 2008. Genetic Analyses of Wild White-Faced Capuchins (*Cebus capucinus*). Ph.D. Thesis, Universität Leipzig.

Muniz, L., Perry, S., Manson, J.H., Gilkenson, H., Gros-Louis, J., Vigilant, L., 2006. Father-daughter inbreeding avoidance in a wild primate population. Curr. Biol. 16, R156–R157.

Muniz, L., Perry, S., Manson, J.H., Gilkenson, J., Gros-Louis, J., Vigilant, L., 2010. Male dominance and reproductive success in wild white-faced capuchins (*Cebus capucinus*) at Lomas Barbudal, Costa Rica. Am. J. Primatol. 72, 1118–1130.

Oppenheimer, J.G., 1968. Behavior and Ecology of the White-Faced Monkey, *Cebus capucinus*, on Barro Colorado Island, C.Z. Ph.D. Thesis, University of Illinois, Urbana.

Oppenheimer, J.G., 1973. Social and communicatory behavior in the *Cebus* monkey. In: Carpenter, C.R. (Ed.), Behavioral Regulators of Behavior in Primates. Associated University Presses, Cranbury, NJ, pp. 251–271.

Oppenheimer, J.G., Oppenheimer, E.C., 1973. Preliminary observations of *Cebus nigrivittatus* (Primates: Cebidae) on the Venezuelan llanos. Folia Primatol. 19, 409–436.

Packer, C., Pusey, A.E., 1982. Cooperation and competition within coalitions of male lions: kin selection or game theory? Nature 296, 740–742.

Panger, M., 1997. Hand Preference and Object-Use in Free-Ranging White-Faced Capuchin Monkeys (*Cebus capucinus*) in Costa Rica. Ph.D. Thesis, University of California, Berkeley.

Perry, S., 1995. Social Relationships in Wild White-Faced Capuchin Monkeys, *Cebus capucinus*. Ph.D. Thesis, University of Michigan, Ann Arbor.

Perry, S., 1996a. Female-female relationships in wild white-faced capuchin monkeys, *Cebus capucinus*. Am. J. Primatol. 40, 167–182.

Perry, S., 1996b. Intergroup encounters in wild white-faced capuchins, *Cebus capucinus*. Int. J. Primatol. 17, 309–330.

Perry, S., 1997. Male-female social relationships in wild white-faced capuchin monkeys, *Cebus capucinus*. Behaviour 134, 477–510.

Perry, S., 1998a. A case report of a male rank reversal in a group of wild white-faced capuchins (*Cebus capucinus*). Primates 39, 51–69.

Perry, S., 1998b. Male-male social relationships in wild white-faced capuchins, *Cebus capucinus*. Behaviour 135, 1–34.

Perry, S., 2003. Coalitionary aggression in white-faced capuchins, *Cebus capucinus*. In: de Waal, F.B.M., Tyack, P. (Eds.), Animal Social Complexity: Intelligence, Culture and Individualized Societies. Harvard University Press, Cambridge, MA, pp. 111–114.

Perry, S., 2011. Social traditions and learning in capuchin monkeys (*Cebus*). Philos. Trans. R. Soc. Lond. B Biol. Sci. 366, 988–996.
Perry, S., Manson, J.H., 2003. Traditions in monkeys. Evol. Anthropol. 12, 71–81.
Perry, S., Manson, J., 2008. Manipulative Monkeys: The Capuchins of Lomas Barbudal. Harvard University Press, Cambridge, MA.
Perry, S., Baker, M., Fedigan, L., Gros-Louis, J., Jack, K., MacKinnon, K., et al., 2003a. Social conventions in wild white-faced capuchin monkeys: evidence for traditions in a neotropical primate. Curr. Anthropol. 44, 241–268.
Perry, S., Panger, M., Rose, L., Baker, M., Gros-Louis, J., Jack, K., et al., 2003b. Traditions in wild white-faced capuchin monkeys. In: Fragaszy, D., Perry, S. (Eds.), The Biology of Traditions: Models and Evidence. Cambridge University Press, Cambridge, UK, pp. 391–425.
Perry, S., Barrett, H.C., Manson, J.H., 2004. White-faced capuchin monkeys exhibit triadic awareness in their choice of allies. Anim. Behav. 67, 165–170.
Perry, S., Manson, J.H., Muniz, L., Gros-Louis, J.G., Vigilant, L., 2008. Kin-biased social behaviour in wild adult female white-faced capuchins, *Cebus capucinus*. Anim. Behav. 76, 187–199.
Perry, S., Godoy, I., Lammers, W., 2011. Alpha male tenures and male reproductive success in wild white-faced capuchin monkeys, *C. capucinus*, at Lomas Barbudal, Costa Rica. Am. J. Primatol. 73 (Suppl. 1), 70.
Perry, S., Godoy, I., Lammers, W., 2012. The Lomas Barbudal Monkey Project: two decades of research on *Cebus capucinus*. In: Kappeler, P., Watts, D. (Eds.), Long-Term Field Studies of Primates. Springer, Heidelberg, pp. 141–164.
Pope, T.R., 1990. The reproductive consequences of male cooperation in the red howler monkey: paternity exclusion in multi-male and single-male troops using genetic markers. Behav. Ecol. Sociobiol. 27, 439–446.
Queller, D.C., Goodnight, K.F., 1989. Estimating relatedness using genetic markers. Evolution 43, 258–275.
Rose, L., 1994. Benefits and costs of resident males to females in white-faced capuchins, *Cebus capucinus*. Am. J. Primatol. 32, 235–248.
Rose, L.M., Fedigan, L.M., 1995. Vigilance in white-faced capuchins (*Cebus capucinus*) in Costa Rica. Anim. Behav. 49, 63–70.
Sheller, C.R., King, Z., Jack, K., 2009. The effects of infant births on male-female relationships in *Cebus capucinus*. Am. J. Primatol. 71, 380–383.
Silk, J.B., Altmann, J., Alberts, S.C., 2006a. Social relationships among adult female baboons (*Papio cynocephalus*). I. Variation in the strength of social bonds. Behav. Ecol. Sociobiol. 61, 183–195.
Silk, J.B., Altmann, J., Alberts, S.C., 2006b. Social relationships among adult female baboons (*Papio cynocephalus*). I. Variation in the quality and stability of social bonds. Behav. Ecol. Sociobiol. 61, 197–204.
Stephan, H., Bauchot, R., Andy, O.J., 1970. Data on size of the brain and of various brain parts in insectivores and primates. Adv. Primatol. 1, 289–297.
Strier, K.B., 1994. The myth of the typical primate. Am. J. Phys. Anthropol. 37, 233–271.
Van Schaik, C.P., 1996. Social evolution in primates: the role of ecological factors and male behaviour. Proc. Brit. Acad. 88, 9–31.
Zahavi, A., 1977. The testing of a bond. Anim. Behav. 25, 246–247.

# Studying Female Reproductive Activities in Relation to Male Song: The Domestic Canary as a Model

GÉRARD LEBOUCHER, ERIC VALLET, LAURENT NAGLE, NATHALIE BÉGUIN, DALILA BOVET, FRÉDÉRIQUE HALLÉ, TUDOR ION DRAGANOIU, MATHIEU AMY and MICHEL KREUTZER

LABORATOIRE D'ETHOLOGIE ET DE COGNITION COMPARÉES, UNIVERSITÉ PARIS OUEST NANTERRE LA DÉFENSE, NANTERRE CEDEX, FRANCE

## I. INTRODUCTION

Birdsong is one of the most complex communication signals in the animal kingdom and offers a unique model to understand animal behavior from a proximal as well as an ultimate perspective. A combination of multiple drifts and selections (Podos et al., 2004) shapes song traits (e.g., phonology, syntax, repertoire size), anatomical, and neurobiological adaptations for both production and reception of song. Therefore, birdsong is studied in various disciplines of sciences, from molecular biology to evolutionary ecology (Naguib and Riebel, 2006).

Within this diversity, birdsong probably represents one of the best models to understand sexual selection (Darwin, 1871); it gave rise to many studies revealing to be fertile. It is commonly recognized that male song in Passerine oscine birds serves two main functions: (a) an intrasexual function: deterring potential competitors and territorial defense; (b) an intersexual function: attracting and stimulating potential mates (Catchpole and Slater, 2008; Kroodsma and Byers, 1991). This question of birdsong functions is the focal point of many bioacoustics studies. About the attractive effects of songs, Møller (1991) suggested that males sing not only to seduce a social partner but also to gain extra-pair copulations (see also Hasselquist et al., 1996).

Female canaries (*Serinus canaria*), whose reproductive activity was the object of the earliest modern studies on animal behavior and hormone–behavior interaction (Hinde, 1958; Warren and Hinde, 1961), proved to be sensitive to the stimulating effect of songs at the onset of the



0065-3454/12 $35.00
DOI: 10.1016/B978-0-12-394288-3.00005-8

reproductive cycle. Pioneering works of Kroodsma (1976) or Hinde and Steel (1976, 1978) pointed out that male canary songs drastically stimulated the setup of female reproductive cycle, increasing plasma luteinizing hormone, follicle recrudescence, egg laying, and nest building. Hinde and Steel (1976) pointed out that only specific songs affected nest building in female canaries; moreover, Kroodsma (1976) demonstrated that, in this regard, large repertoires of male canary songs were more effective than small ones.

Besides that, Spitler-Nabors and Baker (1983) found that female white-crowned sparrow (*Zonotrichia leucophrys*) stimulated by male songs in their natal dialect showed a more intensive nest-building activity than females stimulated by male songs in an alien dialect. This work clearly emphasized the influence of early acoustic environment on female reproductive behaviors.

A study published in 1977 by King and West allowed the scientific community to make further progress in the knowledge of intersexual effects of birdsong. These authors reported that captive brown-headed female cowbirds (*Molothrus ater*), isolated from cage mates, adopted a posture, they termed the "copulatory response" when conspecific songs were broadcast in their cages (King and West, 1977). The female's response, a courtship display that normally precedes copulation in many birds, is better known as the copulation solicitation display (CSD). Many works focusing on CSD responses of captive female generally treated with estradiol were published in the subsequent years (Searcy, 1992a; Searcy and Marler, 1981; West et al., 1981). In these studies, female birds proved to be very sensitive to song quality, for instance to large repertoires (Catchpole et al., 1986; Searcy, 1992b). Moreover, as stated above concerning stimulating effects of songs, the attractive effect of male vocal stimulation appeared to be largely dependent of early experience (Baker et al., 1981b; Casey and Baker, 1992).

Female canaries also display CSDs when stimulated by conspecific male songs (Kreutzer and Vallet, 1991). Moreover, further studies pointed out that special song phrases containing bipartite syllables composed of abrupt frequency falls and short silences appeared to be particularly efficient to elicit sexual responses in female canaries; such phrases were called "A phrases" or "sexy phrases" (Vallet and Kreutzer, 1995; Vallet et al., 1998).

This type of syllable was found by Leitner and coworkers in the non-domesticated, free-living relative of the domesticated canary, the island canary (*S. canaria*) (Leitner et al., 2001a,b). Leitner and coworkers point out that "in wild canaries, rapid frequency-modulated syllable types are likely to occur in the context of female choice. They resemble very much the 'sexy syllables' described previously for domesticated canaries /.../our playback experiments with bipartite syllables of wild canaries suggest that some wild canary syllables have a high reproductive potential, i.e. are 'sexy' syllables" (Leitner et al., 2001b, p. 900). This finding gave us confidence to the adaptive value of this type of signal.

The discovery of sexy syllables in domestic canaries strongly guided further attempts, in our laboratory, to decipher the effects of male canary song, particularly with regard to females' attraction and stimulation, and probably led us to underestimate the intrasexual function of the male canary song. The results of these studies, mainly on the effect of male song on female reproductive activity in the domestic canary of the common strain, will be presented throughout the following sections. To begin, let us get back to the methods used to estimate female responses to male songs and particularly, the A or sexy phrases.

## II. Female Canaries' Preferences—The Sexy Phrases

In male and female song birds, the ability to discriminate and to recognize acoustic stimuli like songs has been usually measured using their responses to song playback like CSDs or attraction. We cannot rule out that recognition may exist in males or females as a cognitive process working in silence; nevertheless, the cognitive processes of song discrimination and recognition are not directly quantifiable and can only be estimated using observable behaviors.

### A. Female Preferences for Male Song: How to Appraise Them? With or Without Estradiol Treatment?

Studies on song recognition in songbirds were male biased for a long period because no reliable method existed to test female response to song in the field, such the territorial playbacks for males (Searcy, 1992a). In 1977, King and West reported that captive North American cowbird *M. ater* females performed CSDs in response to the song of conspecific males: females lowered their wings, arched their bodies and heads, and separated the feathers of the cloacal region (King and West, 1977). A few years later Searcy and Marler developed a method enabling to test sexual preferences of wild song sparrows *Melospiza melodia* in captivity, by using estradiol implants and subsequently measuring CSDs (Searcy and Marler, 1981).

During the next decade, this method was applied to at least 20 different bird species and was frequently used to test female preferences for different classes of songs (conspecific vs. heterospecific, home vs. alien dialects) or different attributes of male song such as repertoire size or song length (Searcy, 1992b). Some species such as blackbirds *Turdus merula* or red-winged blackbirds *Agelaius phoeniceus* perform CSDs only when estradiol implanted (Dabelsteen, 1982; Searcy and Capp, 1997). However, this method has its limits on the practical side as wild birds have to be brought in captivity and to undergo a surgery in order to be implanted. Moreover, some species such as willow warblers *Phylloscopus trochilus* or cirl buntings *Emberiza cirlus* (Michel Kreutzer, unpublished data) do not display even when estradiol implanted (see review by Searcy, 1992a).

We first used the implant method in order to test female domestic canaries' preferences; this method allowed us to highlight female canaries' preferences for conspecific songs and for the songs of their own breed (Kreutzer and Vallet, 1991; Kreutzer et al., 1992; Vallet et al., 1992). In these experiments, females were treated with crystalline 17-β estradiol contained in a Silastic tube. Subsequently, we observed that nonimplanted females naturally performed CSDs in response to song playback between 22 and 27 days after their first offspring had hatched, period that corresponds to their second reproductive cycle (Nagle et al., 1993).

Increasing day length and increasing temperature (e.g., Schaper et al., 2012) can be used by female birds as cues for the onset of the laying. In female canaries under natural conditions, ovarian activity and reproductive behavior are known to be triggered by photoperiod changes from a winter schedule to a spring–summer one (Hinde and Steel, 1978). A further experiment supported the idea that female canaries could display CSDs without any estradiol treatment: two groups of females were tested with male song 3 weeks after a photoperiod change from short days (8 h of light:16 h of darkness, LD 8:16) to long days (LD 16:8): implanted and control females equally responded with copulation displays with a CSD peak at the early beginning of egg laying (Leboucher et al., 1994). As a consequence, a new and less invasive method to test sexual responses of female canary was born using the timing of the female natural reproductive cycle.

The concentrations of female sexual hormones change during the female reproductive cycle. These changes were studied to understand the relationships between hormones and reproductive behavior. Estradiol concentrations are the highest at the beginning of the reproductive cycle before the laying of the first egg; during this period, females actively display CSDs. Leboucher et al. (1994) showed that an artificial increase of estradiol concentrations did not affect the magnitude of the sexual response. Using fadrozole (an inhibitor of the aromatization of androgens to estrogens) as a tool for limiting an individual's exposure to estradiol allowed us to emphasize the crucial role of estradiol in triggering sexual responses: fadrozole treatment at the beginning of the reproductive period significantly delayed the onset of these sexual responses (Leboucher et al., 1998a). When females were injected later, after they were sexually active, the

Fadrozole treatment did not affect sexual displays. These results are consistent with the hypothesis that a threshold level of estradiol is critical to activate the neural circuitry, mediating the CSD response in the female canary. They also give added strength to the idea that the magnitude of sexual response is not related in a dose-dependent manner to estrogen concentrations observed during the period of sexual responsiveness.

Plasmatic progesterone also plays a role in sexual displays. Plasmatic progesterone concentrations change throughout the female reproductive cycle: a plasmatic progesterone surge follows the plasmatic estradiol increase when the females are sexually receptive (Sockman and Schwabl, 1999). Leboucher et al. (2000) studied the influence of systemic injections of progesterone on female CSDs and found that progesterone treatment provoked an inhibition of females' CSDs as soon as 48 h after the beginning of the treatment. Progesterone plays a key role in mediating the transition from active female courtship behavior to sexual refractoriness in this species. Suppressive effects of progesterone on female sexual behavior have been previously described in lizards (Godwin et al., 1996) as well as in rodents (Blaustein and Wade, 1977; Gonzalez-Mariscal et al., 1993). Data on female canaries are consistent with the hypothesis of Godwin et al. (1996) which proposed that the decrease in sexual behavior following plasma progesterone increase represents an evolutionarily conserved mechanism in the regulation of female sexual behavior.

To sum up, under a stimulating light schedule, female domestic canaries naturally exhibit sexual responsiveness during a short period bounded by estradiol and progesterone surges around the laying period. Females' choice and selectivity are also related to the development of the reproductive cycle: females were found to be more discriminative during the 3 days preceding the laying of the first egg when sexual motivation is high (Amy et al., 2008).

Testing females for sexual preferences during their receptive phase of the reproductive cycle reasonably increases the likelihood that the observed responses are adaptive. Between 1994 and 2010, 25 papers were published using timing of the female natural reproductive cycle described above (Leboucher et al., 1994) that enabled us to study in a noninvasive manner female preferences for different aspects of male song and making the domestic canary one of the two most investigated species in this regard, together with the zebra finch *Taeniopygia guttata* (Riebel, 2009).

Besides the classical components of the CSD (lowered wings vibrating, arching their heads and tails), we observed that females also called during song playbacks. We therefore wondered if this behavior could also be used as a measure of female sexual preferences. Two groups of females kept in long days photoperiod were tested with the same stimuli: for one group we measured CSDs during 3 weeks and for the other group female contact calls during 4 days. Female response patterns were identical for the two groups, calling responses matching CSDs responses, suggesting that female calling could be potentially used to test females' preferences (Nagle et al., 2002).

## B. General Female Preferences for Song Types, Phonology, Syntax, and Acoustic Context

In our first experiments, females were tested for general preferences. Conspecific songs elicited stronger responses (CSDs) compared to heterospecific ones such as winter wren *Troglodytes troglodytes* or greenfinch *Chloris chloris* songs. Nevertheless, some conspecific songs coming from other strains than domestic canaries, such as the wild canary or the harzer roller, elicited only weak responses (Kreutzer and Vallet, 1991).

When investigating the salience of song features on common female canary song recognition, we demonstrated that females used both phonology and syntax not only to discriminate between heterospecific and conspecific songs but also to discriminate the songs of their own versus an alien strain (Vallet et al., 1992).

The salience of these two acoustic parameters was confirmed by demonstrating that females clearly prefer crystallized songs from adults compared to plastic ones, with a relaxed phonology and syntax, emitted by subadults' males of the common domesticated canaries (Kreutzer et al., 1996).

The acoustic context of songs playback is also an important cue in eliciting female responses; in this way, a chorus adding two songs appeared to be more efficient in eliciting CSDs than each of these songs heard separately (Nagle et al., 1997). When investigating in more details female responses to song organization, we discovered that the salience of some song phrases might be affected by the preceding or following song phrases, which we called the song bout organization. From these results, one could easily imagine that female needs to hear long strings of songs in order to evaluate their relevance. But surprisingly, females may respond at the onset of some song stimuli without waiting for the end of a song (Kreutzer et al., 1992). These results on the general female preferences led us to evaluate the relative effect of each phrase composing a song and to the discovery of the A or sexy phrases.

## C. A Sexually Attractive Phrase: The A or Sexy Phrase

Male canaries have song repertoires composed of more than a dozen of different phrase types, each of them made of the repetition of a particular note type. Early on, we observed that females preferentially performed CSDs when hearing a particular phrase made of the repetition of a brief two-note syllable with a high-frequency bandwidth (1–3.5 kHz for the low-pitched note and 3.7–5.8 kHz for the high-pitched note) emitted at a sustained rhythm: 17 syllables per second (Vallet and Kreutzer, 1995; Vallet et al., 1998; Fig. 1)

At the same time, Podos suggested the existence of a trade-off relationship between the frequency bandwidth and the syllable repetition rate within the song of several songbirds of the Emberizidae family (Podos, 1997). It is difficult to produce songs with both high bandwidth and high repetition rate. So, when bandwidth increases, repetition rate generally

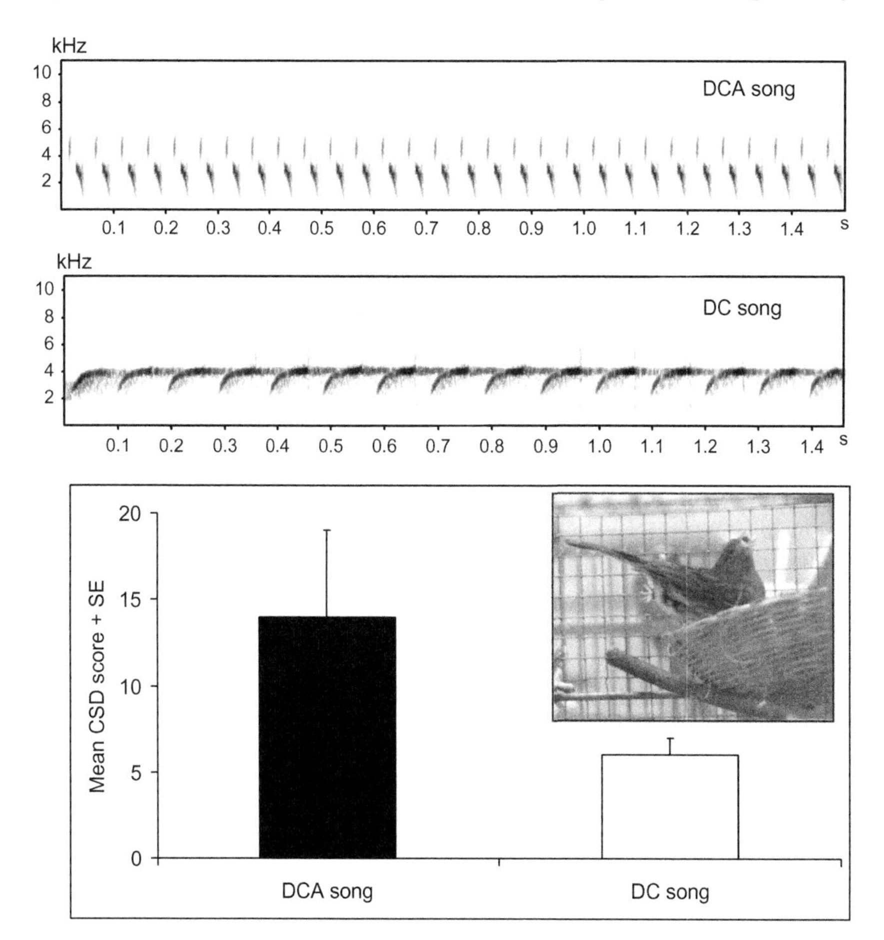


Fig. 1. The mean CSD (copulation solicitation display) number ± SE of nine female canaries for two types of songs: domesticated common canary with A phrase—DCA (dark columns) and domesticated common canary without A phrase—DC (open columns). (For color version of this figure, the reader is referred to the online version of this chapter.)

decreased, and conversely, when repetition rate increases, bandwidth generally decreases. This trade-off relationship is verified within the song of the domestic canary (Draganoiu et al., 2002). In this regard, sexy phrases associating large bandwidth and high repetition rate are particular song elements and are rather rare in male repertoires.

In a further experiment, we tested the importance of both the frequency bandwidth and the syllable rate, and we found that females pay attention to both cues: for a same syllable rate, they prefer phrases with a larger bandwidth, and for a same frequency bandwidth, they prefer phrases with a higher syllable rate. Consequently, females prefer male songs maximizing both the syllable rate and the frequency bandwidth (Draganoiu et al., 2002). Two experiments also investigated female preferences for the frequency range of sexy phrases, and in both cases, females privileged phrases with the lowest minimum frequencies (Pasteau et al., 2007). Females also pay attention to the length (Pasteau et al., 2009a) and the intensity (Pasteau et al., 2009b) of sexy phrases (see Section III below for more detailed data).

As previously mentioned, these phrases have also been described in wild canaries, and they elicit an equally high number of CSDs when broadcast to domesticated female canaries as sexy phrases from domestic canaries (Leitner et al., 2001a,b).

### D. The Sexy Phrase Related to Male Condition and Social Contexts

Since it seems very difficult for a common domesticated male canary to sing syllables with a large frequency bandwidth at a high repetition rate, it should be tempting to regard the sexy phrases as honest signals of male quality (Zahavi, 1975).

The observations of social interactions confirm the importance of A or sexy phrases both for emitters and for receivers. On the production side, males sing longer bouts of sexy phrases in front of both males and females when compared to a situation where they sing alone (Kreutzer et al., 1999). On the perception side, sexy phrases elicit high levels of CSDs in females as previously mentioned, whereas males call less when hearing sexy phrases versus other conspecific or heterospecific phrases, suggesting that the hearing of sexy phrases provokes a social inhibition of male vocal behavior (Parisot et al., 2002).

Such types of sexy phrases are rarely produced by nondomesticated island male canaries (wild canaries) outside the reproductive period, but more frequently during the reproductive period (Leitner et al., 2001b). This may suggest that in this strain as well as in the common domesticated strain, such sound productions are dedicated to social interactions during the breeding season.

Similar findings were reported in a couple of wild songbirds. For example, female swamp sparrows *Melospiza georgiana* display more in response to songs with a high-frequency bandwidth and trill rate that are close to vocal performance limits in this species (Ballentine et al., 2004). In this species, the ability to perform physically challenging songs predicts both age and size in males (Ballentine, 2009). A similar kind of vocalization is an indicator of male mating status and male dominance in water pipits *Anthus spinoletta* (Rehsteiner et al., 1998).

In canaries, we further investigated if the production of this particular phrase type may be an indicator of males' condition. We found that the sexy phrases are produced by a larger proportion of dominants than subordinates and could thus be seen as signal of dominance in male canaries (Parisot, 2004). However, an experimental direct test of the developmental stress hypothesis (Buchanan et al., 2003; Nowicki et al., 1998) in domestic canaries could not find any effect of early conditions, either brood size or the food quality postfledging, on the proportion on males singing sexy phrases or on the length of these phrases (Müller et al., 2010). Another study using parasite injection (*Plasmodium relictum*) during the postfledging stage failed to find any difference in the proportion of adult males singing sexy phrases, but low occurrence of sexy phrases in this study makes it difficult to draw a firm conclusion (Spencer et al., 2005). It should be noted that adult male canaries infected as juveniles develop simpler songs as adults compared to uninfected individuals.

We know that the production of these phrases needs particular peripheral motor skills to coordinate the two parts of the syrinx and the coordination of the right and left HVC at the central level (see Section IV), but for time being, unequivocal proofs revealing the sexy phrase as an indicator of male condition remain to be discovered.

In the previous paragraphs, we gave strong emphasis to responsiveness of female canaries to sexy phrases. One may wonder to what extent the effect of sexy phrases on females' sexual responses is the consequence of early learning. This point is addressed in the next section.

## III. Early Influences and Predisposition

### A. Learning Influences Female Song Preferences

In birdsong species, males learn their songs. Auditory experience plays a vital role in the male song development (Beecher, 2008; Catchpole and Slater, 2008; Marler, 1970; Thorpe, 1958, 1961). For example, young male white-crowned sparrows, *Z. leucophrys nuttalli* (Marler, 1970), or canaries

(Mundinger, 1995) learn songs by copying models heard during their first weeks of life. Male songs convey sexual information and can influence the mate choice of females (Andersson, 1994). Thus, the females can choose a male partly using his song (Payne, 1983); this choice is often based on different male song characteristics (Catchpole and Slater, 2008; Zann, 1996).

The importance of these different song characteristics can vary across species (Riebel, 2003). In addition, in some species, different females can demonstrate different sexual preferences (Burley and Foster, 2006; Holveck and Riebel, 2010; Lerch et al., 2011). Until recently, little was known on the establishment development of females' song preferences. Several authors have proposed that imprinting-like processes could play an important role in shaping female preferences (Owens et al., 1999; Riebel, 2003; ten Cate and Vos, 1999). Different studies have thus demonstrated that preferences in adult females for songs were experienced in early age. In zebra finches, young females can learn songs from their father. In their adult age, when the authors tested these females, birds approached the loudspeaker playing back their father's song rather than the one playing an unfamiliar song (Clayton, 1988; Miller, 1979a), or they preferentially pressed the key which emitted playback of their father's song in an operant task (Riebel et al., 2002). Different studies have recently observed that song learning in birds is influenced by social factors like eavesdropping on singing interactions of adults (Templeton et al., 2009).

## B. Song Tutoring and Sexual Preferences in Female Canaries

In canaries, Nagle and Kreutzer (1997a) and Depraz et al. (2000) observed that young females can learn their sexual song preferences by being tutored from different tape tutors; an indication of female song preferences was obtained from their CSDs.

Female domesticated canaries exposed to wild canary songs during their first months of life developed sexual preferences for these tape tutor songs (Nagle and Kreutzer, 1997a). The wild canary songs have not the same song characteristics (syntax and phonology) as the domesticated canary songs. Depraz et al. (2000) found somewhat different results: they observed that females early exposed to wild canary failed to prefer wild canary songs over domestic canary songs. In comparison, females raised hearing domestic canary songs were reluctant to display CSDs hearing wild canary songs and preferentially responded to domesticated canaries (Fig. 2; Depraz et al., 2000; Nagle and Kreutzer, 1997a). The clear preference of females early tutored by domestic songs seemed to be unrelated to their responsiveness toward sexy phrases: the same results were obtained whether sexy phrases were present in either domestic and wild canary songs or not (Depraz et al., 2000).

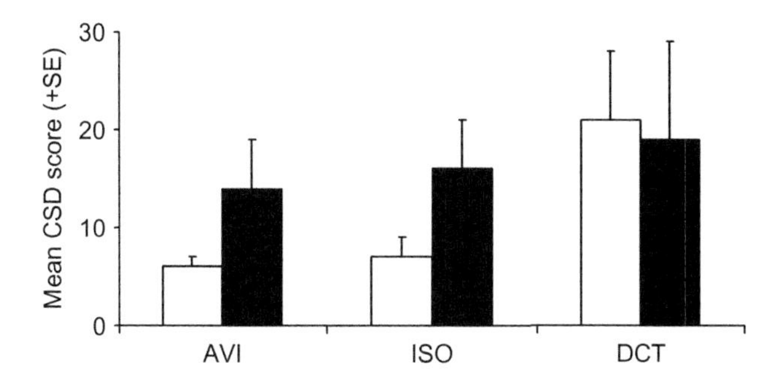


Fig. 2. The mean display scores (±SE) measure the number of copulation solicitation displays (CSD) of the female canaries for two types of songs: domesticated common canary without A phrase (DC, open columns) and domesticated common canary with A phrase (DCA, dark columns) in different groups (AVI = aviary group, 9 females reared in aviary environment exposed to different adult domesticated common canary songs; ISO = isolated group, 10 females reared acoustically isolated from adult songs; DCT = domesticated canary tutored, 6 females tutored with DC song).

Taken together, these results indicate that females can learn their sexual song preferences and that early exposure to song is important in learning processes.

Nevertheless, females do not develop strong sexual preferences to every song proposed by the tape tutor. Nagle and Kreutzer (1997a) observed that females exposed to heterospecific songs (Pine siskin, *Carduelis spinus*) responded less to these tutor songs. Such a weak influence of heterospecific songs was also demonstrated in male song learning (Marler and Peters, 1988). In addition, females reared hearing three different song categories (domesticated canary songs, wild canary songs, and pine siskin songs) develop strong sexual preferences for the two first types. These results show that female domesticated canaries can develop strong preferences to conspecific songs but not, or with more difficulty, to heterospecific songs. In these experiments, with tape tutor, the learning processes seem to have been limited to conspecific songs, but we cannot rule out that the females do memorize and learn heterospecific songs but that these songs do not trigger CSD as conspecific songs do.

## C. Predispositions, Learning, and Song Preferences

Nagle and Kreutzer (1997a) also demonstrated in these experiments that females isolated from live males or tutoring songs in soundproof chambers during their early life responded significantly more to the songs of their own species when adult; in this experiment, the domestic canary songs contained sexy phrases. Similar results were obtained by Nelson and Marler (1993) and Dooling and Searcy (1980), respectively, in white-crowned sparrows and swamp sparrows, *M. georgiana*: these authors observed that hand-reared females prefer conspecific songs without previous exposure. In addition, Nagle and Kreutzer (1997a) observed that females tutored with heterospecific songs respond to songs containing sexy phrases (and not to heterospecific songs). All these results support the hypothesis that female canaries possess a perceptual predisposition (Dooling and Searcy, 1980) or sensory bias (Ryan et al., 1990) for certain parts of the song, in particular to the sexy phrase. This hypothesis should explain why adult females present sexual preferences for this phrase without previous exposure. In addition to these predispositions, females can learn songs, or part of songs, if they are tutored during early age with conspecific songs containing or not the sexy phrase. This early song exposure can impede or add up to the female's predispositions (Depraz et al., 2000; Nagle and Kreutzer, 1997a).

### D. Adult Females Can Modify Their Song Preferences

In addition, Nagle and Kreutzer (1997b) noticed that females were able to develop new song preferences in adult life. Females tutored with wild canary songs in early age responded preferentially to this type of songs in adult age; however, after the breeding period where these females were placed in aviary conditions with male domesticated canaries, they responded in a similar way to the domesticated canary and to the wild canary songs. Thus female song experience can modify song preference. Nevertheless, their preference for the tutor song remains, and they can recognize their taped tutoring songs 10 months later, without hearing these songs during this period. Different authors obtain similar results concerning the modifications of sexual preferences in adult age. For example, passerine females develop sexual preferences to the mate songs (Béguin et al., 1998; Miller, 1979b; O'Loghlen and Beecher, 1997; see Section V) or to the song of a male that they were housed with (Clayton, 1988). In the same way, late song exposure without social reinforcement modifies acoustic preferences in female canaries (Depraz et al., 2004). In contrast, Riebel (2000, 2003) reports that a later tape tutoring did not influence adult female's preferences in zebra finches. Later song exposure could be not efficient if not combined with social or visual stimulation (Riebel, 2003).

### E. Predispositions, Learning, and Sexy Phrases

Predisposition and learning can influence female song recognition and sexual preferences. These both processes are not mutually incompatible and may be complementary. Female domesticated canaries seem predisposed to discriminate a number of phrases of their own strain (like the A phrase) or species. Early or adult song learning can secondarily influence these predispositions.

As previously mentioned, sexy phrases are composed of complex syllables (two notes or more) with a high-frequency bandwidth (4 kHz) emitted at a high repetition rate (minimum 17 syllables per second). Males produce also this sexy phrase with a high amplitude level (in dB), a variable frequency level (low-, median-, or high-pitched phrase), different durations, and a particular amplitude modulation that works as an acoustic parameter implicated in vocal signature (Lehongre et al., 2008). Many experiments in our laboratory have been realized to test the functions of these different acoustic parameters on female sexual preferences (see Section II). For each of these experiments, two groups of females have been constituted: the first was raised in partial acoustic isolation, which means that no male vocalization was heard by these females, and the second was raised in normal acoustic conditions in aviaries containing also male canaries. The influences of predispositions and learning processes on female sexual preferences have been evaluated for each acoustic parameter with CSDs. We observed that females reared in acoustic isolation or in aviary acoustic conditions have a similar preference for broad bandwidth songs ($>4$ kHz) with high syllable rates ($>16$ syllables per second). This result indicates that a learning process seems not to be necessary to develop sexual preferences for these two acoustic parameters (Draganoiu et al., 2002).

We also noticed that females responded to supernormal stimuli, that is, phrases with a broad bandwidth emitted at an artificially increased rate, an exaggerated song male trait beyond the limits of vocal production not naturally produced by male canaries (Draganoiu et al., 2002); in this chapter, we proposed that production of song phrases maximizing both bandwidth and syllable rate might be a reliable index of males' quality in accordance with the honest-signaling hypothesis (Zahavi, 1975). Our results (Draganoiu et al., 2002) suggest that female canaries seem to use at least two cues of males' vocal production (frequency bandwidth and high syllable rates) in order to assess the quality of their future partner.

Nevertheless, other acoustic parameters may be used by the female canaries in this aim. We observed that females, raised in acoustic isolation or in normal conditions, preferred intrasyllabic diversity (three notes in each syllable) in comparison with intrasyllabic simplicity (a syllable constituted by one note) (Pasteau et al., 2004) and that females showed sexual preferences to sexy phrases emitted at normal and low pitch levels versus high pitch (Pasteau et al., 2007). The low-pitched sounds could be correlated with anatomical and physiological condition in males, that is, body size and syrinx size (Bertelli and Tubaro, 2002; Wallschläger, 1980) and high testosterone level (Beani and Dessi-Fulgheri, 1995; Fusani et al., 1994). The preference for lower pitch is also found in other taxa as insect, anuran, fish (Ryan and Keddy-Hector, 1992), and our own species (Feinberg et al., 2005). Nevertheless, different authors showed that females responded more to songs with high pitch in female blackbirds, *T. merula* (Dabelsteen and Pedersen, 1993); white-throated sparrows (Kroodsma and Miller, 1982); and serins, *Serinus serinus* (Cardoso et al., 2007).

Pasteau et al. (2009b) also observed that female canaries give more CSDs in response to loud and normal intensity levels (song amplitude in dB) in comparison with weak intensity level. Searcy (1996) obtained a similar result with female red-winged blackbirds, *A. phoeniceus*, and Ritschard et al. (2010) with zebra finches. In addition, female canaries responded more to the longest sexy phrases (2 s) in comparison with shorter sexy phrases (1 or 1.5 s; Pasteau et al., 2009a). The preference for longer sexy phrases is in agreement with results obtained in other species (Clayton and Pröve, 1989; Nolan and Hill, 2004; Nowicki and Searcy, 2005; Wasserman and Cigliano, 1991). Song duration is often correlated with morphology and physiological conditions of males and could be an indicator of male quality (Kempenaers et al., 1997).

If we look at these experiments overall, the key point to emphasize is that females raised in acoustic isolation presented the same preferences as females raised in normal acoustic conditions. Learning, through acoustic experience, seems not to be necessary for developing sexual preferences concerning these different acoustic parameters. Intrasyllabic diversity, normal or low pitch, loud or normal intensity levels, amplitude modulation, and the duration of sexy phrases are important acoustic cues for female domesticated canaries. These different acoustic parameters should be reliable indicators of male physical or behavioral qualities and should be used by female canaries in mate choice, even though, as previously mentioned, experimental proofs revealing sexy phrases as an indicator of male condition remain to be discovered. However, the next section will emphasize that the emission of sexy phrases involves the two parts of the syrinx and requires particular neuronal pathways and motor skills.

## IV. Some Neuronal and Neuromuscular Aspects of Sexy Phrase Production in Canaries

### A. Bilateral Song Production in Male Common Domesticated Canaries: Neural Correlates at the Central Level

Song production is controlled by a network of discrete interconnected brain nuclei in songbirds including canaries (Nottebohm et al., 1976). A key nucleus of this network is the bilateral nucleus HVC (*nucleus hyperstriatalis ventrale pars caudale* or High Vocal Center), a cortex-analogous sensory-motor integration area, which is present in both right and left telecephalon (Nottebohm et al., 1976; Wild et al., 2000).

Neurobiological studies pointed out that HVC volume is associated with song repertoire size both between songbird species (DeVoogd et al., 1993; Székely et al., 1996) and between males within species (Garamszegi and Eens, 2004). For instance, in song sparrows (*M. melodia*), males with large repertoires had larger HVCs and were in better body condition (Pfaff et al., 2007). Taking into account that HVC is a bilateral structure, another way to decipher the relationship between neural anatomy and song quality was to investigate the influence of both right and left HVC on song organization and complexity, particularly with regard to sexy phrases.

To investigate the possibility of neural specializations in males, we studied the differential role of the male left and right HVC for the control of different features of the song (Hallé et al., 2003a). Songs recorded before and after partial or complete unilateral excitotoxic lesions of the HVC were thoroughly analyzed to correlate lesion size with different features of the male song. Just after the right HVC has been lesioned, birds sang unstable syllables. The phonology of a syllable could vary in the same phrase, with long intersyllables and intersequences pauses, but the overall song syntax was maintained. The songs had low amplitude. Following lesions of the left HVC, birds emitted very monotonous songs, consisting of unstable succession of sound elements rising and falling in pitch. Most of the songs had very low amplitude. The songs of left lesioned birds resembled the immature songs of young birds during development (Güttinger, 1981), lacking a clear phrase organization (Fig. 3).

Whatever the side of the lesion, treated birds failed to produce phrases containing bipartite syllables composed of abrupt frequency falls and short silences; in other words, they failed to produce attractive sexy phases.

As a whole, the experiments suggest a lateralized specialization of both left and right vocal control pathways for particular features of the song. It appears a general right-sided specialization for high frequencies and widest frequency band and a general left-sided specialization for the control of low frequencies. In agreement with a previous study on HVC lesions in Wasserschläger canaries (Nottebohm et al., 1976), we found in the common domesticated canaries that left-sided HVC lesions have a more drastic effect on song than right-sided lesions.

A longitudinal behavioral analysis (Hallé et al., 2003b) suggested that domesticated canaries are not or are only partially able to recover their previous repertoire size of complex syllables following right or left HVC lesions. The loss of most complex syllables and of all complex syllables of the sexy phrases following either left or right HVC lesions was permanent, while on average 56% of the size of simple syllable repertoire was

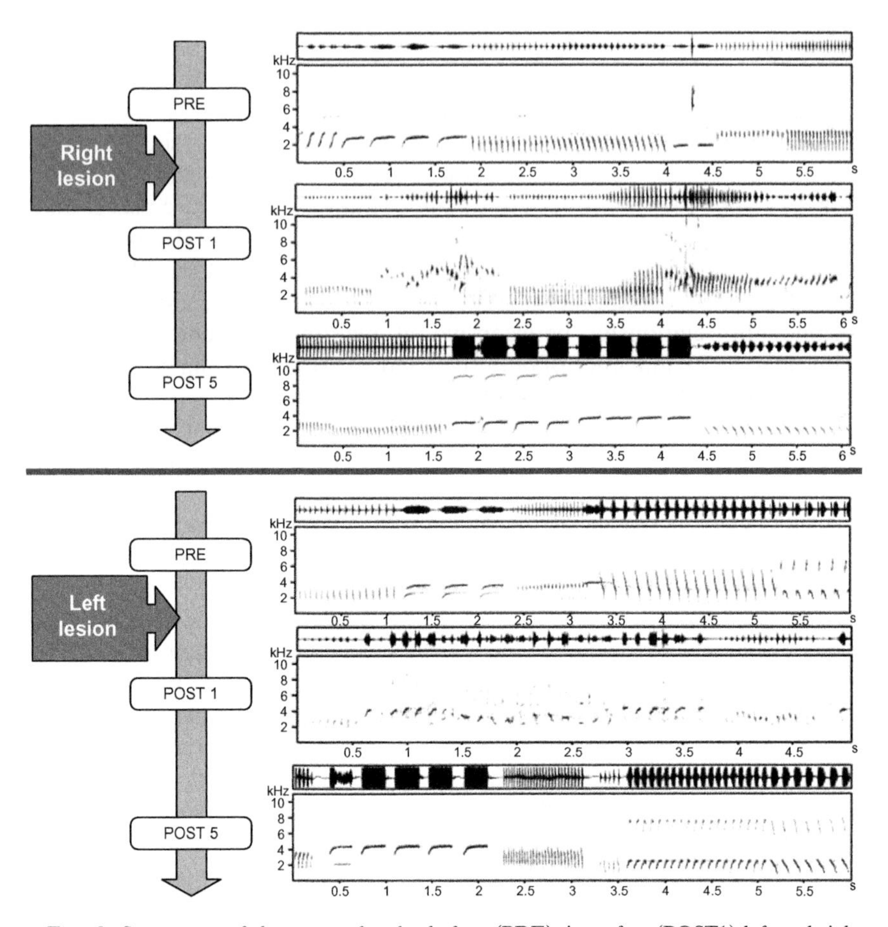


FIG. 3. Sonograms of the songs of males before (PRE), just after (POST1) left and right lesions, and 10 months later (POST5). Following lesions of the left HVC, none, or at most one, of the syllables were produced as they were preoperatively. Just after right lesions, birds sang unstable syllables, but the overall song syntax, that is, the succession of song phrases was maintained. Ten months later (POST5), all males recovered a normal canary typical song structure, but only simple syllables are recovered, no sexy syllables were recovered or only the complex syllables with a low repetition rate.

recovered. This behavioral deficit correlates with the magnitude of the permanent loss of HVC neurons following the lesions. Ten months after surgery, the male canaries in contrast to Wasserschläger canaries (Nottebohm, 1977; Nottebohm et al., 1976) were unable to recover their repertoire (Fig. 3). Moreover, domestic canaries have no significant regeneration of the HVC; the lack of recovery of the repertoire appears to be due to the permanent loss of neuron pools induced by the lesion.

### B. Bilateral Song Production in Male Common Domesticated Canaries: The Peripheral Level

We also studied the mechanisms of song production in the domestic canary at the peripheral level; toward these ends, we monitored airflow through each side of the syrinx, together with subsyringeal pressure, during spontaneous song (Suthers et al., 2004; Vallet et al., 2006). The contribution of each side of the syrinx to song was investigated by observing the effect of unilaterally occluding either the left or the right bronchus, followed by section of the ipsilateral branch of the tracheosyringeal nerve.

We found that some syllables were produced entirely on the left or right side of the syrinx, whereas others contained sequential contributions from each side (Fig. 4). Low fundamental frequencies were produced with the left syrinx and high frequencies by the right syrinx. Syllables at repetition rates below 25 $s^{-1}$ were accompanied by minibreaths, usually bilateral. At higher syllable repetition rates, minibreaths were replaced by another respiratory pattern: the pulsatile expiration.

Seven types of sexy phrases were recorded in three different males. All these phrases included contributions from both sides of the syrinx. Among these seven types of phrases, six were accompanied by minibreaths. The bilateral nature of these syllables is largely responsible for their increased spectral complexity, temporal complexity, and frequency bandwidth.

To conclude, data obtained at the central level show that the production of high attractive sexy phrases by male domestic canaries requires the contribution of both sides of the brain and particularly the integrity of the two HVC nuclei. At the peripheral level, such complex syllables involve the contribution of both sides of the syrinx and the subtle coordination with special temporal dynamics of air sac pressure at the abdominal level. Bilateral motor coordination of the syrinx at high syllable repetition rates, together with the respiratory motor pattern, must require excellent motor skills. This supports the hypothesis that a male's ability to produce such sexy phrases may convey information about his fitness. From an evolutionary point of view, this might explain why all females show a predisposition to prefer this type of syllable patterns.

### C. Central Structures Involved in the Production of Sexy Phrases in Males Are Also Involved in Females' Responses to These Phrases

In receptive female canaries, studies on bilateral lesions of the HVC pointed out the importance of this bilateral nucleus for the discrimination between heterospecific and conspecific songs (Brenowitz, 1991; Del Negro

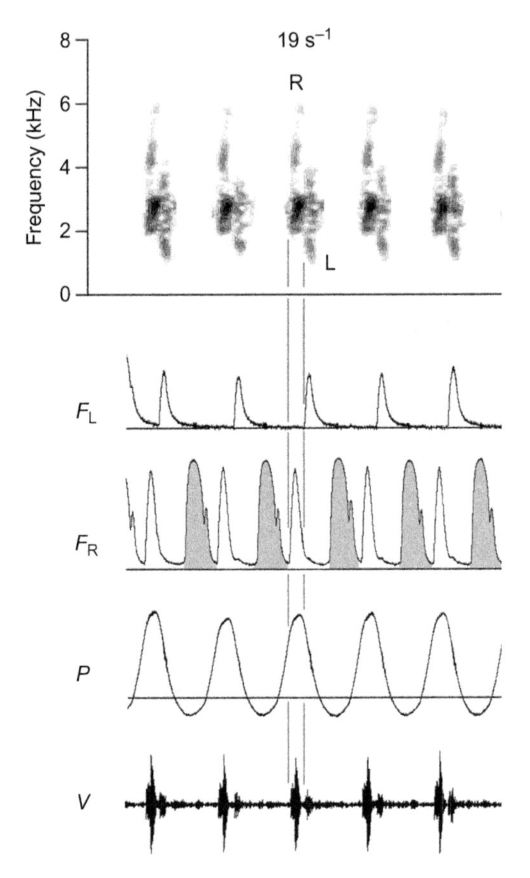


FIG. 4. Description and production of sexy phrases (picture from Rod Suthers). Sonogram of a "A" phrase (top panel). The bipartite syllables are emitted at the repetition rate of 19 $s^{-1}$; R and L indicate the side of the syrinx used to produce the syllable; $F_R$ and $F_L$ are the air flows measured in each bronchus (right or left), the inspirations are in gray, and the expirations are in white; *P* is the abdominal pressure and *V* the oscillogram. This picture clearly demonstrates that the use of the right and left sides of the syrinx enlarges the bandwidth of the syllables and that such syllables are accompanied by minibreathes, requiring special coordination of different motor skills.

et al., 1998) as well as for the discrimination of sexually attractive and less attractive conspecific songs (Del Negro et al., 1998). Electrophysiological studies further emphasized the role of HVC for song discrimination (Del Negro et al., 2000). These data are consistent with the work of Leitner and Catchpole (2002) on female canaries, showing a positive relationship between females' ability to discriminate songs and the size of their HVC.

Further, we showed that the integrity of one of the two HVCs is required in receptive female canaries to recognize sexually attractive phrases (Hallé et al., 2002). Further, we showed that animals recover their song preferences over a period of several months after the lesion. This functional recovery does not involve anatomical recovery of the HVC. These results suggest that the functional recovery was based on a neuronal reorganization of the HVC.

Such functional plasticity of HVC might be involved in the changes of auditory preferences related to reproductive experience observed in normal female canaries (Béguin et al., 1998; Nagle and Kreutzer, 1997b) and presented in the following section.

## V. Reproductive Experience and Female Preferences

As described in Section III, acoustic experience in early life influences song preferences in female domesticated canaries. These preferences can also be modified in adulthood by different reproductive experiences throughout their life.

### A. Choosing Between Mate and Nonmate

In the wild, females are often constrained by the mate choice of other females and are not always able to pair with the mate who should be their optimal choice. An association with a particular male is an important cue for adult females' maintenance of sexual preferences (O'Loghlen and Beecher, 1997).

An experiment conducted in our laboratory with female and male canaries revealed that a single reproductive experience with a particular male could be memorized by females even if this pairing was short lived (only 3 days). Indeed, female canaries displayed more sexual responses (CSDs) to the songs of their previous mate than to the songs of nonmate even after being separated, more than 1 month, from this partner (Béguin et al., 1998). Other studies in different species found females to be more responsive to mate's vocalizations (Eens and Pinxten, 1995; Miller, 1979b; Mundinger, 1970; O'Loghlen and Beecher, 1997; Robertson, 1996; Wiley et al., 1991), strongly suggesting that females are capable of recognizing their mates' song. We concluded from this experiment that a reproductive experience with a particular male could influence female's sexual preferences (Béguin et al., 1998).

## B. Previous Reproductive Experience, Reproductive Success, and Females' Preference

Clearly, the previous study indicated that a reproductive experience influences female preferences for male songs, suggesting an adjustment of their sexual preferences for males emitting such songs. This question was addressed in a further experiment in which we appraised the effect of reproductive success on such a preference. Females previously paired and having reared young were placed in a choice test situation: they were allowed to choose between their previous mate and a familiar male (a neighbor male during the breeding period). During these choice tests, females stayed longer near their previous mate than near a neighbor male when their reproductive success was "good" (at least two chicks). In contrast, females with "poor" reproductive success (one chick) did not show this preference for their previous mate (Béguin et al., 2006; Fig. 5). Further, we observed that during choice tests males responded to the presence of their previous mate in a particular way: they gathered nest material. This behavior was scarcely observed in neighbor males which behaved differently and sang significantly more than previous mates did (Béguin et al., 2006).

These two experiments highlight the influence of reproductive experience in males and females. Mate recognition occurs in both sexes which adapt their reproductive strategies according to their reproductive experience with a particular partner. This mates' recognition has been described

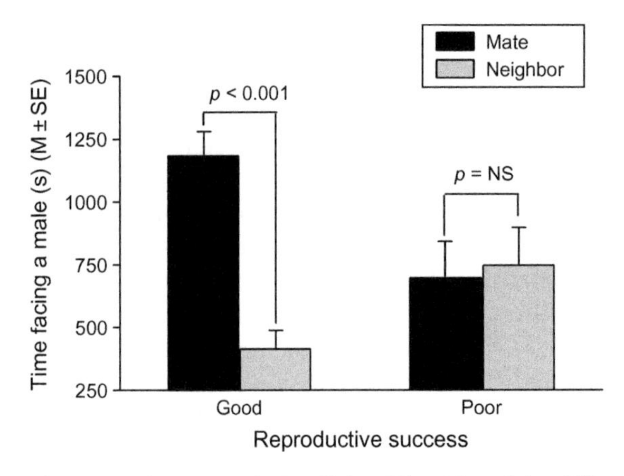


Fig. 5. Effect of reproductive success ("Good" ($n=23$) or "Poor" ($n=15$)) on time spent by females when facing their previous mate (M) versus a neighbor male (N) during choice tests.

in other species (Black, 1996; Vignal et al., 2004, 2008) and has been shown to be important for pair-bond maintenance in birds or for saving time and energy in future reproductive attempts.

Mate switching in bird species often occurs because pairs failed to brood at least one chick during the previous breeding season. Divorce and remating in birds have been described as strategies used to enhance reproductive success by seeking better quality partner (Choudhury, 1995; Lindén, 1991).

Taken together, these data match the hypothesis that adaptive mate preference is strongly constrained by previous experience (Fawcett and Bleay, 2008).

### C. Influence of Physical Condition on Female Preferences

The expression of male secondary sexual traits had been proved to be dependent of their physical condition (David et al., 2000; Hedrick, 2005; McGraw et al., 2002; Møller, 1992; Wagner and Hoback, 1999). The males are often chosen by females on the basis of their secondary sexual traits, which reveal their quality (Andersson, 1994). Less is known about female physical condition and its influence on females' choosiness. Nevertheless, one study in zebra finches *T. guttata* has shown that females are less selective toward males' secondary sexual traits when their flight ability is experimentally suppressed (Burley and Foster, 2006). We altered physical condition of female domestic canaries and studied their selectivity toward attractive songs. In a first experiment, the female physical condition was modified by shortening the wing feathers of some females as described in Burley and Foster (2006). Females with intact wings were more selective than females with altered wing feathers. These results should indicate that females in "better" condition try to mate with "better" males, whereas females in "poor" condition are less selective in their mate choice (Lerch et al., submitted for publication). A similar research (Holveck and Riebel, 2010) found that low-quality female zebra finches prefer low-quality males when choosing a male. In the study of Holveck and Riebel (2010), females reared in poor condition preferred the song of males reared in poor condition rather than those of males reared in good condition.

In a different experiment (Lerch et al., 2011), the physical condition of females was modified by feeding two groups of females with different diets (highly or poorly diversified food during a short period). Females were exposed to recorded highly attractive sexy songs or to less attractive ones. The two groups of females displayed more to sexy songs; nevertheless, the difference between the number of sexual responses toward highly attractive and less attractive songs was weaker among females with a poorly diversified food diet in comparison with females with a highly diversified food diet.

A short-term modification of condition is enough to decrease females' selectivity toward high-value stimuli and increase their response rates toward low-value signals. A similar result was found by Holveck and Riebel (2010).

These results indicate that female domestic canaries can evaluate their own condition and can produce adaptive responses to male song. Individuals with better condition were found to be choosier. If different diets can change female sexual selectivity, we can suppose that the access to food could also modify this choosiness. Access to food is a mean to determine the social status of a bird in several species. The more the bird has access to food, the more dominant she or he is considered (Baker et al., 1981a; Delestrade, 1999; Parisot et al., 2004) that can impact on the reproductive behavior of birds.

Individual's reproductive experiences, environmental, or physical conditions of females are some of numerous parameters that can shape female preferences for different males or male song types. Female preferences for particular acoustic elements of songs have been largely studied in our laboratory, but others' breeding parameters like egg quality should be influenced by song quality. This point is developed in the next section.

## VI. Consequences of the Exposure to Male Sexy Phrases on Females' Reproduction

### A. Sexy Phrases Do not Affect the Development of Nest Building or Egg Laying

As mentioned in Section I, pioneering works of Hinde or Kroodsma (Hinde, 1958; Hinde and Steel, 1976, 1978; Kroodsma, 1976; Warren and Hinde, 1961) emphasized the crucial role of male stimulation, particularly vocal stimulation on female reproductive development in the canary. Paradoxically, studies on physiological consequences of canary song on reproductive activity failed to develop during the following years. So, little was known of the song features that stimulate female reproductive activity apart from the facts that songs should be conspecific (Hinde and Steel, 1976) and that large repertoires were more efficient than small ones (Kroodsma, 1976). In a closely related species, the serin (*S. serinus*), females exposed daily to playbacks of songs during the nest-building stage were found to spend more time nest building than females not exposed to such additional songs (Mota and Depraz, 2004). The discovery of the stimulating activity of sexy phrases on sexual activity (see Section II) led us to wonder whether these sexy phrases should also hasten reproductive development of females, particularly with regard to nest building and egg laying. We studied the effect of continuous exposure to sexy or nonsexy phrases on CSDs and on reproductive activity estimated using nest building and egg laying in sexually inexperienced and experienced females (Leboucher et al., 1998b). Whereas sexy songs significantly elicited more CSDs than nonsexy ones in inexperienced as well as in experienced females, we failed to observe any significant influence of sexy songs on females' reproductive activity: no difference was found between groups exposed to either sexy songs or to nonsexy songs. We concluded that the efficiency of songs in promoting nest building or egg laying appeared to be unrelated to their efficiency in eliciting sexual responses (Leboucher et al., 1998b). Further works, described below, have led us to reconsider the effects of sexy phrases on reproduction.

### B. But Sexy Phrases Affect Egg Quality

Reproduction involves transferring genetic and nongenetic resources from parents to offspring. Parental investment in terms of breeding and transmission of nongenetic resources is assumed to be costly (e.g., Charnow and Krebs, 1974; Hanssen et al., 2005; Williams, 1966). As a consequence, depending on breeding conditions, females may modulate their reproductive investment. In the differential allocation hypothesis, Burley (1988) postulated that females paired with attractive mates should invest more in their young than females paired with poorly attractive ones; variations in the allocation of nongenetic resources affect the development and survival of the young.

In 1993, Schwabl demonstrated that yolk is a source of maternal testosterone (T) for developing birds, notably in canaries; he stated that yolk T was exclusively provided by the female and was probably the result of follicular activity. He also demonstrated that T contained in eggs has various effects on young development. For instance, domestic canary chicks born out of eggs containing high concentrations of T beg more often (Schwabl, 1996a); they are also more aggressive and more dominant later in their lives (Schwabl, 1993); however, negative effects of high concentrations of T in yolks, particularly on immune response, were also suggested (e.g., Andersson et al., 2004). Some studies pointed out that breeding conditions might affect androgen deposition in yolks (Schwabl, 1996b, 1997); in this way, Gil et al. (1999) showed that female zebra finches paired to attractive, red-banded, males deposited higher quantities of T in their eggs than females paired to nonattractive, green-banded, males.

Using a similar paradigm, we wondered whether attractive songs including sexy phrases should increase T deposition in yolks and should affect other aspects of egg quality. In two consecutive studies (Gil et al., 2004;

Tanvez et al., 2004), female domestic canaries were chronically exposed to either attractive sexy songs or unattractive ones during the pre-laying period. In summary, it can be said that these two experiments emphasized the stimulating effect of sexy songs on yolk T concentration. In contrast, sexy songs failed to modify other aspects of egg production like laying latency, clutch size, egg volume, and yolk mass. Though in some species female preferences can influence egg size (e.g., in the mallard, *Anas platyrhynchos* (Cunningham and Russel, 2000), this is less obvious in the canary. Neither Tanvez et al. (2004) nor Gil et al. (2004) found an effect of song attractiveness on egg mass contrary to the findings of Leitner et al. (2006). For their part, Marshall, et al. (2005) found that attractive songs failed to induce a significant increase of T concentration in egg yolks but provoked an increase of circulating T detectable into feces.

Our studies showing a significant effect of male attractiveness on androgen deposition in eggs are consistent with data obtained in other species. For instance, in barn swallows, *Hirundo rustica*, females paired to more colorful males laid eggs with greater concentrations of androgens (Safran et al., 2008). In the same line, in great tits, *Parus major*, yolk androgen concentration in eggs was found to be related to male attractiveness, breeding density, and territory quality (Remeš, 2011).

The neuroendocrine pathway underlying the translation of song audition to changes in T deposition in yolks is far to be clearly understood. Neurobiology studies using the expression of immediate early genes (IEGs) revealed that the *caudomedial mesopallium* (CMM), a brain region outside the song control system, was differentially stimulated by sexy songs (Barker et al., 2010; Leitner et al., 2005). In female canaries, CMM is involved in female perception and discrimination of male song quality (Leitner et al., 2005). Moreover, hearing sexy song enhanced IEGs' expression in the medial basal hypothalamus, a brain region involved in neuroendocrine regulation, more than in birds hearing a white noise. Thus differential auditory processing in the CMM may be an initial stage in the assessment of song quality by a female and the translation of this information to differential testosterone deposition in the egg (Barker et al., 2010).

It is noteworthy that egg mass, volume, or composition, on the one hand, and T deposition, on the other hand, seem to be controlled by different mechanisms in female canaries: situations affecting egg volume do not affect testosterone deposition and vice versa (see experiments above and see below Section VII. B, Garcia-Fernandez et al., 2010, next section). Moreover, Tanvez et al. (2008) found that dominant females laid eggs with higher T concentrations in the yolk than subordinate ones; this could be a mechanism of dominance inheritance, leading chicks from dominant females to be more aggressive and dominant later. In contrast, in line of previous findings, no relationship between mothers' rank number and their clutch size or egg mass was found. In great tits too, androgen deposition and egg size appeared to be controlled by different factors: yolk mass increase is related to the area of male black breast band, whereas androgens increase with carotenoid chroma of male yellow breast feathers (Remeš, 2011).

On the whole, stimulating configurations efficient to induce a rise of T deposition appeared to be inefficient to provoke an increase of egg volume, mass, or composition; in contrast, stimulating configurations efficient to bring an increase of egg volume, mass, or composition failed to heighten T deposition. As far as the relative egg mass, volume, or composition are classically considered indexes of maternal investment (Christians, 2002), the question arises of whether T deposition can also be considered in the same way.

Singing sexy phrases allows males to entice females (Section II) and to affect egg quality, particularly with regard to hormone deposition in the yolk (this section). However, emitting sexy phrases is not the only mean the male canary can influence activities of conspecifics; communication in social networks offers other possibilities, as developed in the next section.

## VII. Canaries' Communication in Social Networks

Average distance between individuals is often shorter than the range of signals. Therefore, several individuals can communicate with one another, forming a social environment with multiple senders and receivers (McGregor, 2005; McGregor and Dabelsteen, 1996). Such communication networks have consequences for both senders and receivers. A third party can witness conspecifics' interactions. Therefore, individuals can gather relative information about the motivation, status, or quality of conspecifics by witnessing their interactions, that is, they eavesdrop (McGregor and Dabelsteen, 1996). At the opposite, controlling the information in the presence of potential eavesdroppers should be as important as obtaining information.

### A. Eavesdropping on Male–Male Singing Interactions by Females

During mate choice, females may assess male quality sequentially and choose the individual of highest quality (Otter and Ratcliffe, 2005), but studies also revealed that females may use male–male signaling interactions to assess the relative quality of males and to direct their sexual behavior accordingly (e.g., Mennill et al., 2002; Otter et al., 1999). In this way, females may assess directly the relative quality of the two interacting males and, therefore, may identify mates of higher quality faster than by sequential sampling, thereby reducing time, energy, and other costs associated with mate searching (Otter and Ratcliffe, 2005). As stated above, this process of gathering information from signaling interactions of others has been called eavesdropping (McGregor and Dabelsteen, 1996).

In songbirds, neighboring territorial males often interact through countersinging, in which birds sing in response to the singing of neighbors. During countersinging, males can use different strategies by singing the same song as their opponent (song matching), by increasing the rate of singing or the rate at which song types are switched, or by overlapping the opponent's song (Gil and Gahr, 2002; Vehrencamp, 2000). These different strategies can reflect differences in status, motivation, or quality between the singers (Naguib, 2005). Studies have highlighted the importance of the timing of singing in vocal interactions, whereby song overlapping is perceived as an antagonistic signal (e.g., Naguib, 1999; see also Naguib and Mennill, 2010; Searcy and Beecher, 2009, 2011 for a recent debate). In other words, song overlapping has been considered by some authors as a more aggressive strategy than song alternating (Todt and Naguib, 2000).

First studies investigating how females use information gathered by eavesdropping have been carried out in the field in Paridae. In two different studies, experimenters have simulated two kinds of singing interactions with territorial males using interactive playbacks: an overlapping interaction where the resident male was overlapped by a simulated intruder and an alternating interaction where the song of a simulated intruder alternated with the song of the resident male. In the great tits, *P. major*, females paired to a male who was overlapped by a simulated intruder were more likely to intrude into neighboring territories after the experiment than females paired to males with whom the intruder alternated (Otter et al., 1999). It suggests that females pay attention to the singing interactions of their mate, and that they use this information to direct their extra-pair behaviors. In black-capped chickadees, *Poecile atricapillus*, females paired to high-ranking males that lost song contests during countersinging (males which were overlapped by the simulated intruder) have more extra-paired young in their nest (Mennill et al., 2002).

### B. Female Canaries Eavesdrop

We investigated eavesdropping in several experiments under laboratory controlled conditions. Overall, these experiments were designed with a similar protocol with a presentation phase and a test phase. During the presentation phase, females were exposed to a simulated overlapping interaction twice daily during five consecutive days. During the test phase, songs previously heard in the overlapping interaction were broadcast twice daily to females until they stopped to elicit CSDs; songs were broadcast separately one after the other.

We first investigated if female canaries can eavesdrop on male–male singing interactions. In a series of two experiments, we assessed females' sexual preferences for male songs previously heard during a simulated singing interaction during which one song overlapped another one. In the first experiment, male songs contained sexy phrases, whereas in the second, sexy phrases were removed. In the first experiment, where the sexy phrases were included in the songs, the females did not prefer the overlapping song, whereas in the second experiment, where songs did not contained sexy phrases, females preferred the songs previously heard as the overlapping songs (Leboucher and Pallot, 2004). In the first experiment, we had probably given the females contradictory information: on the one hand, some males seemed to be of poorer quality than others, as they were losing the contest, but, on the other hand, these same males were producing songs with a sexy phrase that was probably regarded as a signal of quality. Nevertheless, the second experiment shows that female songbirds can also use eavesdropping in initial stage of mate choice.

Therefore, following experiments were designed by discarding sexy phrases and using less attractive elements. We investigated female canaries' preferences toward male songs previously heard in different types of simulated interaction. A first group of females heard simulated overlapping interactions; in the other group, females heard alternated interactions. Overlapping interactions provided information to females about the relative quality or motivation of the singers, whereas alternated interactions did not allow females to identify hierarchical relationships between the two singers (Peake, 2005). Results showed that females preferred the song previously heard as the overlapping one (Fig. 6); when females heard an alternated interaction, they subsequently did not show a preference for either of the two songs (Amy et al., 2008).

A further experiment was designed to assess whether female investment in eggs (see also Section VI) should be affected by the information gathered by eavesdropping. Contrarily to the previous experiment, during the test phase, females were only exposed to *one of the two songs*. During 2 h, twice daily, they were exposed either to the song previously heard as the overlapping one or to the song previously heard as the overlapped one. This testing phase ended when the female laid its last egg. Eggs were collected and analyzed. Females exposed to overlapping songs laid eggs with greater yolk ratio than females exposed to overlapped song (Garcia-Fernandez et al., 2010). Egg yolk influences neonatal quality as a nutritional supplement, and hatchlings keep the remaining yolk sac after hatching as an

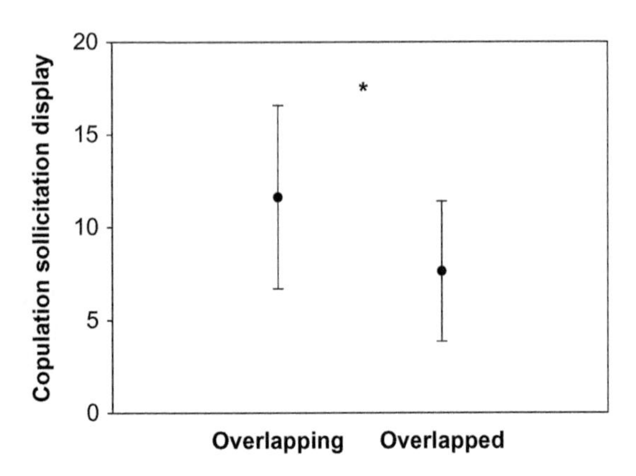


Fig. 6. Number of copulation solicitation displays of female canaries ($n=11$) to the overlapping song or the overlapped song. Means±SE are given. *$p=0.018$, paired $t$ test.

energy reserve during the first few days after hatching. Therefore, our results suggest that female exposed to overlapping songs would have chicks with more survival probability than females exposed to overlapped songs (Garcia-Fernandez et al., 2010).

Finally, female canaries gather information from male–male singing interactions and used the information to direct their preferences and their maternal investment.

### C. Male Canaries Eavesdrop Too

Male songbirds sing to defend territory against rivals and to attract potential partners (Gil and Gahr, 2002). Territories of songbirds often are close to each others forming a communication network. By eavesdropping on conspecifics' singing interactions, males can obtain information about relative motivation, status, or quality of their neighbors and of potential intruders. Even, if the intrasexual function of the song in canaries is poorly understood yet, we investigated eavesdropping in male domestic canaries with a protocol similar to that used for females and described above. To sum up, we found that signaling behavior of male canaries was affected by the songs they were listening to: they emitted less calls during the song previously heard as the overlapping song than during all other songs (Amy and Leboucher, 2009). This result is not consistent with previous studies on eavesdropping in which males responded more intensely to the more threatening song, the overlapping song (Mennill and Ratcliffe, 2004;

Naguib and Todt, 1997), or less intensely to the song previously heard as the overlapped song (Peake et al., 2001, 2002). In these studies, males were defending their territories, whereas in our study male canaries were placed in an unknown room, creating an inhibiting situation, and probably driving male canaries to subaverage responses when facing threatening signals (Amy and Leboucher, 2009). Overall, eavesdropping on singing interactions seems widespread among songbirds, thus highlighting the need to study birdsong by taking into account the social environment.

### D. Visual Eavesdropping

Individuals can also obtain information by paying attention not only to singing interactions but also to other types of interactions. Female fighting fish (*Betta splendens*) gather information about relative male quality from male–male visual displays (Doutrelant and McGregor, 2000) and female Japanese quails (*Coturnix japonica*) pay attention to male–male physical interaction and use previously observed aggressive interactions between two males to direct their sexual preferences (Ophir and Galef, 2003; Ophir et al., 2005). Two studies on canaries showed that both female and male canaries proved to be able to extract information from a competition for food between two males to adjust their behavior.

Females who had previously seen a competition between two males spent more time near the loser of the competition than near the winner (Amy et al., 2008). These results are congruent with those in the Japanese quails, where females that visually "eavesdropped" on fighting males prefer losers to winners (Ophir and Galef, 2003). This preference is supposed to be due more to an avoidance of aggressive males than to a sexual preference for the loser (Ophir et al., 2005).

With regard to male canaries, we found that males behaved differentially to winners and losers of competition for food they had witnessed, initiating fewer attacks against the winner and spending less time foraging. In contrast, we found no such effect when males had no prior knowledge of the relative competitive ability of their opponents (Amy and Leboucher, 2007).

Overall, these studies show that domestic canaries are able to extract information from different types of interactions between two males, from singing interactions as well as competition for food. They not only gather information, but they also use the information to direct their subsequent behaviors. The effect of eavesdropping on female investment described above emphasizes that in canaries as well as in black-capped chickadees, eavesdropping during playback sessions "can have long-lasting and far-reaching effects on individual fitness" (Mennill et al., 2002).

### E. Audience Effects

Individuals can modify their signals according to the presence and the characteristics of the receiver. In the domestic chicken (*Gallus gallus domesticus*), production of food calls emitted by males is modulated by the social context: males do not emit food calls in the presence of another male, whereas they emit these calls in the presence of a female mostly if the latter is not paired (Marler et al., 1986). Others studies in the domestic chicken show that males modify their alarm calls (Evans and Marler, 1991, 1992, 1994; Gyger et al., 1986) according to the social context. As previously mentioned in Section II, male domestic canaries sing longer bouts of sexy phrases in front of both males and females when compared to a situation where they sing alone (Kreutzer et al., 1999).

Animals can obtain information about competitors or potential mates by observing or listening to conspecifics' interactions. In such a social context, interacting individuals may benefit by changing their signaling behavior when a third conspecific is present. This phenomenon is better known as the "audience effect." Audience effects have been observed during various types of interaction and vary according to the social context in different species (e.g., Baltz and Clark, 1997; Doutrelant et al., 2001; Price and Rodd, 2006; Vignal et al., 2004).

We performed an experiment to assess how the presence of an audience influences the behavior of male canaries during an interaction with an extra-pair female and during a competition for food with another male. Males' behaviors were observed in the absence of an audience, in the presence of their mate or of a familiar female. Male canaries, when engaged in an extra-paired interaction, courted less in the presence of their mate than in the presence of a familiar female. They mostly courted more in the absence of an audience than in the presence of their mate or a familiar female. Moreover, they also adjusted their responses with regard to the behavioral activity of the interacting female by courting more winsome females than aggressive ones (Ung et al., 2011). When engaged in an intrasexual competition for food, male canaries were more aggressive in the presence of an audience than without (Ung et al., 2011). Males also adjusted their agonistic behavior toward their competitor according to the behavior of the audience by significantly attacking less the competitor if the female displayed affiliative behaviors. Therefore, male domestic canaries modify their courting and agonistic behaviors according to the type of the audience and to the behaviors of both the audience and the interactants. These results highlight the ability of domestic canaries to adjust their behavior according to their social environment when interacting with a conspecific.

## VIII. Summary and Conclusion

### A. From Findings to Questions for the Future

The aim of this chapter was to contribute to a better understanding of how birdsong is involved in birds' reproductive activity; at the same time, it was an opportunity to summarize 20 years of studies, from our group "Comparative Ethology and Cognition" in Nanterre. Most of our works focused on acoustical communication and reproduction in domestic canaries and were strongly influenced by the finding, published in 1995, of what we called the "A" or "sexy" phrases, phrases composed of syllables with a large frequency bandwidth and sung at a high repetition rate. These acoustical elements were found to be particularly efficient to elicit females' sexual responses.

While females' responses toward songs are known to be constrained by auditory experience at the beginning of life, the ability of sexy phrases to elicit females' sexual displays appears to be rather independent of the early acoustic milieu. Females reared in acoustic isolation present a sexual preference for broad bandwidth songs with high syllable rates; so, learning processes seem not to be necessary to develop sexual preferences for these acoustic parameters. Several hypotheses may account for such a result; the sensory bias hypothesis (Ryan, 1990) and the honest-signaling hypothesis (Zahavi, 1975) were mentioned above. We should also cite Darwin's hypothesis of a latent aesthetic mate preferences for particular traits (Burley and Szymanski, 1998; Darwin, 1871). These hypotheses are not mutually exclusive, and at the present time, none of them can be discarded. Here is a nice challenge for future research.

Sexy phrases associating large bandwidth and high repetition rate are clearly difficult to produce. Neuronal and neuromuscular studies indicated that sexy songs are produced using both left- and right-sided specializations of the brain pathway. Sexy songs are not only efficient in eliciting sexual responses but can also modify egg quality, namely yolk testosterone concentration, thus influencing the development and survival of the young. The neuroendocrine pathway underlying the translation of song audition to changes in androgen deposition in yolks is not still understood, and this field remains to be cleared. Moreover, it is not obvious whether yolk testosterone deposition should be considered as a maternal investment, and this question remains a topic of debate.

The discovery of acoustic elements stimulating sexual responses in female birds probably led us to underestimate the intrasexual function of the male canary song. A study pointed out that the hearing of sexy phrases provokes an inhibition of male vocal behavior (Parisot et al., 2002); however, a clear evidence of the effect intrasexual function of sexy phrases remains to be found. In the serin (*S. serinus*), a closely related species, Cardoso et al. (2007) found sex differences in the discrimination of broadcast songs. Because bird song serves both intra- and intersexual functions, it is possible that a song trait adaptive to one function trades off with the other (Cardoso et al., 2007); in the chaffinch (*Fringilla coelebs*) fast trills affect male–male interactions but are not attractive to females (Leitão and Riebel, 2003).

It must be born in mind that sexy phrases are sung by domestic as well as by free-living island canaries, so one may reasonably build on the adaptive value of this signal and, consequently, on the relevance of our studies with regard to the evolutionary mechanisms. We must also remember that there are males who may not be able to sing sexy phrases; consequently, when hearing songs from potential mates, females have to take information from other sources. Females' previous reproductive experience and reproductive success have been found to affect their preferences towards songs. Moreover, female canaries eavesdrop to male–male signaling interactions in order to assess the relative quality of males and to direct their sexual behavior accordingly. In such a situation, the strategic response of male canaries is to adjust their display according to the audience. The study of reproductive strategies developed by males who do not sing sexy phrases certainly represents a new way of research.

Finally, one of the most puzzling findings is that theories and concepts issued from behavioral ecology could be relevant for a domestic species. This clearly indicates that artificial selection during many generations failed to shake off the effects of natural and sexual selection and this should be questioned too.

**References**

Amy, M., Leboucher, G., 2007. Male canaries can visually eavesdrop in conspecific food interactions. Anim. Behav. 74, 57–62.

Amy, M., Leboucher, G., 2009. Effects of eavesdropping on subsequent signalling behaviours in male canaries. Ethology 115, 239–246.

Amy, M., Monbureau, M., Durand, C., Gomez, D., Théry, M., Leboucher, G., 2008. Female canary mate preferences: differential use of information from two types of male–male interactions. Anim. Behav. 76, 971–982.

Andersson, M., 1994. Sexual Selection. Princeton University Press, Princeton, NJ.

Andersson, S., Uller, T., Lõhmus, M., Sundström, F., 2004. Effects of egg yolk testosterone on growth and immunity in a precocial bird. J. Evol. Biol. 17, 501–505.

Baker, M.C., Belcher, C.S., Deutsch, L.C., Sherman, G.L., Thompson, D.B., 1981a. Foraging success in Junco flocks and the effects of social hierarchy. Anim. Behav. 29, 137–142.

Baker, M.C., Spitler-Nabors, K.J., Bradley, D.C., 1981b. Early experience determines song dialect responsiveness of female sparrows. Science 214, 819–820.

Ballentine, B., 2009. The ability to perform physically challenging songs predicts age and size in male swamp sparrows, *Melospiza Georgiana*. Anim. Behav. 77, 973–978.
Ballentine, B., Hyman, J., Nowicki, S., 2004. Vocal performance influences female response to male bird song: an experimental test. Behav. Ecol. 15, 163–168.
Baltz, A.P., Clark, A.B., 1997. Extra-pair courtship behaviour of male budgerigars and the effect of an audience. Anim. Behav. 53, 1017–1024.
Barker, J.M., Monbureau, M., Leboucher, G., Balthazart, J., 2010. Auditory forebrain activation in the female canary is modulated by male song quality. In: Communication at the 40th Annual Meeting of the Society for Neuroscience (SfN 2010), 13–17 November 2010, San Diego, California, USA.
Beani, L., Dessi-Fulgheri, F., 1995. Mate choice in the grey partridge *perdix perdix*: role of physical and behavioural traits. Anim. Behav. 49, 347–356.
Beecher, M.D., 2008. Function and mechanisms of song learning in song sparrows. Adv. Stud. Behav. 38, 167–225.
Béguin, N., Leboucher, G., Kreutzer, M., 1998. Sexual preferences for mate song in female canaries (*Serinus canaria*). Behaviour 135, 1185–1186.
Béguin, N., Leboucher, G., Bruckert, L., Kreutzer, M., 2006. Mate preferences in female canaries (*Serinus canaria*) within a breeding season. Acta Ethol. 9, 65–70.
Bertelli, S., Tubaro, P.L., 2002. Body mass and habitat correlates of song structure in a primitive group of birds. Biol. J. Linn. Soc. 77, 423–430.
Black, J.M., 1996. Partnerships in Birds: The Study of Monogamy. Oxford University Press, Oxford.
Blaustein, J.D., Wade, G.N., 1977. Sequential inhibition of sexual behavior by progesterone in female rats: comparison with a synthetic anti-estrogen. J. Comp. Physiol. Psychol. 91, 752–760.
Brenowitz, E.A., 1991. Altered perception of species-specific song by female birds after lesions or a forebrain nucleus. Science 251, 303–305.
Buchanan, K.L., Spencer, K.A., Goldsmith, A.R., Catchpole, C.K., 2003. Song as a honest signal of past developmental stress in the European starling (*Sturnus vulgaris*). Proc. R. Soc. Lond. B 270, 1149–1156.
Burley, N., 1988. The differential-allocation hypothesis: an experimental test. Am. Nat. 132, 611–628.
Burley, N.T., Foster, V.S., 2006. Variation in female choice of mates: condition influences selectivity. Anim. Behav. 72, 713–719.
Burley, N.T., Szymanski, R., 1998. "A taste for the Beautiful": latent aesthetic Mate preferences for white crest in two species of Australian grassfinches. Am. Nat. 152, 792–802.
Cardoso, G., Gama Mota, P., Depraz, V., 2007. Female and male serins (*Serinus serinus*) respond differently to derived song traits. Behav. Ecol. Sociobiol. 61, 1425–1436.
Casey, R.M., Baker, M.C., 1992. Early social tutoring influences female sexual response in white-crowned sparrows. Anim. Behav. 44, 983–986.
Catchpole, C.K., Slater, P.J.B., 2008. Bird Song: Biological Themes and Variations, second ed. Cambridge University Press, New York.
Catchpole, C.K., Leisier, B., Dittami, J., 1986. Sexual differences in the responses of captive great reed warblers, *Acrocephalus arundinaceus*, to variation in song structure and repertoire size. Ethology 73, 69–77.
Charnow, E.L., Krebs, J.R., 1974. On clutch size and fitness. Ibis 116, 217–219.
Choudhury, S., 1995. Divorce in birds: a review of the hypotheses. Anim. Behav. 50, 412–429.
Christians, J.K., 2002. Avian egg size: variation within species and inflexibility within individuals. Biol. Rev. Camb. Philos. Soc. 77, 1–26.
Clayton, N.S., 1988. Song discrimination learning in zebra finches. Anim. Behav. 36, 1016–1024.
Clayton, N.S., Pröve, E., 1989. Song discrimination in female Zebra Finches and Bengalese Finches. Anim. Behav. 38, 352–362.

Cunningham, E.J.A., Russel, A.F., 2000. Egg investment is influenced by male attractiveness in the mallard. Nature 404, 74–76.

Dabelsteen, T., 1982. Variation in the response of freeliving blackbirds (*Turdus merula*) to playback of song: I. Effect of continuous stimulation and predictability of the response. Z. Tierpsychol. 58, 311–328.

Dabelsteen, T., Pedersen, S.B., 1993. Song-based species discrimination and behaviour assessment by female blackbirds, *Turdus merula*. Anim. Behav. 45, 759–771.

Darwin, C., 1871. The Descent of Man and Selection in Relation to Sex. John Murray, London.

David, P., Bjorksten, T., Fowler, K., Pomiankowski, A., 2000. Condition-dependent signalling of genetic variation in stalk-eyed flies. Nature 406, 186–188.

Del Negro, C., Gahr, M., Leboucher, G., Kreutzer, M., 1998. The selectivity of sexual responses to song displays: effects of partial chemical lesion of the HVC in female canaries. Behav. Brain Res. 96, 151–159.

Del Negro, C., Gahr, M., Vallet, E., Kreutzer, M., 2000. Sexually stimulating signals of canary (*Serinus canaria*) songs: evidence for a female-specific auditory representation in the HVC Nucleus during the breeding season. Behav. Neurosci. 114, 526–542.

Delestrade, A., 1999. Foraging strategy in a social bird, the Alpine Chough: effect of variations in quantity and distribution of food. Anim. Behav. 57, 299–305.

Depraz, V., Leboucher, G., Kreutzer, M., 2000. Early tutoring and adult reproductive behaviour in female domestic canary (*Serinus canaria*). Anim. Cogn. 3, 45–51.

Depraz, V., Kreutzer, M., Leboucher, G., 2004. Sexual preferences for songs in female domestic canaries (*Serinus canaria*): can late song exposure, without social reinforcement, influence the effects of early tutoring? Acta Ethol. 6, 73–78.

DeVoogd, T.J., Krebs, J.R., Healy, S.D., Purvis, A., 1993. Relations between song repertoire size and the volume of brain nuclei related to song: comparative evolutionary analyses amongst oscine birds. Proc. R. Soc. Lond. B 254, 75–82.

Dooling, R., Searcy, M., 1980. Early perceptual selectivity in the swamp sparrow. Dev. Psychobiol. 13 (5), 499–506.

Doutrelant, C., McGregor, P.K., 2000. Eavesdropping and mate choice in female fighting fish. Behaviour 137, 1655–1669.

Doutrelant, C., McGregor, P.K., Oliveira, R.F., 2001. The effect of an audience on intrasexual communication in male Siamese fighting fish *Betta splendens*. Behav. Ecol. 12, 283–286.

Draganoiu, T., Nagle, L., Kreutzer, M., 2002. Directional female preference for an exaggerated male trait in canary (*Serinus canaria*). Proc. R. Soc. Lond. B 269, 2525–2531.

Eens, M., Pinxten, R., 1995. Inter-sexual conflicts over copulations in the European starling: evidence for the female mate-guarding hypothesis. Behav. Ecol. Sociobiol. 36, 71–81.

Evans, C.S., Marler, P., 1991. On the use of video images as social stimuli in birds: audience effects on alarm calling. Anim. Behav. 41, 17–26.

Evans, C.S., Marler, P., 1992. Female appearance as a factor in the responsiveness of male chickens during anti-predator behaviour and courtship. Anim. Behav. 43, 137–145.

Evans, C.S., Marler, P., 1994. Food calling and audience effects in male chickens, *Gallus gallus*. Their relationships to food availability, courtship and social facilitation. Anim. Behav. 47, 1159–1170.

Fawcett, T.W., Bleay, C., 2008. Previous experiences shape adaptive mate preferences. Behav. Ecol. 20, 68–78.

Feinberg, D., Jones, B., Burt, D., Perett, D., 2005. Manipulations of fundamental and formant frequencies influence the attractiveness of human male voices. Anim. Behav. 69, 561–568.

Fusani, L., Beani, L., Dessi-Fulgheri, F., 1994. Testosterone affects the acoustic structure of male call in the grey partridge *Perdix perdix*. Behaviour 128, 301–310.

Garamszegi, L.Z., Eens, M., 2004. Brain space for a learned task: strong intraspecific evidence for neural correlates of singing behaviour in songbirds. Brain Res. Rev. 44, 187–193.

Garcia-Fernandez, V., Amy, M., Lacroix, A., Malacarne, G., Leboucher, G., 2010. Eavesdropping on male singing interactions leads to differential allocation in eggs. Ethology 116, 662–670.

Gil, D., Gahr, M., 2002. The honesty of bird song: multiple constraints for multiple traits. TREE 17 (3), 133–141.

Gil, D., Graves, J., Hazon, N., Wells, A., 1999. Male attractiveness and differential testosterone investment in Zebra Finch eggs. Science 286, 126–128.

Gil, D., Leboucher, G., Lacroix, A., Cue, R., Kreutzer, M., 2004. Female canaries produce eggs with greater amounts of testosterone when exposed to attractive male song. Horm. Behav. 45, 64–70.

Godwin, J., Hartman, V., Grammer, M., Crews, D., 1996. Progesterone inhibits female-typical receptive behavior and decreases hypothalamic estrogen and progesterone receptor messenger ribonucleic acid levels in whiptail lizards (genus *Cnemidophorus*). Horm. Behav. 30, 138–144.

Gonzalez-Mariscal, G., Melo, A.I., Beyer, C., 1993. Progesterone, but not LHRH or prostaglandin E2, induces sequential inhibition of lordosis to various lordogenic agents. Neuroendocrinology 57, 940–945.

Güttinger, H.R., 1981. Self-differentiation of song organization rules by deaf canaries. Zf. Tierpsychol. 56, 323–340.

Gyger, M., Karakashian, S.J., Marler, P., 1986. Avian alarm calling: is there an audience effect? Anim. Behav. 34, 1570–1572.

Hallé, F., Gahr, M., Pieneman, A.W., Kreutzer, M., 2002. Recovery of song preferences after excitotoxic HVc lesion in female canaries. J. Neurobiol. 52, 1–13.

Hallé, F., Gahr, M., Kreutzer, M., 2003a. Effects of unilateral lesions of HVC on song patterns of male domesticated canaries. J. Neurobiol. 56, 303–314.

Hallé, F., Gahr, M., Kreutzer, M., 2003b. Impaired recovery of syllable repertoire after unilateral lesions of the HVC of male domesticated canaries. Anim. Biol. 53, 113–128.

Hanssen, S.A., Hasselquist, D., Folstad, I., Erikstad, K.E., 2005. Cost of reproduction in a long-lived bird: incubation effort reduces immune function and future reproduction. Proc. R. Soc. Lond. B 272, 1039–1046.

Hasselquist, D., Bensch, S., Von Schantz, T., 1996. Correlation between male song repertoire, extra-pair paternity and offspring survival in the great reed warbler. Nature 381, 229–232.

Hedrick, A., 2005. Environmental condition-dependent effect on a heritable, preferred male trait. Anim. Behav. 70, 1121–1124.

Hinde, R.A., 1958. The nest-building behaviour of domesticated canaries. Proc. Zool. Soc. Lond. 131, 1–48.

Hinde, R.A., Steel, E., 1976. The effect of male song on an estrogen-dependent behaviour pattern in the female canary (*Serinus canaria*). Horm. Behav. 7, 293–304.

Hinde, R.A., Steel, E., 1978. The influence of daylength and male vocalizations on the estrogen-dependant behavior of female canaries and budgerigars, with discussion of data from other species. Adv. Stud. Behav. 8, 39–73.

Holveck, M.J., Riebel, K., 2010. Low-quality females prefer low-quality males when choosing a mate. Proc. R. Soc. Lond. B 277, 153–160.

Kempenaers, B., Verheyen, G.R., Dhondt, A.A., 1997. Extra-pair paternity in the blue tit (*Parus caeruleus*): female choice, male characteristics and offspring quality. Behav. Ecol. 8, 481–492.

King, A.P., West, M.J., 1977. Species identification in the North American cowbird: appropriate responses to abnormal song. Science 195, 1002–1004.

Kreutzer, M., Vallet, E., 1991. Differences in the response of captive female canaries to variation in conspecific and heterospecific songs. Behaviour 117, 106–116.
Kreutzer, M., Vallet, E.M., Doucet, S., 1992. Sexual responses of female canaries at the onset of song stimuli. Experientia 48, 679–682.
Kreutzer, M., Vallet, E., Nagle, L., 1996. Female canaries display to songs of early isolated males. Experientia 52, 277–280.
Kreutzer, M., Beme, I., Vallet, E., Kiossova, L., 1999. Social stimulation modulates the use of the "A" phrase in male canary songs. Behaviour 136, 1325–1334.
Kroodsma, D.E., 1976. Reproductive development in a female song bird: differential stimulation by quality of male song. Science 192, 574–575.
Kroodsma, D.E., Byers, B.E., 1991. The function(s) of bird song. Am. Zool. 31, 318–328.
Kroodsma, D.E., Miller, D.B., 1982. Acoustic Communication in Birds. Academic Press, New York, NY.
Leboucher, G., Pallot, K., 2004. Is he all he says he is? An investigation into inter-sexual eavesdropping in the domestic canary (*Serinus canaria*). Anim. Behav. 68, 957–963.
Leboucher, G., Kreutzer, M., Dittami, J., 1994. Copulation solicitation displays in female canaries (Serinus canaria): are oestradiol implants necessary? Ethology 97, 190–197.
Leboucher, G., Béguin, N., Mauget, R., Kreutzer, M., 1998a. Effects of Fadrozole on sexual displays and reproductive activity in the female canary. Physiol. Behav. 65 (2), 233–240.
Leboucher, G., Depraz, V., Kreutze, M., Nagle, L., 1998b. Male song stimulation of female reproduction in canaries: features relevant to sexual displays are not relevant to nest-building or egg-laying. Ethology 104, 613–624.
Leboucher, G., Béguin, N., Lacroix, A., Kreutzer, M., 2000. Progesterone inhibits courtship behavior in domestic canaries (*Serinus canaria*). Horm. Behav. 38, 123–129.
Lehongre, K., Aubin, T., Robin, S., Del Negro, C., 2008. Individual signature in canary songs: contribution of multiple levels of song structure. Ethology 114, 425–435.
Leitão, A., Riebel, K., 2003. Are good ornaments bad armaments? Male chaffinch perception of songs with varying flourish length. Anim. Behav. 66, 161–167.
Leitner, S., Catchpole, C.K., 2002. Female canaries that respond and discriminate more between male songs of different quality have a larger song control nucleus (HVC) in the brain. J. Neurobiol. 52, 294–301.
Leitner, S., Voigt, C., Garcia-Segura, L.M., Van't Hof, T., Gahr, M., 2001a. Seasonal activation and inactivation of song motor memories in wild canaries is not reflected in neuroanatomical changes of forebrain song areas. Horm. Behav. 40, 160–168.
Leitner, S., Voigt, C., Gahr, M., 2001b. Seasonal changes in the song pattern of the non-domesticated island canary (*Serinus canaria*), a field study. Behaviour 138, 885–904.
Leitner, S., Voigt, C., Metzdorf, R., Catchpole, C.K., 2005. Immediate early gene (ZENK, Arc) expression in the auditory forebrain of female canaries varies in response to male song quality. J. Neurobiol. 64, 275–284.
Leitner, S., Marshall, R.C., Leisler, B., Catchpole, C.K., 2006. Male song quality, egg size and offspring sex in captive canaries (*Serinus canaria*). Ethology 112, 554–563.
Lerch, A., Rat-Fischer, L., Gratier, M., Nagle, L., 2011. Diet quality affects mate choice in domestic female canary *Serinus canaria*. Ethology 117, 769–776.
Lerch, A., Rat-Fischer, L., Nagle., L., Submitted to Ethology. Condition-dependent choosiness for highly attractive songs in female canaries.
Lindén, M., 1991. Divorce in great tits: chance or choice? An experimental approach. Am. Nat. 138, 1039–1048.
Marler, P., 1970. A comparative approach to vocal learning: song development in white-crowned sparrows. J. Comp. Physiol. Psychol. 71, 1–25.

Marler, P., Peters, S., 1988. The role of song phonology and syntax in vocal learning preferences in the song sparrow, *Melospiza melodia*. Ethology 77, 125–149.
Marler, P., Dufty, A., Pickert, R., 1986. Vocal communication in the domestic chicken. II. Is a sender sensitive to the presence and nature of a receiver? Anim. Behav. 34, 194–198.
Marshall, R.C., Leisler, B., Catchpole, C.K., Schwabl, H., 2005. Male song quality affects circulating but not yolk steroid concentrations in female canaries (*Serinus canaria*). J. Exp. Biol. 209, 4593–4598.
McGraw, K.J., Mackillop, E.A., Dale, J., Hauber, M.E., 2002. Different colors reveal different information: how nutritional stress affects the expression of melanin and structurally based ornamented plumage. J. Exp. Biol. 205, 3747–3755.
McGregor, P.K., 2005. Bird song, sexual selection and female mating strategies. Introduction In: McGregor, P.K. (Ed.), Animal Communication Networks. Cambridge University Press, New York, pp. 1–6.
McGregor, P.K., Dabelsteen, T., 1996. Communication networks. In: Kroodsma, D.E., Miller, E.H. (Eds.), Ecology and Evolution of Acoustic Communication in Birds. Cornell University Press, Ithaca, New York, pp. 409–425.
Mennill, D.J., Ratcliffe, L.M., 2004. Overlapping and matching in the song contests of black-capped chickadees. Anim. Behav. 67, 441–450.
Mennill, D.J., Ratcliffe, L.M., Boag, P.T., 2002. Female eavesdropping on male song contests in songbirds. Science 296, 873.
Miller, D.B., 1979a. Long-term recognition of father's song by female zebra finches. Nature 280, 389–391.
Miller, D.B., 1979b. The acoustic basis of mate recognition by female zebra finches (*Taeniopygia guttata*). Anim. Behav. 27, 376–380.
Møller, A.P., 1991. Why mated songbirds sing so much: mate guarding and male announcement of mate fertility status. Am. Nat. 138, 994–1014.
Møller, A.P., 1992. Parasites differentially increase the degree of fluctuating asymmetry in secondary sexual characters. J. Evol. Biol. 5, 691–699.
Mota, P.G., Depraz, V., 2004. A test of the effect of male song on female nesting behaviour in the Serin (*Serinus serinus*): a field playback experiment. Ethology 110, 841–850.
Müller, W., Vergauwen, J., Eens, M., 2010. Testing the developmental stress hypothesis in canaries: consequences of nutritional stress on adult song phenotype and mate attractiveness. Behav. Ecol. Sociobiol. 64, 1767–1777.
Mundinger, P.C., 1970. Vocal imitation and individual recognition in finch calls. Science 168, 480–482.
Mundinger, P.C., 1995. Behaviour–genetic analysis of canary song: inter-strain differences in sensory learning, and epigenetic rules. Anim. Behav. 50, 1491–1511.
Nagle, L., Kreutzer, M., 1997a. Song tutoring influences female song preferences in domesticated canaries. Behaviour 134, 89–104.
Nagle, L., Kreutzer, M., 1997b. Adult female domesticated canaries can modify their song preferences. Can. J. Zool. 75, 1346–1350.
Nagle, L., Kreutzer, M., Vallet, E.M., 1993. Obtaining copulation solicitation displays in female canaries without estradiol implants. Experientia 49, 1022–1023.
Nagle, L., Kreutzer, M., Vallet, E.M., 1997. A chorus song style influences sexual responses in female canaries. Can. J. Zool. 75 (4), 638–640.
Nagle, L., Kreutzer, M., Vallet, E., 2002. Adult female canaries respond to male song by calling. Ethology 108, 463–472.
Naguib, M., 1999. Effects of song overlapping and alternating on nocturnally singing nightingales. Anim. Behav. 58, 1061–1067.

Naguib, M., 2005. Singing interactions in songbirds: implications for social relations and territorial settlement. In: McGregor, P.K. (Ed.), Animal Communication Networks. Cambridge University Press, Cambridge.

Naguib, M., Mennill, D.J., 2010. The signal value of bird song: empirical evidence suggests song overlapping is a signal. Anim. Behav. 80, e11–e15.

Naguib, M., Riebel, K., 2006. Bird song: a key model in animal communication. Brown, K. (Ed.), second ed. In: Encyclopedia for Language and Linguistics, vol. 2. Elsevier, Oxford, pp. 40–53.

Naguib, M., Todt, D., 1997. Effects of dyadic vocal interactions on other conspecific receivers in nightingales. Anim. Behav. 54, 1535–1543.

Nelson, D.A., Marler, P., 1993. Innate recognition of song in white-crowned sparrows: a role in selective vocal learning? Anim. Behav. 46, 806–808.

Nolan, P.E., Hill, G.E., 2004. Female choice for song characteristics in the House Finch. Anim. Behav. 67, 403–410.

Nottebohm, F., 1977. Asymmetries in neural control of vocalization in the canary. In: Harnard, S., Doty, R.W., Goldstein, L., Jaynes, J., Krauthamer, G. (Eds.), Lateralization of the Nervous System. Academic Press, New York, pp. 23–44.

Nottebohm, F., Stokes, T.M., Leonard, C.M., 1976. Central control of song in the canary, *Serinus canarius*. J. Comp. Neurol. 165, 457–486.

Nowicki, S., Searcy, W.A., 2005. Song and mate choice: how the development of behavior helps us understand function. Auk 122, 1–14.

Nowicki, S., Peters, S., Podos, J., 1998. Song learning, early nutrition and sexual selection in songbirds. Am. Zool. 38, 179–190.

O'Loghlen, A.L., Beecher, M.D., 1997. Sexual preferences for mate song types in female song sparrows. Anim. Behav. 53, 835–841.

Ophir, A.G., Galef, B.G.J., 2003. Female Japanese quail that 'eavesdrop' on fighting males prefer losers to winners. Anim. Behav. 66, 399–407.

Ophir, A.G., Persaud, K.N., Galef, B.G.J., 2005. Avoidance of relatively aggressive male Japanese quail (Coturnix japonica) by sexually experienced conspecific females. J. Comp. Psychol. 119, 3–7.

Otter, K.A., Ratcliffe, L.M., 2005. Enlightened decisions: female assessment and communication networks. In: McGregor, P.K. (Ed.), Animal Communication Networks. Cambridge University Press, Cambridge.

Otter, K., McGregor, P.K., Terry, A.M.R., Burford, F.R.L., Peake, T.M., Dabelsteen, T., 1999. Do female great tits (*Parus major*) assess males by eavesdropping? A field study using interactive song playback. Proc. R. Soc. Lond. B 266, 1305–1309.

Owens, I.P., Rowe, C., Thomas, A.L., 1999. Sexual selection, speciation and imprinting: separating the sheep form the goats. TREE 14, 131–132.

Parisot, M., 2004. Le statut social et le choix du partenaire chez le canari domestique commun (Serinus canaria): indices comportementaux, hormonaux et vocaux. Unpublished PhD thesis, Université Paris Ouest Nanterre La Défense.

Parisot, M., Vallet, E., Nagle, L., Kreutzer, M., 2002. Male canaries discriminate among songs: call rate is a reliable measure. Behaviour 139, 55–63.

Parisot, M., Nagle, L., Vallet, E., Kreutzer, M., 2004. Dominance-related foraging in female domesticated canaries under laboratory conditions. Can. J. Zool. 82, 1246–1250.

Pasteau, M., Nagle, L., Kreutzer, M., 2004. Preferences and predispositions for intra-syllabic diversity in female canaries (*Serinus canaria*). Behaviour 141, 571–583.

Pasteau, M., Nagle, L., Kreutzer, M., 2007. Influences of learning and predispositions on frequency level preferences on female canaries (*Serinus canaria*). Behaviour 144, 1103–1118.

Pasteau, M., Nagle, L., Monbureau, M., Kreutzer, M., 2009a. Aviary experience has no impact on predispositions of the female canary to prefer longer sexy phrases. Auk 126, 383–388.

Pasteau, M., Nagle, L., Kreutzer, M., 2009b. Preferences and predispositions of female canaries (*Serinus canaria*) for loud intensity of male sexy phrases. Biol. J. Linn. Soc. 96, 808–814.

Payne, R.B., 1983. Wasser, S.K. (Ed.), Social Behavior of Female Vertebrates. Academic Press, New York, pp. 108–110.

Peake, T.M., 2005. Eavesdropping in communication networks. In: McGregor, P.K. (Ed.), Animal Communication Networks. Cambridge University Press, Cambridge.

Peake, T.M., Terry, A.M.R., McGregor, P.K., Dabelsteen, T., 2001. Male great tits eavesdrop on simulated male-to-male vocal interactions. Proc. R. Soc. Lond. B 268, 1183–1187.

Peake, T.M., Terry, A.M.R., McGregor, P.K., Dabelsteen, T., 2002. Do great tits assess rivals by combining direct experience with information gathered by eavesdropping? Proc. R. Soc. Lond. B 269, 1925–1929.

Pfaff, J.A., Zanette, L., MacDougall-Shackleton, S.A., MacDougall-Shackleton, E.A., 2007. Song repertoire size varies with HVC volume and is indicative of male quality in song sparrows (*Melospiza melodia*). Proc. R. Soc. Lond. B 274, 2035–2040.

Podos, J., 1997. A performance constraint on the evolution of trilled vocalizations in a songbird family (Passeriformes: *Emberizidae*). Evolution 51, 537–551.

Podos, J., Huber, S.K., Taft, B., 2004. Bird song: the interface of evolution and mechanism. Annu. Rev. Ecol. Evol. Syst. 35, 55–87.

Price, A.C., Rodd, F.H., 2006. The effect of social environment on male–male competition in guppies (*Poecilia reticulata*). Ethology 112, 22–32.

Rehsteiner, U., Geisser, H., Reyer, H.U., 1998. Singing and mating success in water pipits: one specific song element makes all the difference. Anim. Behav. 55, 1471–1481.

Remeš, V., 2011. Yolk androgens in great tit eggs are related to male attractiveness, breeding density and territory quality. Behav. Ecol. Sociobiol. 65, 1257–1266.

Riebel, K., 2000. Early experience leads to repeatable preferences for male song in female zebra finches. Proc. R. Soc. Lond. B 267, 2553–2558.

Riebel, K., 2003. The 'mute' sex revisited: vocal production and perception learning in female songbirds. Adv. Stud. Behav. 33, 49–86.

Riebel, K., 2009. Song and female mate choice in zebra finches: a review. Adv. Stud. Behav. 40, 197–239.

Riebel, K., Smallegange, I.M., Terpstra, N.J., Bolhuis, J.J., 2002. Sexual equality in zebra finch song preference: evidence for a dissociation between song recognition and production learning. Proc. R. Soc. Lond. B 269, 729–733.

Ritschard, M., Riebel, K., Brumm, H., 2010. Female zebra finches prefer high-amplitude song. Anim. Behav. 79, 877–883.

Robertson, B.C., 1996. Vocal mate recognition in a monogamous, flock-forming bird, the silvereye (*Zosterops lateralis*). Anim. Behav. 51, 303–311.

Ryan, M.J., 1990. Sexual selection, sensory system and sensory exploitation. Oxford Surv. Evol. Biol. 7, 157–195.

Ryan, M.J., Keddy-Hector, A., 1992. Directional patterns of female mate choice and the role of sensory biases. Am. Nat. 139, 4–35.

Ryan, M.J., Fox, J.H., Wilczynski, W., Rand, A.S., 1990. Sexual selection for sensory exploitation in the frog *Physalemus pustulosus*. Nature 343, 66–67.

Safran, R.J., Pilz, K.M., McGraw, K.J., Correa, S.M., Schwabl, H., 2008. Are yolk androgens and carotenoids in barn swallow eggs related to parental quality? Behav. Ecol. Sociobiol. 62, 427–438.

Schaper, S.V., Dawson, A., Sharp, P.J., Gienapp, P., Caro, S.P., Visser, M.E., 2012. Increasing temperature, not mean temperature, is a cue for avian timing of reproduction. Am. Nat. 179, E55–E69.

Schwabl, H., 1993. Yolk is a source of maternal testosterone for developing birds. Proc. Natl. Acad. Sci. USA 90, 11446–11450.

Schwabl, H., 1996a. Maternal testosterone in the avian egg enhances postnatal growth. Comp. Biochem. Phys. A 114, 271–276.

Schwabl, H., 1996b. Environment modifies the testosterone levels of a female bird and its eggs. J. Exp. Zool. 276, 157–163.

Schwabl, H., 1997. The content of maternal testosterone in House Sparrow *Passer domesticus* eggs vary with breeding conditions. Naturwissenschaften 84, 406–408.

Searcy, W.A., 1992a. Measuring responses of female birds to male songs. In: McGregor, P.K. (Ed.), Playback and Studies of Animal Communication. Plenum, New York, pp. 175–182.

Searcy, W.A., 1992b. Song repertoires and mate choice in birds. Am. Zool. 32, 71–80.

Searcy, W.A., 1996. Sound-pressure levels and song preferences in female red-winged blackbirds (*Aegelaius phoeniceus*). Ethology 102, 187–196.

Searcy, W.A., Beecher, M.D., 2009. Song as an aggressive signal in songbirds. Anim. Behav. 78, 1281–1292.

Searcy, W.A., Beecher, M.D., 2011. Continued scepticism that song overlapping is a signal. Anim. Behav. 81, e1–e4.

Searcy, W.A., Capp, M.S., 1997. Estradiol dosage and the solicitation display assay in red-winged blackbirds. Condor 99, 826–828.

Searcy, W.A., Marler, P., 1981. A test for responsiveness to song structure and programming in female sparrows. Science 213, 926–928.

Sockman, K.W., Schwabl, H., 1999. Daily estradiol and progesterone levels relative to laying and onset of incubation in canaries. Gen. Comp. Endocrinol. 114, 257–268.

Spencer, K.A., Buchanan, K.L., Leitner, S., Goldsmith, A.R., Catchpole, C.K., 2005. Parasites affect song complexity and neural development in a songbird. Proc. R. Soc. Lond. B 272, 2037–2043.

Spitler-Nabors, K.J., Baker, M.C., 1983. Reproductive behavior by a female songbird: differential stimulation by natal and alien song dialects. Condor 85, 491–494.

Suthers, R.A., Vallet, E., Tanvez, A., Kreutzer, M., 2004. Bilateral song production in domestic canaries. J. Neurobiol. 60, 381–393.

Székely, T., Catchpole, C.K., DeVoogd, A., Marchl, Z., DeVoogd, T.J., 1996. Evolutionary changes in a song control area of the brain (HVC) are associated with evolutionary changes in song repertoire among European warblers (*Sylvidae*). Proc. R. Soc. Lond. B 263, 607–610.

Tanvez, A., Béguin, N., Chastel, O., Lacroix, A., Leboucher, G., 2004. Sexually attractive phrases increase yolk androgens deposition in Canaries (*Serinus canaria*). Gen. Comp. Endocrinol. 138, 113–120.

Tanvez, A., Parisot, M., Chastel, O., Leboucher, G., 2008. Does maternal social hierarchy affect yolk testosterone deposition in domesticated canaries? Anim. Behav. 75, 929–934.

Templeton, C.N., Akçay, C., Campbell, S.E., Beecher, M.D., 2009. Juvenile sparrows preferentially eavesdrop on adult song interactions. Proc. R. Soc. Lond. B 277, 447–453.

ten Cate, C., Vos, D.R., 1999. Sexual imprinting and evolutionary processes in birds: a reassessment. Adv. Stud. Behav. 28, 1–31.

Thorpe, W.H., 1958. The learning of song patterns by birds, with special reference to the song of the chaffinch *Fringilla coelebs*. Ibis 100, 535–570.

Thorpe, W.H., 1961. Bird Song. Cambridge University Press, London and New York.

Todt, D., Naguib, M., 2000. Vocal interactions in birds: the use of songs as a model in communication. Adv. Stud. Behav. 29, 247–296.

Ung, D., Amy, M., Leboucher, G., 2011. Heaven it's my wife! Male canaries conceal extra-pair courtships but increase aggressions when their mate watches. PLoS One 6, e22686.

Vallet, E., Kreutzer, M.L., 1995. Female canaries are sexually responsive to special song phrases. Anim. Behav. 49, 1603–1610.

Vallet, E.M., Kreutzer, M.L., Richard, J.-P., 1992. Syllable phonology and song segmentation: testing their salience in female canaries. Behaviour 121 (3–4), 155–167.

Vallet, E., Beme, I., Kreutzer, M.L., 1998. Two-note syllables in canary songs elicit high levels of sexual display. Anim. Behav. 55, 291–297.

Vallet, E., Suthers, R.-A., Kreutzer, M., Tanvez, A., 2006. Bilateral motor skills in domestic canary song. Acta Zool. Sin. 52 (Suppl.), 475–477.

Vehrencamp, S.L., 2000. Handicap, index, and conventional signal elements of bird song. In: Espmark, Y., Amundsen, T., Rosenqvist, G. (Eds.), Animal Signals: Signalling and Signal Design in Animal Communication. Tapir Academic Press, Trondheim, Norway, pp. 277–300.

Vignal, C., Mathevon, N., Mottin, S., 2004. Audience drives male songbird response to partner's voice. Nature 430, 448–451.

Vignal, C., Mathevon, N., Mottin, S., 2008. Mate recognition by female zebra finch: analysis of individuality in male call and first investigations on female decoding process. Behav. Proc. 77, 191–198.

Wagner Jr., W.E., Hoback, W.W., 1999. Nutritional effects on male calling behaviour in the variable field cricket. Anim. Behav. 57, 89–95.

Wallschläger, D., 1980. Correlation of song frequency and body in passerine birds. Experientia 36, 412.

Warren, R.P., Hinde, R.A., 1961. Does the male stimulate oestrogen secretion in female canaries? Science 133, 1354–1355.

Wasserman, F.E., Cigliano, J.A., 1991. Song output and stimulation of the female in White-throated Sparrows. Behav. Ecol. Sociobiol. 29, 55–59.

West, M.J., King, A.P., Eastzer, D.H., 1981. Validating the female bioassay of cowbird (*Molothrus ater*) song: relating differences in song potency to mating success. Anim. Behav. 29, 490–501.

Wild, J., Williams, M., Suthers, R., 2000. Neural pathways for bilateral vocal control in songbirds. J. Comp. Neurol. 423, 413–426.

Wiley, R.H., Hatchwell, B.J., Davies, N.B., 1991. Recognition of individual mates' songs by female dunnocks: a mechanism increasing the number of copulatory partners and reproductive success. Ethology 88, 145–153.

Williams, G.C., 1966. Natural selection, the cost of reproduction, and a refinement of Lack's principle. Am. Nat. 100, 687–690.

Zahavi, A., 1975. Mate selection: a selection for a handicap. J. Theor. Biol. 53, 205–214.

Zann, R., 1996. The Zebra Finch. Oxford University Press, Oxford.

ADVANCES IN THE STUDY OF BEHAVIOR, VOL. 44

# Causes and Consequences of Differential Growth in Birds: A Behavioral Perspective

Mark C. Mainwaring and Ian R. Hartley

LANCASTER ENVIRONMENT CENTRE, LANCASTER UNIVERSITY, LANCASTER, UNITED KINGDOM

## I. Introduction

Evolutionary processes, such as natural and sexual selection, tend to favor organisms that maximize their evolutionary fitness, which itself depends on the allocation of resources to both reproduction and the growth, maintenance, and survival necessary to achieve it (Darwin, 1859). This resource allocation depends on the kind of environment in which an organism lives and how reproductive opportunities vary with age, size, social status, and a variety of other factors. The outcome of the subsequent allocation of resources constitutes the organism's "life history" (Roff, 1992; Stearns, 1992). The challenge for behavioral and evolutionary biologists is to understand the processes that lead to variation in life history traits, both within and between species.

Here, we focus on the variation in early development and growth in birds and how this relates to the ecology and behavior of the parents and offspring. Birds are an ideal taxonomic group in which to investigate the interactions between parental investment and offspring. They are readily observable and individuals can be marked and followed; they have discrete breeding events and the units of investment, such as eggs and nestlings, can be counted and measured; and they provide parental care that can be quantified.

### A. Differential Growth and Trade-Offs

Ontogeny is an important stage of an individual's life history, and variation in growth trajectories and juvenile phenotypes has important consequences for a wide range of traits including behavioral traits that



0065-3454/12 $35.00
DOI: 10.1016/B978-0-12-394288-3.00006-X

subsequently influence fitness, because they directly influence mating success or competitive interactions over food (Arendt, 1997; Dmitriew, 2011; Kilner and Hinde, 2008; Lessells et al., 2011; Lindström, 1999; Metcalfe and Monaghan, 2001, 2003; Ricklefs, 1968; Schew and Ricklefs, 1998; West-Eberhard, 2003). Ontogeny encompasses both growth, which refers to increases in body size over time, and development, which refers to the allocation of resources to cell differentiation for specialized systems (Arendt, 1997). There is widespread evidence that early developmental conditions have important long-term consequences for an individual's fitness (Lindström, 1999). For example, the body condition of juvenile great tits (*Parus major*) during the nestling period was found to be positively correlated with the quality of the breeding habitat that the birds later occupied during adulthood (Verhulst et al., 1997). Additionally, female nestlings from experimentally enlarged broods of collared flycatchers (*Ficedula albicollis*) were found to lay smaller clutches than control females in their first breeding season (Gustafsson and Sutherland, 1988), presumably as a direct effect of the increased competition for limited resources during their development.

Despite the importance of their rearing circumstances, offspring raised in competitive and limiting environments are not necessarily at the mercy of their parents. Developmental plasticity enables offspring to adjust their morphology, physiology, and behavior to maximize their fitness within the confines of their environmental conditions. External cues, such as competition within the nest or other evolutionary pressures, can influence rates of development and lead to the production of distinct phenotypes from the same genotype (Arendt, 1997; Beldade et al., 2011; Dmitriew, 2011; Schew and Ricklefs, 1998; West-Eberhard, 2003). So the final body structure of a bird, and how it reaches that end point, will depend on its behavioral and physiological responses to its rearing environment, as well as its genetics (Schew and Ricklefs, 1998; van Noordwijk and Marks, 1998).

One aspect of developmental plasticity which has received a lot of attention has been that of compensatory growth. If developing offspring experience a period of nutritional deficit, they can subsequently show accelerated growth, should conditions improve, apparently compensating for the initial setback (reviews in Metcalfe and Monaghan, 2001, 2003). However, empirical studies suggest that although compensatory growth can bring benefits in the short term, it is also associated with a variety of costs that are often not evident until much later in adult life.

For example, a study of green swordtail fish (*Xiphophorus helleri*) found that adult body size and sexual ornament size were not affected by a period of experimentally reduced food availability during development, but, as adults, experimental fish were found to be behaviorally subordinate to controls of similar size (Royle et al., 2005). In guppies (*Poecilia reticulata*), another experimental study found that, while adult body size was not affected by periods of food shortage during development, subsequent fecundity was reduced by 20% (Auer et al., 2010). In birds, an experimental study of carrion crows (*Corvus corone*) found that following a period of reduced food availability during the nestling phase, experimental nestlings remained smaller than control birds into adulthood and also had lower social status (Richner et al., 1989).

Compensatory growth has been shown to affect a number of behavioral traits during the development period and into adulthood. A study of southern gray shrikes (*Lanius meridionalis*) found that when nests were experimentally divided into zones of differing profitability in terms of food acquisition, nestlings were able to detect differences in zone profitability and position themselves so that they spent most time in the most profitable areas of the nest. This suggests that nestlings are able to learn about their environment and adjust their solicitation behaviors accordingly (Budden and Wright, 2005). In experimental studies of domesticated zebra finches (*Taeniopygia guttata*), being raised on a nutritionally poor diet did not affect final body size of individuals but resulted in long-term costs during adulthood, such as reduced learning performance (Fisher et al., 2006), significantly higher resting metabolic rate (Criscuolo et al., 2008), and reduced exploratory behavior (Krause and Naguib, 2011). In house sparrows (*Passer domesticus*), nestlings incurring an experimentally induced period of reduced food availability showed no evidence of compensatory changes in digestive efficiency when compared to control nestlings. This suggests that the gut has little spare capacity to deal with increased food intake following a period of food shortage (Lepczyk et al., 1998). Therefore, conditions experienced during early life are likely to have significant fitness costs for individuals during adulthood.

Another aspect of developmental plasticity, which has received less attention to date, is that of differential growth, where resources are directed toward some traits, at the expense of others, according to their importance in maximizing fitness (Coslovsky and Richner, 2011; Gil et al., 2008; Mainwaring et al., 2010a). Differential growth occurs because resources are limited and so some aspects of growth may be prioritized or traded-off against others (Mock and Parker, 1997; Roff, 1992; Stearns, 1992). Additionally, the development of different traits may be under antagonistic selection due to shifting pressures between periods of the life history. For example, it may be beneficial in the short term to invest a disproportionately large amount of limited resources into traits that enable success in scramble or signaling contests between siblings (e.g., Gil et al., 2008; Leonard and Horn, 1996), but this might be in a trade-off with investment in traits that enable efficient flight immediately after fledging or influence fitness later in life. Developing a large gape area during the nestling phase may enable nestlings to procure more food from their parents, as it is part of the multicomponent behavioral display that is used to solicit food during begging contests. Allocating resources toward flight feather development, meanwhile, is adaptive in the medium term as it facilitates fledging, especially as it might be important to fledge with nest-mates to avoid being ignored by provisioning parents once the majority of the brood fledge (Glassey and Forbes, 2002; Kilner and Hinde, 2008; Rosivall et al., 2005). Flight feather development is also likely to have a longer-term impact as it determines flight maneuverability, and so foraging success and predation risk, so there is the potential for a trade-off between growth speed and trait quality (Mainwaring et al., 2009).

### B. Developmental Mode

Birds exhibit a broad range of interspecific variation in their life histories. Accordingly, there is also a large amount of variation in developmental modes which occur along the altricial–precocial developmental continuum (Bennett and Owens, 2002; Gebhardt-Henrich and Richner, 1998; Martin, 1995; Roff, 1992; Stearns, 1992). Some groups, such as ducks, gamebirds, waders, and rails, have precocial, also called nidifugous, chicks which are mobile, feathered, and able to feed themselves within hours of hatching. The young of the mallee fowl (*Leipoa ocellata*) are even able to fly within a day of hatching (Marchant and Higgins, 1993). At the other end of the spectrum are groups such as the pigeons, swifts, and passerines, which have altricial, also called nidicolous, young which hatch blind and virtually naked, and which are fully dependent upon parents for warmth and food for a period of several weeks. Between the two extremes, there are groups of species which can be classed as semi-precocial or semi-altricial, depending on their relative independence after hatching (Bennett and Owens, 2002; Harrison, 1975; O'Connor, 1984).

Developmental mode in birds is largely divided along taxonomic lines. Comparative analyses suggest that, across bird families, the evolution of developmental mode is driven by the risk of predation (Lack, 1968; Martin et al., 2011b; Owens and Bennett, 1995). So, species which nest on the ground are likely to have precocial chicks which develop their senses very soon after hatching, whereas species which use nests or cavities to raise their young can delay the development of traits such as vision and escape capabilities in favor of growth in other traits. Moreover, the freedom to move around and independently forage for food means that the levels of sibling competition are much lower than those experienced within those nests of altricial species.

The nestlings of altricial species are confined to a period of time in the nest when they are totally reliant on their parents for food (Bennett and Owens, 2002). However, the family members have divergent evolutionary interests which lead to intrafamilial conflict, which itself can impact upon the growth of individual nestlings. Therefore, we briefly summarize the conflicts of interest that occur within families before reviewing the empirical studies of differential growth. Evolutionary conflicts of interest occur because resources are often limited (Kilner and Hinde, 2008; Trivers, 1974), and therefore, it can be expected that life history components will be traded off against each other to maximize the individual's lifetime reproductive success (Clutton-Brock, 1988; Newton, 1989). Individuals are expected to act selfishly to maximize their own fitness (Dawkins, 1999), yet individuals of species that reproduce sexually must collaborate to breed (Chapman et al., 2003; Clutton-Brock, 1991; Mock and Parker, 1997).

Both reproduction and parental care are costly to parents, so sexual conflict occurs over the level of parental care that each parent should provide to the current offspring (Andersson, 1994; Arnqvist and Rowe, 2005; Chapman et al., 2003; Parker et al., 2002a). In anisogamous reproduction, the asymmetry in gamete size means that females put more resources into each offspring than do males. Furthermore, the maximum number of independent offspring that parents can produce per unit time is potentially much higher in males than it is in females (Clutton-Brock and Vincent, 1991), although there are species in which the potential rate of reproduction is higher in females (Owens et al., 1994; Székely et al., 1999). Consequently, the sex with the lower potential rate of reproduction courts or competes for the other sex (Andersson, 1994; Chapman et al., 2003). In species with investment beyond the gamete stage, which includes nearly all bird species, each sex should be expected to exploit the other by reducing its own share in the investment. In altricial species, where the nestlings are entirely dependent on the parents for providing food until fledging, reproductive success is often limited by the amount of food that parents can deliver to the brood. We should expect, therefore, that when investing a similar amount of parental effort, two parents should be able to raise more offspring than a single parent (Royle et al., 2006; Webster, 1991; but see Skutch, 1949).

The theory of parent–offspring conflict suggests that, on an evolutionary time scale, offspring are generally under selection to demand more parental investment than it is in each parent's interest to provide (Godfray, 1991, 1995; Johnstone and Godfray, 2002; Kilner and Hinde, 2008; Lessells et al., 2011; Trivers, 1974). Therefore, in altricial species, the total amount of food provided is less than the offspring demand, and this evolutionary conflict of interest will favor those offspring that are able to manipulate their parents into providing more food, either by physically outcompeting nest-mates for access or by signaling more effectively to parents through begging displays (Kilner and Hinde, 2008; Parker, 1985; Parker et al., 2002a,b; Saino et al., 2000). Given that it is the parents who provide all of the resources required for offspring development in altricial species, it could follow that parent birds control both the level of food provisioning to the brood and its distribution between nestlings (Kilner and Johnstone, 1997; Royle et al., 2002a). Female parents can assert some control by determining the brood size and manipulating the competitive hierarchy among nestlings through hatching asynchrony, which subsequently reduces the benefits that offspring may gain through intense begging behavior and/or rapid growth (Ricklefs, 2002). Sibling competition within a brood favors nestlings whose rapid development increases their competitive status (Miller, 2010; Nilsson and Svensson, 1996; Ricklefs, 2002; Royle et al., 1999), which may subsequently reduce parental fitness by increasing the food requirements of individual nestlings, and therefore the brood as a whole. The benefit of slower nestling growth for parents is that the immediate food requirements of individual nestlings and the brood as a whole are reduced (Ricklefs, 2002), and for the nestlings themselves, benefits include a more responsive immune system and delayed senescence (Dmitriew, 2011; Gebhardt-Henrich and Richner, 1998; Martin et al., 2011a; Metcalfe and Monaghan, 2001, 2003). Note that while this pattern reflects evolutionary relationships, it may not necessarily reflect proximate relationships. This is because slower growth caused by poor feeding conditions, which is a proximate cause, may result in nestlings in poorer, and not better, condition (Hoi-Leitner et al., 2001).

Therefore, brood demand exceeds parental supply and sibling competition will occur over the within-brood division of that investment (Mock and Parker, 1997). The intensity and ferocity of sibling competition varies widely among species. For example, a relatively peaceful form of sibling competition is that shown by nestlings upon the arrival of a parent bird with food, when they reveal brightly colored gapes, assume a range of begging postures, beg loudly, and scramble or jostle for position within the nest (e.g., Dearborn, 1998; Leonard and Horn, 1996; Smith and Montgomerie, 1991; see review in Kilner, 2002). The majority of empirical studies have found that eldest, largest, and tallest nestling gets the most food (Mock and Parker, 1997), although it is not a universal pattern (Ardia, 2007; Low et al., 2012). A more forceful form of sibling competition is exhibited by the facultative brood reducers, where the smallest nestling/s are much more likely to die than the oldest nestlings, but only when food is in short supply (Drummond, 2002). For example, cattle egrets (*Bubulcus ibis*) usually have three eggs which hatch asynchronously, so creating a competitive hierarchy due to the size variation (Mock, 1985, 2004). A series of observations and experiments have shown that the eldest nestling is the largest and tallest and is the first to intercept the food from the provisioning parent; once it has had its fill, the rest of the food is divided between the middle and smallest nestlings. The middle nestling does little to interfere with the largest nestling's feeding, but instead concentrates on ensuring that it is second to feed by ferociously pecking the smallest nestling until it retreats to the back of the nest (Mock, 1985; Mock and Ploger, 1987). Perhaps, the most aggressive form of sibling aggression is exhibited by obligate brood reducers such as the black eagle (*Ictinaetus malayensis*). Of 58 nests where a clutch of two eggs were laid, there was only one case where both nestlings fledged because siblicidal aggression by the larger nestling almost universally resulted in the death of the smaller nestling (Brown et al., 1977).

Despite these examples, however, siblings should not be seen solely as competitors, because there are several reasons to expect them to cooperate (Forbes, 2007; Wilson and Clark, 2002). First, siblings share a relatively high proportion of genes by common descent and so could potentially gain indirect fitness from each other (Mock and Parker, 1997). Second, siblings are beneficial for heat conservation and thermoregulation, especially during the first few days after hatching, because the effective surface area to volume ratio increases with brood size (Royama, 1966; Visser, 1998). Third, in some species, siblings associate with each other after they have fledged, which has been shown to enhance foraging skills (e.g., in ospreys (*Pandion haliaetus*); Edwards, 1984) and effective learning of social behavior and, for example, helping in cooperatively breeding species (Dickinson and Hatchwell, 2004). Fourth, parents may desert a brood that is deemed too small, either to conserve time, energy, and risk in favor of another breeding attempt during the same season or to enhance their survival and condition for another breeding attempt the following season (Mock and Parker, 1997). Finally, it has been suggested that in some species, for example, parrots (Krebs, 2002), that food allocation within broods is purely parentally determined, which makes direct sibling competition redundant.

Therefore, different selection pressures at different taxonomic levels have important consequences for the development of avian offspring. In this review, we examine the causes and consequences of differential growth and growth trade-offs in birds, although we also examine the influence the behavior of individual nestlings, as growth patterns are inextricably linked to the different selection pressures at different taxonomic levels. We begin by examining interspecific patterns of developmental resource allocation and the evolutionary pressures which have selected for such patterns. For example, varying levels of predation risk, sibling competition, and parasitism all exert strong selection pressures on growth and developmental rates. We then examine the causes and consequences of differential growth and growth trade-offs that occur both between- and within-broods. Variation in growth patterns between broods within a single species may occur, for example, due to variation in the environmental conditions experienced by the nestlings while in the nest. For example, levels of sibling competition, ectoparasites, and/or predation may cause variation in the conditions experienced by different broods. Variation in growth patterns within broods results from variation between individual nestlings with respect to genotypic factors such as relatedness and gender, and/or phenotypic factors such as laying order and hatching asynchrony. These five factors commonly result in asymmetric sibling competition and have important consequences for patterns of developmental resource allocation. We then review the long-term consequences of size variation, differential growth, and growth trade-offs. This is important as while allocating resources toward gape area may enable nestlings to procure more food from their parents during the period spent in the nest, such a resource allocation pattern is unlikely to be beneficial after fledging. We then finish by suggesting promising avenues for future research.

## II. Interspecific Patterns of Differential Growth

A number of comparative studies have examined interspecific patterns of differential growth and have tried to explain the reasons why growth and developmental rates vary widely among species. For example, while the precocial chicks of gamebirds and megapodes grow rapidly and quickly acquire independent foraging skills, the altricial nestlings of passerine birds are born naked and helpless, grow relatively slowly, and take a relatively long time to develop independent foraging skills (Bennett and Owens, 2002; Gebhardt-Henrich and Richner, 1998; Martin and Briskie, 2009; Martin and Scwabl, 2008). Initially, Lack (1968) noted the higher predation rates upon species constructing open nests but later argued that differences in number of offspring and their development duration was actually driven by food limitation. However, there is evidence from comparative studies that suggests the risk of losing eggs or nestlings to predators exerts strong selective pressures of developmental mode (Bosque and Bosque, 1995; Lack, 1968; Owens and Bennett, 1995), and so ground-nesting taxa tend to have precocial young, whereas cavity or off-ground nesters have altricial young. In general, we should expect that selection pressure from predation risk should drive the evolution of faster development in nestlings and, even within altricial species, there is evidence that this is the case (Martin and Briskie, 2009). For example, the nestlings of hole-nesting blue tits (*Cyanistes caeruleus*) remain in the nest for

16–22 days and fledge at approximately 90% of their final body mass (Cramp and Perrins, 1990; Mainwaring et al., 2011a), whereas nestlings of the ground-nesting corn bunting (*Miliaria calandra*) fledge after 9–13 days in the nest, and at 65% of their final body mass (Hartley and Shepherd, 1994; Hartley et al., 2000). However, it is important to remember that in non-avian taxa, increased predation risk commonly favors slower development in order to reduce both food requirement and feeding activity (Dmitriew, 2011). It is also important to clarify that these patterns have evolved at taxonomic scales above the species, yet, within species, predation risk may be negatively correlated with developmental speed (e.g., East African stonechats (*Saxicola torquata axillaris*); Scheuerlein and Gwinner, 2006).

The earliest comparative studies (Ricklefs, 1969; Ricklefs et al., 1998) found no effect of possible environmental influences on growth rates, suggesting that nestlings grow at the maximum rates allowed within physiological constraints (Arendt, 1997; Dmitriew, 2011). A more recent study, however, considered that estimates of growth rates may have been confounded by variation in the length of nestling periods and therefore reestimated growth rates to take this into account (Remeš and Martin, 2002). In a sample of 115 North American passerines, the growth rates of altricial nestlings were strongly positively correlated to daily nest predation rates, even after controlling for adult body mass and phylogeny. Additionally, nestlings of species under stronger predation pressure remained in the nest for a shorter period, and they left the nest at lower body mass relative to adult body mass. Thus nestlings both grew faster and left the nest at an earlier developmental stage in species at higher risk of predation. On a wider geographic scale, another study found similarly that predation risk was the main driver of growth rate evolution between species, but that within species, growth rates were limited by food in temperate regions, but not in tropical regions (Martin et al., 2011b).

Remeš (2007) tested the theory that as annual adult mortality rates increase, then parental investment in a single reproductive event should increase and hence, the growth rates of the nestlings should also increase. In a sample of 84 North American passerines, it was shown that the growth rates of altricial nestlings were strongly positively correlated with annual adult mortality rates, even after controlling for other covariates, including nest predation rates and phylogeny. Thus nestlings grew faster in species with higher levels of annual adult mortality rates and higher levels of parental care (Remeš, 2007). More broadly, these results (Martin et al., 2011b; Remeš, 2007; Remeš and Martin, 2002) support a view that, between species, growth and developmental rates of altricial nestlings are influenced by their environmental rearing conditions. They show that nest predation is an agent of selection on nestling growth rates, but also that a number of other factors affect developmental rates, including food availability, biome type, and ambient temperatures.

The extent and form of competition among siblings for limited food resources may also be an important determinant of growth patterns over evolutionary time scales (Royle et al., 1999; Werschkul and Jackson, 1979). Theoretical models of sibling competition that are based upon kin selection suggest that sibling rivalry should be greatest when average relatedness within broods is lowest (Bonabeau et al., 1998; Cockburn, 2006; Parker et al., 1989). Although in birds the most common social mating system is monogamy, we know from studies that have used DNA to identify parentage that mixed parentage, and especially mixed paternity, is common in many species but rare in others (Birkhead and Møller, 1992; Griffith et al., 2002). So, for example, in species where offspring are often sired through extra-pair copulations, the average relatedness within broods is lower than in species where extra-pair parentage is rare. Interspecific studies have shown that levels of extra-pair paternity are negatively correlated with male provisioning rates, so that in species where extra-pair paternity is widespread, the male contribution toward parental care of the growing offspring is relatively low (Møller, 2000; Møller and Birkhead, 1993). Therefore, interspecific patterns of paternity may further intensify sibling competition and thereby strongly influence offspring development. Royle et al. (1999) predicted that this variation in within-brood relatedness between species should influence competition for resources such as food delivered to the nest by parents. In such competitions, faster growing nestlings would be at an advantage over slower growing nest-mates, because in general, larger nestlings are able to dominate the competitive environment of the nest. In a comparative study across species, Royle et al. (1999) found a strong, positive correlation between the mean nestling growth rate and the incidence of extra-pair paternity. This result even held after controlling for such confounding variables as body size, brood size, mating system, and the form of parental care. This suggests that variation in growth rate among bird species is not only simply dependent on ecological and developmental factors but is also influenced by interactions, over an evolutionary timescale, among kin (Royle et al., 1999; Werschkul and Jackson, 1979). Note that a number of intraspecific studies demonstrate costs of faster growth (Metcalfe and Monaghan, 2001) and that these patterns are occurring at the interspecific level. It is not clear, however, whether faster growth at the interspecific level comes at a cost to the nestlings, through trade-offs against other life history traits or from their parents through greater investment in parental care. Evidence from a similar comparative study, which explored the relationship across species between nestling begging intensity and relatedness within broods, found that nestlings begged with louder calls in species where extra-pair parentage was more common (Briskie et al., 1994), so it is possible that the parents are paying for the extra growth speed as a response to nestling begging signals. Alternatively, it is possible that the nestlings begged louder in order to compensate for a reduction in other species-specific, that is, visual, cues.

Given that interactions with family members are an important selective pressure on the evolution of growth and developmental rates, we might expect that brood parasites, which lay their eggs in the nests of other species, have an even greater influence than the presence of half-siblings in mixed parentage broods (Davies, 2000; Davies et al., 1998; Dearborn and Lichtenstein, 2002; Rivers et al., 2010). Brood parasites, such as cuckoos and cowbirds, are unrelated to their host parents and nest-mates, so we should expect no kin-selected benefits to hold them back from begging as hard as they need to and growing at their optimum pace. The evidence suggests that cuckoos, cowbirds, and other brood parasites do beg more intensely for a given level of need (Davies, 2000; Dearborn and Lichtenstein, 2002; Redondo and Zuñiga, 2002). Ortega and Cruz (1992) examined how the nestlings of brood parasitic brown-headed cowbirds (*Molothrus ater*), which are much smaller and lighter than the nestlings of their yellow-headed blackbird (*Xanthocephalus xanthocephalus*) hosts, allocated resources during development. Brown-headed cowbird nestlings were experimentally cross-fostered into yellow-headed blackbird nests, and it was found that the yellow-headed blackbird nestlings were heavier and had larger gapes than the brown-headed cowbird nestlings throughout the nestling period. However, brown-headed cowbird nestlings had significantly larger gapes, relative to their weight, than yellow-headed blackbird nestlings throughout the nestling period (Ortega and Cruz, 1992). The suggestion is that the disproportionately large gape area gives the brown-headed cowbird an advantage in competition against the hosts' nestlings and manipulates the host parents provision and allocation of food. Host–parasite systems are, however, coevolutionary arms races involving evolutionary advances by the parasite and evolutionary counter-advances by the hosts (Davies, 2000; Davies et al., 1998; Dearborn and Lichtenstein, 2002). Remeš (2006) used a comparative analysis across 134 North American passerine species to test the hypothesis that risk of brood parasitism was related to growth and development speed in the potential hosts. The analysis showed that species at greater risk of brood parastism developed faster, fledged at a lower proportion of the final body size, and had shorter nestling periods than species at lower risk, thereby alleviating the negative impacts of brood parasitic nestlings on host young.

Together, these studies across species show that competition within the nest, whether caused by brood parasites or kin, can have strong effects on the evolution of resource allocation and growth strategies of nestlings (Boncoraglio et al., 2008; Møller, 2000; Møller and Birkhead, 1993; Ortega and Cruz, 1992; Remeš, 2006; Royle et al., 1999). More specifically, they show that nestlings can allocate resources toward traits which facilitate effective nest-mate competition and potentially manipulate the parents' provisioning behavior to the advantage of the nestling.

Variation in avian growth rates has therefore been explained by both predation risk and food availability. The relative influence of predation risk and food availability in determining avian growth rates was tested by Martin et al. (2011b) using data from 64 species of passerine birds from 3 continents, including tropical and temperate regions. The growth rates of nestlings were found to increase strongly with increasing nest predation rates, which in turn were strongly associated with reduced feeding rates (Martin et al., 2011b). In other words, parents provisioning nestlings in risky environments did so at a lower rate than parents provisioning nestlings in relatively safe environments. Therefore, even if parents were fully compensating for the reduced number of trips by provisioning larger quantities of food, this still provides support for the idea that predation rates, and not food availability, explain variation in growth rates in birds (Martin et al., 2011b). Consequently, the increase in growth rates despite lower food delivery rates suggests an antagonistic interaction between nest predation and food limitation on growth rates (Martin et al., 2011b).

In addition to predation risk and food availability, it has also been suggested that parasitism by ectoparasites, as opposed to brood parasites, may be an important natural selection pressure on reproductive periods and developmental rates. Using a comparative analysis involving 43 bird species, Møller (2005) examined the relative importance of daily nest predation rates and daily nest mortality rates due to parasites in determining the relative duration of incubation and nestling periods in birds. Daily nest predation rates were found to be significantly negatively correlated with daily mortality rates due to parasitism. It was concluded that the relative duration of nestling periods is influenced by the effects of parasites rather than nest predators and that parasites constitute a neglected selective force affecting the evolution of nestling periods in birds (Møller, 2005). Further empirical support for this idea comes from a study of house finches (*Carpodacus mexicanus*) which aimed to test the theory that the presence of parasites selects for a shorter period in the nest (Badyaev et al., 2006). The presence of a hematophagous ectoparasitic nest mite, *Pellonyssus reedi*, affected the survival of male nestlings much more than daughters. Consequently, it was shown that in nests containing large numbers of mites, laying females laid male eggs last and female eggs first which resulted in a reduction in the sons' exposure to the mites. Males also grew faster than females, which meant that a simultaneous adjustment of laying order and sex-specific growth patterns enabled shorter exposure to nest ectoparasites and a 10% reduction in mite-induced nestling mortality in both sexes (Badyaev et al., 2006).

There may also be an interspecific trade-off between the growth of various morphological characters and the development of an immune system. Long embryonic periods are associated with slower intrinsic development, which is thought to allow enhanced physiological systems, such as immune function, to develop (Ricklefs, 1992, 2002; Ricklefs et al., 1998). Consequently, it should be expected that the length of embryonic periods is positively associated with immune function, yet two tests for such a trade-off provided no support for such a trade-off (Palacios and Martin, 2006; Tella et al., 2002). However, the extrinsic effects of temperature were ignored in those studies, and a later study (Martin et al., 2011a) showed that when using data from 34 passerine birds from tropical Venezuela and north temperate Arizona, the immune function of offspring was positively correlated. Therefore, the length of embryonic periods seemed to have been traded-off against intrinsic rates of embryonic development once the intrinsic rates of embryonic temperatures were taken into account (Martin et al., 2011a).

In summary, if we examine patterns across species of birds, we see that there are ecological and behavioral correlates with the speed of growth and development mode. The risk of egg or nestling loss through predation seems to have driven the evolution of development mode and especially the main altricial and precocial divide, which is largely split along taxonomic lines at the order and family level (Owens and Bennett, 1995). There is also clear evidence that food limitation is also a major influence on the evolution of developmental strategies among species (Lack, 1968). However, Lack (1968) sometimes confused proximate and evolutionary pressures upon growth rates, which is important as there is evidence that food limitation influences growth rate variation within species (Martin, 2004). Furthermore, there is also evidence that at the species level, a high predation risk is associated with faster growth, nest departure at earlier stages of development, and shorter periods in the nest (Remeš and Martin, 2002), but there are also other factors which are important at this taxonomic level. In altricial species, competition from siblings, half-siblings, and unrelated brood parasites, for limited resources such as parentally provisioned food, has been shown to impact on growth speed and time spent in the nest as a nestling. Similarly, across bird species, ectoparasites also have a positive impact on nestling speed of growth and development. However, it is important to remember that this is in contrast to intraspecific studies which commonly show that parasites lower the mass of nestlings (Richner et al., 1993). The general pattern is that as nests become riskier or more competitive environments, nestlings have evolved to leave them more quickly, either by growing faster or leaving at an earlier stage of development. Together, such studies go a long way toward explaining why growth and developmental rates vary widely among species, although further studies which examine the relative contribution of predation risk, sibling competition, and parasitism in explaining variation in growth rates would be useful. Most comparative studies use change in mass as the measure of growth and developmental, presumably because it is the trait most often reported in growth studies. We would encourage the measurement and reporting of other morphological traits that give further information on the structure of growing individuals in empirical studies. In particular, biometrics which partition skeletal and feather development would be useful in addition to mass, as they would enable the inclusion of potential trade-offs in future comparative analyses.

## III. Intraspecific Variation in Growth: Between Broods

The variation in growth and development that we see between orders, families, and species of birds has evolved over long evolutionary timescales. In contrast, variation between individuals within species can arise over comparatively short timescales and can occur in direct response to the conditions experienced during the nestling period. In this section, we consider the variation in growth between broods, but within a species. Factors such as habitat quality, parental quality, brood size, seasonality, and subsequent food availability may all influence the growth and success of broods. When resources, such as parentally delivered food, are scarce, nest-mates compete against each other for the food that is provided (Mock and Parker, 1997). We would predict that variation in nestling growth within a species should follow the general pattern seen across species, namely, that faster growth and earlier development should be found in nests where competition and risk are greatest.

Gil et al. (2008) explored the effect of nestling competition on growth trade-offs in spotless starlings (*Sturnus unicolor*) by experimentally manipulating brood size. When compared to nestlings in control broods, nestlings in experimentally enlarged broods were smaller in terms of overall size, as defined as a composite index of mass, wing length, tarsus length, and bill width, while gape width was larger (Fig. 1).

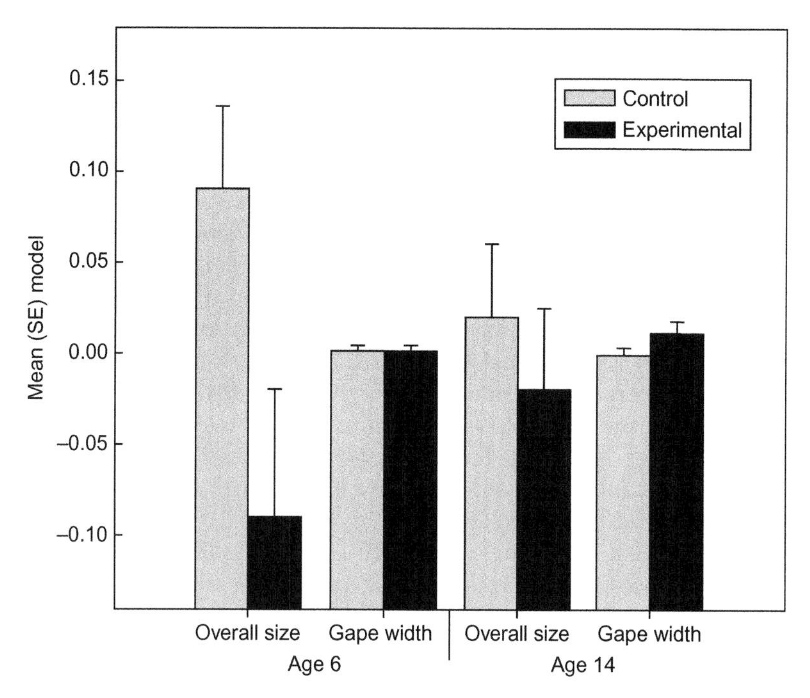


FIG. 1. Sibling competition and differential gape development in spotless starling nestlings. The level of sibling competition experienced by nestlings was experimentally increased (experimental) or left unmanipulated (control). When compared to nestlings from control broods at day 6, nestlings raised in experimental broods had similar gape widths but were smaller in terms of their overall size, as defined as a composite index of mass, wing length, tarsus length, and bill width. At day 14, nestlings raised in experimental broods had wider gapes and had reached a similar size to nestlings raised in control broods. (Data from Gil et al., 2008.)

It was concluded that nestlings which were forced to compete harder for food adaptively allocated resources toward growth of their gapes because this acted as a stronger signal to provisioning parents (Gil et al., 2008). Gapes are an integral part of the multicomponent behavioral display that is used by nestlings to solicit food from parents. Gapes are largest around the middle of the nestling period and get smaller after that, which suggests that they have a temporary function during sibling competition (Clark, 1995). Illustratively, one study showed that the visual gape display of Eurasian reed warbler (*Acrocephalus scirpaceus*) nestlings influenced parental provisioning behaviors, but provisioning behaviors were influenced to a greater extent by a combination of a visual gape display and a vocal begging display (Kilner et al., 1999). Consequently, there is good evidence to suggest that nestlings invest a disproportionately large amount of resources into developing a large gape area which helps them to procure more food from their parents, as part of the multicomponent behavioral display used to solicit food during begging contests.

Similar experimental approaches have been used to explore the effect of increased competition on development in marsh tits (*Poecile palustris*) (Nilsson and Gårdmark, 2001) and mourning doves (*Zenaida macroura*) (Miller, 2010). In both species, experimentally increased brood sizes induced a shift in growth effort toward flight feathers and away from body mass gain. In the marsh tit, the suggestion was that this trade-off was for the short-term gain of fledging more quickly to escape the low-quality environment of the nest (Nilsson and Gårdmark, 2001), with a probable longer-term cost of decreased postfledging survival (Lindström, 1999). In the mourning dove, the trade-off was shown to be adaptive as although nestlings from larger broods fledged at later ages owing to slower overall growth, the allocation of resources toward wing growth reduced this effect by an estimated 11% (Miller, 2010).

In species which lay a single egg, such as many seabirds, overt sibling competition would appear to be absent, but that is not the case. Many seabirds are long-lived species and so the contribution of one breeding event to lifetime reproductive success is proportionately quite small. Nestlings may, therefore, be abandoned during periods of food shortage (Gaston, 2004) because the parents will have opportunities to breed in following years. The single nestlings are effectively competing for parental care with siblings in future broods, rather than directly in the same time and space (Trivers, 1974). Øyan and Anker-Nilssen (1996) studied the effect of experimentally inducing food shortage in Atlantic puffins (*Fratercula arctica*), which lay one egg per breeding event. During periods of food shortage, nestlings allocated resources toward body mass increase and away from other morphological traits, such as gape size, which might facilitate sibling competition in other species which suffer direct competition from simultaneously reared siblings.

It is interesting to note the contrasting patterns of developmental resource allocation between species. First, in the absence of direct simultaneous sibling competition, the puffin nestlings allocated their resources toward body mass increases during times of food shortage. Meanwhile, an increase in direct, simultaneous competition from siblings in the nests of spotless starlings, marsh tits, and mourning doves resulted in nestlings allocating resources toward those morphological traits that facilitate rapid fledging or give an advantage in competitive begging contests. Furthermore, it is interesting to note that spotless starling and marsh tit nestlings employed different growth strategies, despite both species being hole-nesters and presumably being subjected to similar selective pressures.

The differences may lie in how responsive the parents are to visual stimuli presented by the nestlings, and such behavioral interactions could be usefully explored in the future.

Another study examined resource allocation patterns among broods of wild and domesticated zebra finches (Mainwaring et al., 2010b). The zebra finch is native to the inhospitable arid zone of Australia but is also one of the most widely used vertebrate model systems for studying a variety of behavioral and evolutionary questions. Domesticated populations have been maintained for over 100 years, and consequently over a hundred generations, by amateur aviculturists and live in an environment where natural selection pressures such as predation and food acquisition, which are likely to be the strongest selection pressures acting on populations in the wild (Sossinka, 1982; Zann, 1996), have been greatly relaxed through the provision of *ad libitum* food in a benign environment. Nestlings in domesticated environments had significantly higher masses and larger skeletal characters than nestlings in wild broods, while prefledging wing lengths were similar in wild and domesticated broods. Wing growth, at least up until the prefledging stage, may be a priority for wild zebra finch nestlings as it reduces the amount of time spent in the nest and hence minimizes nest predation risk (Zann, 1996). However, it is plausible that nestlings in the wild prioritized wing growth at the expense of other fitness-related traits such as immune function (Eraud et al., 2008). Alternatively, wing growth may be a low priority for domestic birds as simultaneous fledging is not a priority in such a benign environment and they are hardly required to fly and, consequently, wing growth is likely to be under only weak selection (Mainwaring et al., 2010b). Meanwhile, the combination of greater food availability, lower predation risk (Remeš and Martin, 2002), and the selective breeding of heavier birds in captivity is likely to explain why domesticated birds in Europe are approximately 50% heavier than their wild counterparts in Australia (Sossinka, 1982).

Nestlings may also experience adverse conditions while in the nest due to the presence of ectoparasites. A series of studies have examined the effect of a hematophagous ectoparasite, the hen flea (*Ceratophyllus gallinae*), on the development and behavior of great tit nestlings by either experimentally adding or removing fleas from nests. Great tits breeding in nests where ectoparasites were added were found to delay breeding, then laid similarly sized clutches, yet had lower hatching and fledging success when compared to those birds breeding in nests in which the ectoparasites were removed (Oppliger et al., 1994). Furthermore, nestlings raised in nests with high levels of ectoparasite abundance had significantly lower body masses and tarsus lengths compared to nestlings raised in broods where ectoparasites were removed (Richner et al., 1993; Tschirren et al., 2003), and went on to have lower lifetime reproductive success (Fitze et al., 2004).

Meanwhile, other experimental studies have examined patterns of differential growth in alpine swift (*Apus melba*) nestlings that were raised in nests where blood-sucking louse flies (*Crataerina melbae*) were either added or removed (Bize et al., 2003, 2004). Nestlings raised in nests where ectoparasites were added grew their wing feathers at a reduced rate when compared to nestlings raised in nests where ectoparasites were removed, although there was no effect on body mass development. The wings of nestlings in more heavily parasitized nests grew for an additional 3 days when compared to nestlings raised in nests where ectoparasites were removed, and hence were of a similar size at fledging (Bize et al., 2003). This suggests some level of developmental plasticity in alpine swift nestlings, but this may entail fitness costs because the resources required to sustain compensatory growth may be invested at the expense of developmental stability. Indeed, those nestlings raised in nests where ectoparasites were added were found to have wing feathers which were less symmetrical than nestlings raised in nests where ectoparasites were removed (Bize et al., 2004), which suggests there may be a long-term cost to growth plasticity.

In barn swallow (*Hirundo rustica*) nestlings, an observational study showed that the number of louse flies (*Ornithomyia biloba*) found within a nest was positively correlated with the growth rate of feathers (Saino et al., 1998). This suggested that nestlings may allocate resources toward wing development in order to fledge sooner and escape the louse flies. When additional louse flies were experimentally added to nests, nestlings grew their wing feathers at a disproportionately faster rate, compared to other traits, and developed greater immune responses, but at a cost of having slower development of tarsus and body mass (Fig. 2). Therefore, nestlings from broods where the number of louse flies was increased grew their flight feathers significantly faster than nestlings from control broods in order to fledge sooner. However, they also fledged in poorer condition than those nestlings from control broods and so were likely to pay long-term costs for short-term adaptive trade-offs (Saino et al., 1998).

The study by Saino et al. (1998) suggested that there may be a trade-off between growth and immunity, as nestlings from nests where ectoparasite load was increased, developed stronger immune systems at the expense of body mass increases. However, the most compelling evidence of such a trade-off comes from a study of Eurasian magpies (*Pica pica*) (Soler et al., 2003). The trade-off between growth and immune function is an important one, as the development of an immune system is crucial in resisting attacks from pathogens. The development of an immune system is costly, however, and is likely to be involved in a trade-off against the development of other traits (Saino et al., 1998). Eurasian magpie nestlings were provided with

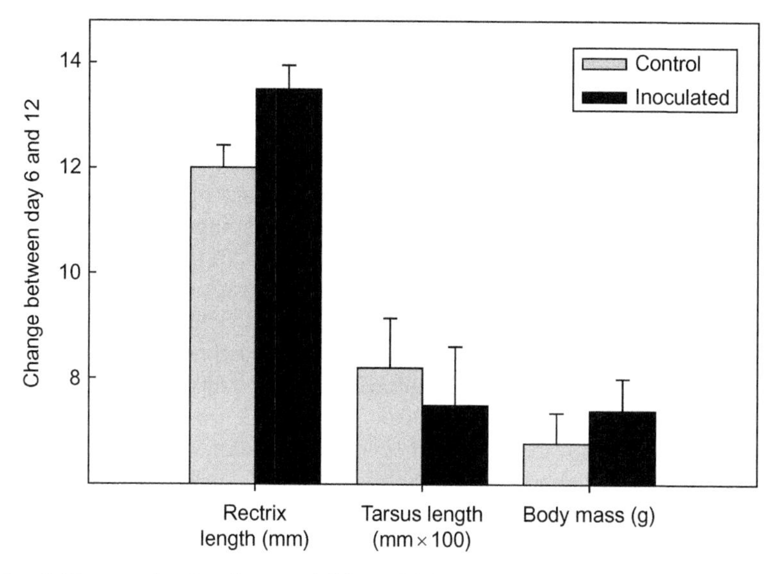


FIG. 2. Ectoparasite abundance and differential feather development in barn swallow nestlings. Ectoparasites were experimentally added to "inoculated" nests, then the growth of nestlings was compared to "control" broods, where ectoparasite loads were not manipulated. Nestlings raised in experimental broods grew their flight feathers at a disproportionately faster rate than nestlings from control broods, but had similar developmental rates for tarsus and body mass. The differential growth of feathers by nestlings raised in experimental broods allowed them to fledge comparatively sooner than nestlings from control broods and hence escape from the low-quality environment of their infested nest. (Data from Saino et al., 1998.)

methionine, a sulfur amino acid that specifically enhances T-cell immune response in chickens, and were found to mount a larger immune response when challenged with phytohemagglutin than nestlings from control broods. However, the nestlings from experimental broods had a slower growth rate, which provides direct evidence for a trade-off between immunity and growth (Soler et al., 2003).

Predation risk seems to have a great effect on developmental mode and growth on a long evolutionary timescale, as evidenced by the strong effect of family and order in the precocial–altricial split, but predation risk has also been shown to have an influence on growth in the short term, within a species. Coslovsky and Richner (2011) examined how great tit mothers may perceive and transfer information about the predation risk within their breeding environment. Breeding pairs of great tits were exposed before and during ovulation to taxidermy mounts of predators in the form of Eurasian sparrowhawks (*Accipiter nisus*) or nonpredatory song thrushes

(*Turdus philomelos*) as controls. Nestlings were then cross-fostered in order to separate maternal effects from posthatching environmental effects, and it was found that when compared to nestlings from control broods, nestlings in experimental broods were smaller and lighter, but had higher growth rates of their flight feathers (Fig. 3). It was concluded that allocating resources to wing growth may enable nestlings from riskier environments to evade predators by both fledging early and escaping from aerial predators once out of the nest (Coslovsky and Richner, 2011). One interspecific study showed that the level of hormone concentrations within eggs was positively correlated with nest predation risk, which therefore provides a possible mechanism for the observation that the nestlings of species which suffer high levels of predation develop relatively quickly and fledge relatively fast (Schwabl et al., 2007).

In summary, a number of studies have considered how patterns of differential growth aimed at mediating adverse conditions experienced during the nestling period may vary between broods, but within a species.

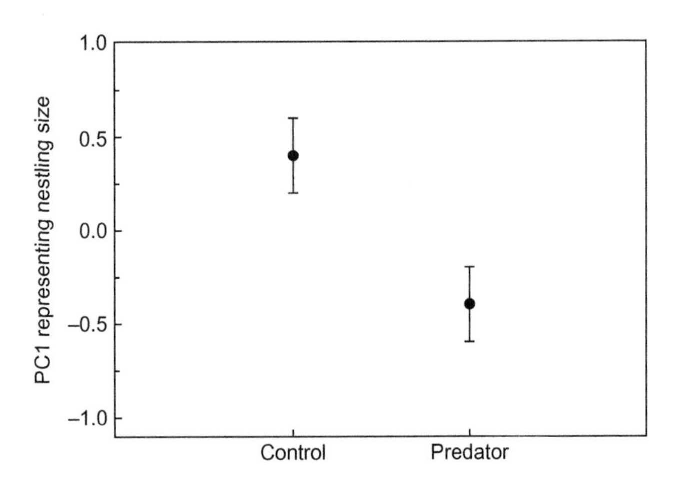


Fig. 3. Predation risk and differential feather growth in great tit nestlings. The perceived predation risk experienced by parents was experimentally manipulated by exposing experimental pairs to taxidermy mounts of predatory Eurasian sparrowhawks before and during ovulation, while control pairs were exposed to nonpredatory song thrush mounts. Nestlings were then cross-fostered in order to separate maternal effects from posthatching environmental effects. When compared to nestlings from control broods, nestlings raised in experimental broods had higher growth rates of their flight feathers, but were smaller and lighter, as illustrated by the first principal component (PC1: mean ± 95% CI) of a principal component analysis combining nestling tarsus, wing and sternum lengths. The figure shows data for nestlings at day 14 after hatching, but the pattern is similar at day 8. (Data from Coslovsky and Richner, 2011.)

Studies have demonstrated adaptive trade-offs between various morphological characters and immune function in relation to levels of sibling competition, ectoparasite abundance, and predation risk. There is also a potential effect of predation risk on parental provisioning rates which can lead to decreased growth rates from reduced food intake. There is widespread evidence that predation risk reduces parental provisioning rates (review in Martin and Briskie, 2009), but few studies have examined the consequences of reduced provisioning rates for patterns of nestling growth (but see Scheuerlein and Gwinner, 2006). Therefore, nestlings respond to adverse conditions in the nest by allocating resources to traits which facilitate success in sibling competition or rapid fledging. While these patterns of differential growth are adaptive in the time spent in the nest, they are likely to be maladaptive once out of the nest, as diverting resources away from body mass increases are likely to have a range of detrimental consequences (Metcalfe and Monaghan, 2001). This is because having a relatively low body mass at fledging is associated with high levels of postfledging mortality and reduced mating success (Both et al., 1999). For example, body size has been shown to be an important determinant of reproductive success in socially monogamous blue tits, as larger males acquire larger total fertilization through extra-pair, as well as within-pair, copulations, when compared to smaller males (Kempenaers et al., 1992). Nevertheless, it is important to consider that low fledging weight may be compensated for in later life. These conclusions are drawn from only a handful of studies and further work is required to assess the generality of these findings.

## IV. Intraspecific Variation in Growth: Within Broods

For a variety of reasons, it is common to see variation in size between nestlings within broods. Size variation can occur because of differences due to the genotype, including sexual size dimorphism (Kalmbach and Benito, 2007) and effects of extra-pair parentage (Ferree et al., 2010; Magrath et al., 2009), and also variation due to maternally induced effects such as hatching asynchrony, and egg quality variation associated with laying order (Glassey and Forbes, 2002; Mock and Parker, 1997; Schwabl, 1996). There will also be variation due to parental and offspring behaviors, such as differential begging and parentally biased favoritism (Dickens and Hartley, 2007a; Draganoiu et al., 2005; Kilner and Hinde, 2008; Mainwaring et al., 2011b). Such size variation within broods has a direct impact on the competitive hierarchy within the nest. Larger nestlings, or those able to produce more effective begging signals, will dominate the allocation of parentally provided resources such as food and individual nestlings should be expected to prioritize the development of those morphological characters that maximize their chances of soliciting food from their parents.

### A. Sexual Size Dimorphism

Sexual size dimorphism refers to sexual differences in body size (Bennett and Owens, 2002; Hartley, 2007). The scale of dimorphism is vast and ranges from both sexes of silver-eyes (*Zosteropidae*) being almost identical through to Western capercaillies (*Tetrao urogallus*) where males can be twice the weight of females. However, males are not always the larger sex and reversed sexual dimorphism often occurs in raptors; for example, female Eurasian sparrowhawks are 60% heavier than males (Moss, 1979). Sexual size dimorphism may have evolved in birds because larger females may be able to produce more eggs and hence have a reproductive advantage and larger males are more likely to win intrasexual competitions to access females and/or size differences may reduce competition between the sexes through niche separation (Hartley, 2007).

Sexual dimorphism can be manifested in size or plumage differences, although size differences are more prominent during growth. Consequently, sexual size dimorphism results in asymmetric sibling competition, and thus has the potential to result in differential growth, whereby nestlings of the smaller sex are expected to prioritize the development of those morphological characters that maximize effective sibling competition (Bortolotti, 1986; Kalmbach and Benito, 2007).

When the relative fitness of male and female offspring varies with environmental conditions, evolutionary theory suggests that avian parents should adjust the sex of their offspring accordingly, as has been demonstrated in some, but not all, empirical studies (review in Hardy, 2002). For example, female tawny owls (*Strix aluco*) had female-biased primary sex ratios in territories with a good food supply, which was found to enhance the subsequent reproductive success of those fledglings as there was a significant correlation between the number of nestlings fledged by adult females and the vole abundance in the territory on which they were reared as nestlings: a relationship that did not hold for males (Appleby et al., 1997).

An observational study showed that there was no variation between the sexes in gape area development in the sexually size dimorphic red-winged blackbird (*Agelaius phoeniceus*) (Clark, 1995). This is surprising, as gapes are part of the multicomponent behavioral display that is used by nestlings to solicit food from parents (Kilner et al., 1999) and reach a maximum width around the middle of the nestling period and regress thereafter, suggesting that they have a temporary function in sibling competition (Clark, 1995;

Gil et al., 2008). However, smaller nestlings may not develop larger gape areas than larger nestlings for two reasons: first, gape color, rather than area, may be the key determinant influencing parental allocation rules (Hunt et al., 2003; Kilner, 1999; Saino et al., 2000) and, second, body size may override any aspect of gape display in influencing parental allocation rules (Dickens and Hartley, 2007a,b; Dickens et al., 2008; Mainwaring et al., 2011b; Slagsvold and Wiebe, 2007).

Corn buntings are among the most sexually dimorphic passerines for size: adult males are approximately 20% larger than females, although they appear similar in plumage, and size dimorphism is also apparent in nestlings in the nest. Within broods, the smaller female nestlings grew differently to males and tended to grow their tarsi at a proportionately faster rate than males (Fig. 4; Hartley et al., 2000). It was concluded that this would facilitate simultaneous fledging with the larger male siblings: corn buntings are ground-nesting birds, with open nests, and so are potentially at a relative high risk in the nest. Unfortunately, wing growth was not measured in that study, and it is equally plausible that the females' disproportionate investment in faster growth of tarsi was a response to competition within the nest from larger male siblings: longer legs being more effective when stretching to receive food delivered by parents.

Blue tit adults are also sexually size dimorphic, although not to the same extent as corn buntings. In blue tit broods, the smaller female nestlings grew flight feathers disproportionately faster than males (Mainwaring et al., 2011a). It was concluded that this trade-off would facilitate simultaneous fledging with the larger male siblings, which is important because provisioning of those nestlings remaining in the nest dramatically decreases once the fledging process has begun (Glassey and Forbes, 2002; Rosivall et al., 2005). A similar conclusion was reached from a study of Eurasian sparrowhawks, which exhibit reversed sexual size dimorphism. The smaller male nestlings grew their outermost primary feathers faster than their larger female siblings, which enabled simultaneous fledging of whole broods (Moss, 1979).

These studies suggest that in a trade-off between the development of the skeleton, muscle mass, and flight feathers, it is flight that is a priority for nestlings of the smaller sex. Allocating resources toward wing development is adaptive in the short term as it facilitates simultaneous brood fledging, but is also likely to have postfledging benefits. For example, Eurasian sparrowhawks prey heavily on newly fledged blue tits and so rapidly gaining a good flight capability is likely to benefit individuals in both species (Cramp and Perrins, 1990).

The allocation of resources away from mass gain and tarsal growth and toward feather development by female blue tit nestlings also appears to put them at a disadvantage in the competition for parentally provided food.

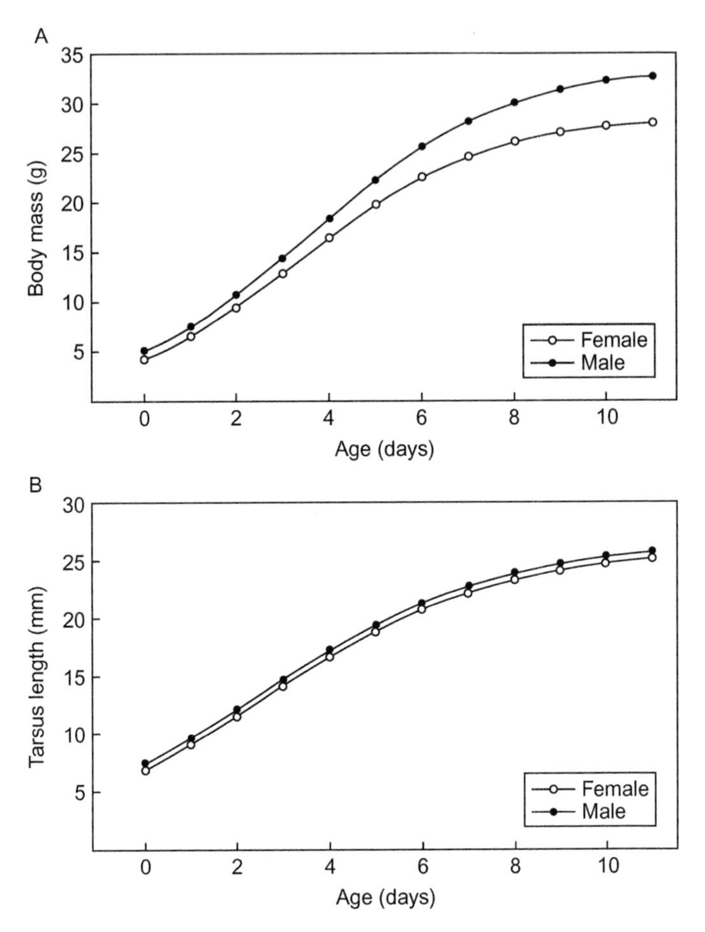


FIG. 4. Sexual dimorphism and differential growth in corn bunting nestlings. Corn buntings are among the most sexually dimorphic passerines for size, with adult males being approximately 20% larger than females. Within broods, the smaller female nestlings grew differently to males: females gained (A) body mass at a lower rate than males, yet grew their (B) tarsi at a similar rate. (Data from Hartley et al., 2000.)

However, it has been shown that blue tit parents respond to competitive interactions between nestlings by allocating more food to those young that begged at higher intensities (Dickens and Hartley, 2007a). Moreover, while the largest nestlings in the brood, irrespective of sex, received more food than their smaller siblings when they were hungry, female parents also showed a preference toward provisioning smaller nestlings when they were further away in the nest (Dickens and Hartley, 2007a). This may reduce the net cost to female offspring of investing in growth of feathers rather than muscle or tarsal length. However, that study did not examine the effect of offspring sex in parental allocation rules, leaving open the possibility that parents may have been provisioning nestlings of either sex preferentially as has been found in zebra finches (Mainwaring et al., 2011b).

In summary, sexual size dimorphism results in asymmetric sibling competition during the nestling period and therefore has important consequences for sex-specific patterns of developmental resource allocation (Kalmbach and Benito, 2007). A number of observational studies show no evidence for nestlings of the smaller sex preferentially allocating resources toward gape development (Clark, 1995), but they do allocate resources toward feather growth in order to facilitate simultaneous fledging with their larger siblings (Mainwaring et al., 2011a; Moss, 1979). This is important because provisioning of those nestlings remaining in the nest dramatically decreases once the fledging process has begun (Glassey and Forbes, 2002; Rosivall et al., 2005). This implies that sex-specific growth patterns may have important consequences for nestlings, which are likely to have long-term effects for individuals. Sex-specific differential growth may also be an important determinant of primary sex ratios and sex-biased laying/hatching orders. However, sex-specific patterns of differential growth are poorly understood at present and further studies could usefully examine sex-specific patterns of developmental resource allocation. Also, there are no studies of precocial species, presumably due to the difficulty of recapturing individual nestlings on a regular basis after they have left the nest site. While this bias is understandable, it may be limiting our understanding of sex-specific patterns of differential growth.

### B. Parentage and Relatedness

The advent of molecular genetic tools for identifying parentage has made it apparent that males and females in socially monogamous species often mate outside of their pair bonds (Burke, 1989; Burke and Bruford, 1987). The original view of monogamous pair bonds (Lack, 1968) has now been replaced by the knowledge that 86% of bird species indulge in extra-pair matings (Griffith et al., 2002). Mixed paternity within broods, due to either extra-pair matings or polygynandry, will mean that at least some of the nestlings are half-siblings. Kin selection theory predicts that in species where progeny compete for limited parent care, individual offspring should be more prone to monopolizing parental resources as their genetic relatedness to brood competitors decreases. For example, in a cross-fostering study with barn swallows, the magnitude of scramble competition for food increased as relatedness among competitors decreased (Boncoraglio et al., 2009). Nestlings may modulate their competitive behavior according to vocal cues that vary with their origin and allow kin recognition (Boncoraglio et al., 2009).

Within-family comparisons of offspring growth rates have provided important tests of genetic benefits of extra-pair mating for females (Ferree et al., 2010; Magrath et al., 2009). Extra-pair nestlings are larger than within-pair nestlings and this pattern has traditionally been attributed to genetic effects as proposed by good genes or genetic compatibility effects (review in Griffith et al., 2002). However, it has recently been shown that extra-pair blue tit (Fig. 5) and tree swallow (*Tachycineta bicolor*) nestlings also hatch earlier in the clutch, thus benefiting from a head start in growth and benefits that come from being the first hatched nestlings in a brood with hatching asynchrony (Ferree et al., 2010; Magrath et al., 2009). By controlling for hatch order and other nongenetic factors and comparing mixed-paternity broods with genetically monogamous broods, it was shown that the extra-pair nestling growth advantage is not genetically based (Ferree et al., 2010; Magrath et al., 2009). However, given that Westneat et al. (1995) found no relationship between hatching order and paternity within red-winged blackbird broods, further studies are required to examine the generality of this pattern.

Nevertheless, in an observational study of collared flycatchers, extra-pair nestlings were found to be in better condition than within-pair nestlings, even when hatching order was taken into account (Krist et al., 2004). Moreover, extra-pair tree swallow nestlings were found to have longer wings than within-pair nestlings, but the same mass, tarsus length, and immune system function. Wing length is a strong predictor of the timing of fledging in tree swallows and therefore has important fitness consequences in the short term (O'Brien and Dawson, 2007).

While half-siblings might be expected to compete harder against each other than full siblings, the occurrence of brood parasitism should be expected to induce even greater competition between individuals sharing the same nest. In the European cuckoo (*Cuculus canorus*), females lay eggs in the nests of species such as reed warblers (*A. scirpaceus*), dunnocks (*Prunella modularis*), or meadow pipits (*Anthus pratensis*), where they are reared by the host (Davies, 2000). In what might be considered to be the ultimate competitive move, a newly hatched European cuckoo chick will push the eggs or young of the host out of the nest, so that it is the only chick left for the parents to provision. In other cuckoo species, however, host young are often reared alongside the brood parasite, and these can have a significant impact upon their growth and development. Ridley and Thompson (2012) investigated the effect of Jacobin Cuckoo (*Clamator jacobinus*) parasitism on the body mass development and survival of

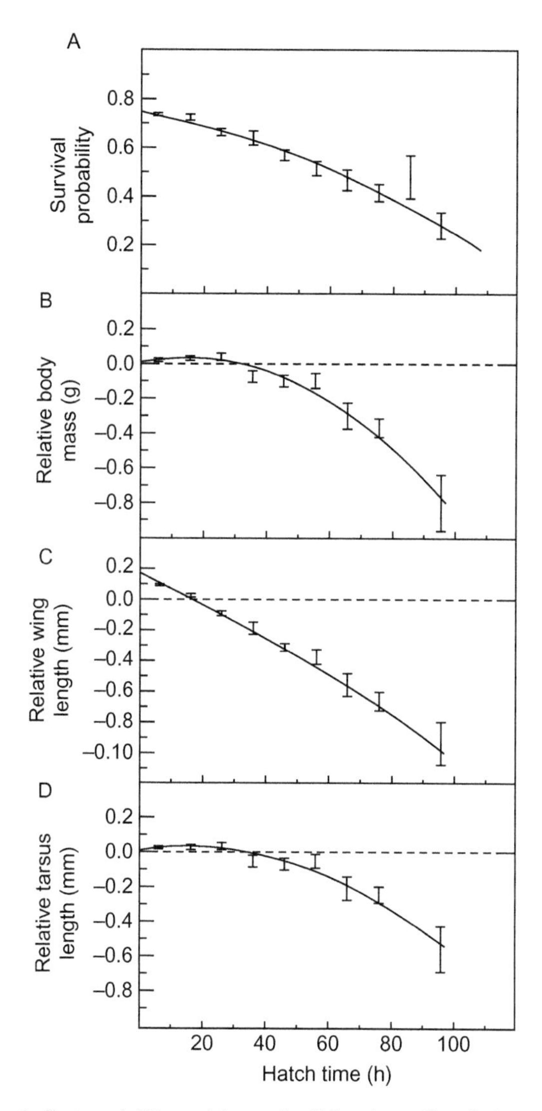


FIG. 5. Maternal effects and differential growth of blue tit nestlings fathered through within-pair and extra-pair matings. Nestlings sired through extra-pair copulations were found to hatch earlier in the clutch. The figure shows hatch time of nestlings within broods relative to the first egg, in relation to measures of performance at brood age 15 days: (A) probability of surviving to brood age 15 days, (B) body mass, (C) wing length, and (D) tarsus length. These morphological measurements have been centered about brood means to correct for between-brood variation. The solid lines show the within-brood model-predicted relationships. The error bars represent the standard error of the model-predicted values for each 10-h interval in hatch time. The horizontal dotted lines in (B–D) represent zero deviation from the brood mean values. Extra-pair chicks benefited from a head start in growth and benefits that come from being the first hatched nestlings in a brood with hatching asynchrony. The data suggest, however, that once hatching order and other nongenetic factors have been controlled for, the extra-pair nestling growth advantage was not genetically based. (Data from Magrath et al., 2009.)

southern pied babbler (*Turdoides bicolor*) host nestlings. They found that host nestlings in parasitized nests survived to fledging just as well as those in unparasitized nests and had similar body masses at day 11 after hatching. Survival to independence, however, was reduced in host chicks raised alongside a brood parasite, which suggests that there was a long-term effect of brood parasitism, and that host nestlings may be allocating resources toward body mass and away from other traits which affect postfledging survival.

To summarize, there have been no studies which have explicitly examined patterns of developmental resource allocation by within- and extra-pair nestlings within mixed-paternity broods, although one study did examine parental provisioning rules in relation to parentage and found no effect (Westneat et al., 1995). This is somewhat surprising as there is a growing body of evidence that relatedness can influence offspring competition, and hence developmental resource allocation patterns, through selection on begging call intensity (Boncoraglio and Saino, 2008; Briskie et al., 1994) and scramble competition (Boncoraglio et al., 2009). Moreover, relatedness is likely to have consequences for family interactions as kin selection theory predicts that offspring behave more selfishly in monopolizing parental care as relatedness with competitors declines (Burke, 1989; Mock and Parker, 1997). Consequently, examining patterns of developmental resource allocation in within- and extra-pair nestlings would be a fruitful area for future research.

### C. Laying Order and Hatching Asynchrony

Phenotypic variation within broods arises due to differential investment by female parents in, for example, egg size, maternally derived embryonic hormones, and the position in asynchronous hatching orders, all of which can influence offspring growth and competitive ability. First, the majority of empirical studies have found that egg size decreases with laying order and that offspring from relatively large eggs initially grow faster (Krist, 2011; Williams, 1994). Second, females can either enhance (Schwabl, 1996; Schwabl et al., 1997) or mitigate (Gil et al., 1999; Lipar et al., 1999; Royle et al., 2001, 2003) the within-brood inequalities by changing the levels of various hormones or vitamins deposited within their eggs prior to laying. Third, many species of altricial birds hatch their young asynchronously, thereby producing an age and size hierarchy within the brood (Lack, 1947, 1954; Magrath, 1990). While egg size and maternally derived hormone levels both influence early growth, they are frequently overridden by nestling size asymmetries caused by hatching asynchrony (Forbes and Wiebe, 2010; Glassey and Forbes, 2002; Krist, 2011). This is aptly illustrated in a study of Atlantic canary (*Serinus canaria*) nestlings, where more frequent begging of later hatched nestlings was associated with higher maternally derived testosterone in later laid eggs. While this resulted in access to more food early in the nestling period and was associated with faster growth, these initial advantages were lost as later hatched nestlings could not overcome size differences associated with hatching asynchrony (Schwabl, 1996). Therefore, we only concentrate on the role of hatching asynchrony and its consequences for asymmetric sibling competition and growth trade-offs.

Many species of altricial birds hatch their young asynchronously, thereby producing an age and size hierarchy within the brood. The evolution and function of hatching asynchrony have received much attention, as hatching asynchrony is seemingly paradoxical, as the age and size hierarchies result in later hatched offspring being at a competitive disadvantage in the competition for parentally provided food (Stoleson and Beissinger, 1995). The adaptive hypotheses viewing the size hierarchy as a selected trait that allows for facilitative adaptive brood reduction when resources are limited (Lack, 1947, 1954) and promotes parental feeding efficiency by spreading out the demands for food by individual offspring (Bryant, 1978; Hussel, 1972). Also, the "hurry-up" hypothesis states that the shorter nestling periods in asynchronously hatching broods are important in those species whose food supplies decline as the season progresses, such as blue tits and pied flycatchers breeding in seasonal environments (Both and Visser, 2001; Perrins, 1991), as it ensures that at least part of the brood fledges (Clark and Wilson, 1981). This hypothesis has been used to explain why some species hatch their young more asynchronously as the season progresses (Perrins, 1991). Meanwhile, the nonadaptive hypotheses view the size hierarchy as a side effect of selection for other traits and involve minimizing the total amount of time that the eggs and nestlings spend in the nest, thereby reducing the likelihood of predation (Clark and Wilson, 1981; Hussel, 1985; Nilsson, 1993; Slagsvold, 1986). The various hypotheses should not be considered equal in terms of importance, as some hypotheses refer only to a subset of species with particular natural history traits (e.g., Stoleson and Beissinger, 1995). Irrespective of the selective forces, hatching patterns in nestling birds are clearly controlled by the parent birds through the timing of the onset of incubation (Amundsen and Slagsvold, 1991; Magrath, 1990; Stoleson and Beissinger, 1995).

Despite the uncertainty regarding the evolutionary causes, it is clear that hatching asynchrony results in a structured kinship, effectively dividing the brood into "early hatched" and "late-hatched" offspring (Glassey and Forbes, 2002). The core brood represents the minimum subset that parents can or will rear across a breeding attempt, and core siblings are largely unaffected by the presence of marginal siblings, although the reverse is not true.

When considering the impact of phenotypic handicaps placed on certain siblings, it is more appropriate to examine the early hatched versus late hatched divide rather than individual nestlings in a staggered hierarchy within broods (Glassey and Forbes, 2002). This maternal manipulation of phenotype results in disparities in growth and mortality rates within broods, and the vast majority of empirical studies find that older progeny enjoy superior prospects, both when competing for parentally provided resources and as adults (Magrath, 1990). A recent study has shown that the dramatic difference in phenotype between early- and late-hatched red-winged blackbird siblings has important consequences for nestling survival, with early hatched nestlings having much higher survival than late-hatched nestlings. Nestlings that lived in the same physical space and reared by the same parents effectively lived in different worlds (Forbes, 2011).

Size differences within broods may strongly influence how parental care is distributed among siblings (e.g., Bengtsson and Rydén, 1983; Leonard and Horn, 1996; McRae et al., 1993; Parker et al., 1989; review in Glassey and Forbes, 2002) and late-hatched offspring rarely grow fast enough to catch up with their siblings. When parents respond to competitive interactions between offspring (Royle et al., 2002a), first hatched and larger nestlings are dominant and are often able to access a greater share of the food delivered by parents, primarily by reaching higher than their younger and smaller siblings during begging contests (Dearborn, 1998; Leonard and Horn, 1996; Smith and Montgomerie, 1991; Teather, 1992). Early hatched nestlings reach key developmental landmarks earlier, such as acquiring sight, having sclerotized (hardened) legs, initiating thermoregulation and gaining greater strength in the neck and gastrocnemius muscles (Glassey and Forbes, 2002). Allocating resources toward body mass increases is adaptive in the short term as early hatched nestlings are able to compete effectively with their siblings within the nest, and thereby gain access to provisioning parents by jostling or reaching (e.g., Dickens et al., 2008; Lotem, 1998; Mainwaring et al., 2011b; McRae et al., 1993). Moreover, early hatched nestlings, by definition, are older than late-hatched nestlings and so are able to initiate a begging event sooner and maintain it for longer than late-hatched nestlings. Body mass increases are also beneficial in the long term as there are fitness consequences for body size.

This phenotypic disadvantage can have important consequences for late-hatched nestlings as they seek to mitigate the adverse conditions they find themselves in soon after hatching. While parents may choose to actively feed later hatched and smaller nestlings (Godfray, 1995; Kilner and Johnstone, 1997), empirical studies report that late hatched and smaller nestlings are subordinate and often receive less food than their early hatched siblings, despite begging more intensively (Dearborn, 1998;

Leonard and Horn, 1996; Mainwaring et al., 2011b; Smith and Montgomerie, 1991; Teather, 1992). However, late-hatched nestlings may benefit from such developmental asymmetry as they become endothermic later, and thus expend less energy on thermoregulation than their early hatched siblings (Forbes, 2007). Nevertheless, late-hatched nestlings usually remain smaller than their early hatched siblings throughout the nestling period. Consequently, hatching asynchrony and parental provisioning rules have important consequences for the developmental resource priorities of late-hatched nestlings (Glassey and Forbes, 2002).

In an observational study, late-hatched barn swallow nestlings had slower wing feather growth than their nest-mates but were found to allocate resources to body mass and skeletal characters similarly (Mainwaring et al., 2009). This indicates either that body mass and skeletal development are of highest importance to late-hatched nestlings or that wing development is relatively of lesser importance during early development. The fact that late-hatched nestlings were able to match body mass and skeletal development may be as a result of barn swallows being among those species considered to employ a "brood-survival" strategy (Slagsvold et al., 1984), which means that they may at least partly compensate for the initial disadvantage of late-hatched nestlings by, for example, increasing egg mass (Saino et al., 2004) and egg albumen content (Ferrari et al., 2006) with laying order. However, the early hatched nestlings would have been larger and heavier than their late-hatched siblings, so throughout the nestling period late-hatched nestlings would have been at a competitive disadvantage when begging to provisioning parents (Mainwaring et al., 2009).

It is interesting to compare the finding of that study (Mainwaring et al., 2009) with the finding of another study involving tree swallows, which found that later hatched nestlings allocated resources toward wings development and away from mass development, in order to fledge simultaneously with their older siblings (Zach, 1982a). This difference is interesting as both species are aerial insectivores and both have a "brood-survival" strategy (Zach, 1982a). However, another study of tree swallows showed that late-hatched nestlings allocated resources toward the development of body mass and tarsus, but grew shorter wings (Johnson et al., 2003). Consequently, the findings of the two studies contradict each other and such spatial variation in patterns of developmental resource allocation may be a fruitful area for future research. Meanwhile, in hole-nesting marsh tits, it was found that late-hatched nestlings allocated resources toward wing development and away from mass development, in order to fledge simultaneously with their older siblings (Nilsson and Svensson, 1996). This pattern of resource allocation was also found in an observational study of house wrens (*Troglodytes aedon*). It was found that late-hatched nestlings allocated resources toward wing development and away from mass and tarsi development, which was associated with simultaneously fledging with their older siblings (Lago et al., 2000).

Another study examined the patterns of developmental resource allocation in early- and late-hatched blue tit nestlings. An observational study showed that while early hatched nestlings allocated resources toward the development of body size and skeletal characters, late-hatched nestlings allocated resources to leg development and away from wing development (Fig. 6; Mainwaring et al., 2010a).

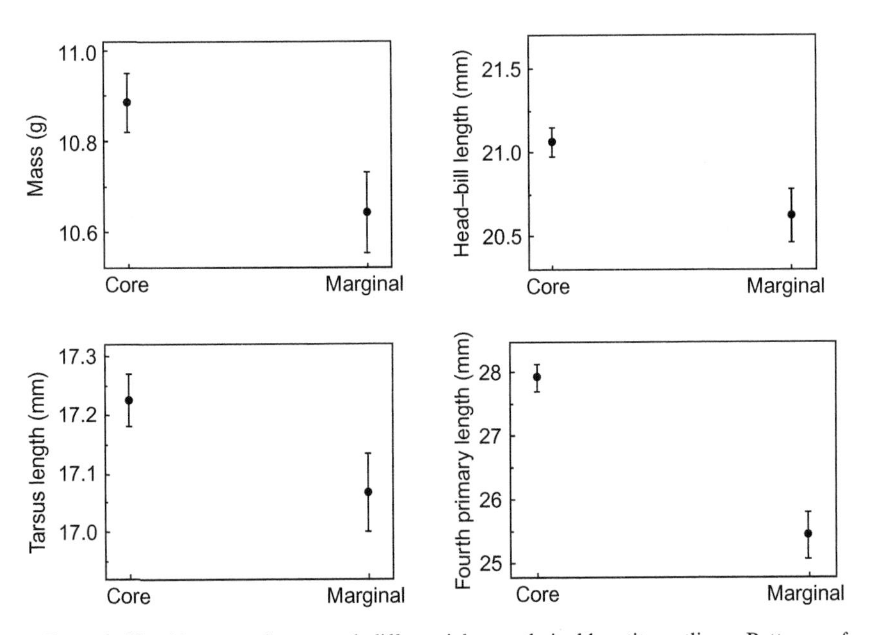


FIG. 6. Hatching asynchrony and differential growth in blue tit nestlings. Patterns of differential growth of four morphological traits: body mass, head-bill length, tarsus length, and fourth primary length, in core (early hatched) and marginal (late-hatched) nestlings. Core nestlings allocated resources toward the development of body mass and skeletal traits, while marginal nestlings allocated resources to tarsus development and away from flight feather development. Selection pressure from sibling competition within the nest probably drives the differences in development rates: marginal offspring prioritize tarsi development to keep up with nest-mates during begging competitions. By allocating resources to their tarsi, marginal nestlings are presumably able to stretch higher, as height within nests has been shown to be an accurate indicator of begging success in blue tits. (Data from Mainwaring et al., 2010a.)

Blue tit parents usually respond to the competitive interactions of their nestlings, which normally results in the largest nestlings receiving most food (Dickens and Hartley, 2007a,b; Dickens et al., 2008) and this suggests that early hatched nestlings should direct resources toward body size as this enables them to monopolize the best positions within the nest cup (Glassey and Forbes, 2002). The selection pressures operating on late-hatched offspring prioritizing tarsi development and not feather growth are probably the need to maintain height within the nest. By allocating resources to their tarsi, late-hatched nestlings are presumably able to stretch higher, as height within nests has been shown to be an accurate indicator of begging success both in blue tits (Dickens et al., 2008) and a range of other passerine birds (Dearborn, 1998; Leonard and Horn, 1996). Moreover, attempts by late-hatched nestlings to maintain comparable masses with early hatched nestlings may have enabled individual nestlings to avoid being the "tasty" chick. The "tasty" chick hypothesis states that within the coevolutionary arms race between hosts and their parasites, parent hosts may sacrifice one or more later hatched nestlings and thereby minimizing parasite attacks on the early hatched nestlings, which maximizes parental breeding success (Christe et al., 1998). Support for this hypothesis comes from a study of Mediterranean blue tits, where the ectoparasite load of blue tit nestlings was found to be negatively correlated with individual mass (Simon et al., 2003).

Having established that early- and late-hatched nestlings have different resource allocation patterns, a hatching order manipulation experiment was performed in order to determine the relative contributions of environmental effects and maternal effects. Such differences could be due to an environmental effect brought on by hatching asynchrony, where the competitive environment directly influences growth, or a maternal effect where the mother routinely supplies the last laid eggs with different hormones or nutrients which influence growth (Mousseau et al., 2009; Uller, 2008). In order to separate the two mechanisms, the order in which the eggs hatched was experimentally disrupted, so that nestlings from first laid eggs actually hatched last. If environmental effects were important, then experimental nestlings should grow similarly to late-hatched nestlings, and if maternal effects were important, then experimental nestlings should grow similarly to early hatched nestlings (Mainwaring et al., 2010a). The growth of experimental nestlings, from first laid but last hatched eggs, had similar patterns of developmental resource allocation to late-hatched nestlings. The experimental manipulation of hatching order in this study revealed that environmental effects were more important than maternal effects in determining phenotypic variation among nestlings. This is surprising as a large number of empirical studies have reported differential investment of egg size and mass (Williams, 1994) and various hormones and vitamins (Lipar et al., 1999; Royle et al., 2001; Schwabl, 1996) through the laying sequence and found strong effects on avian growth patterns. This suggests that while egg size and maternally derived hormone levels both influence early growth, they can be overridden by nestling size asymmetries caused by hatching asynchrony (Glassey and Forbes, 2002). The selection pressures operating on late-hatched, experimental nestlings to allocate resources toward tarsus development and away from feather growth was probably the need to maintain height within the nest. In this way, experimental nestlings would have been able to stretch higher within the nest, which is important as height within nests has been shown to be an accurate indicator of begging success in blue tits (Dickens et al., 2008).

However, offspring growth consists of the development of a number of morphological characters and an immune system. Therefore, in a study which examined resource allocation between growth, immunity, and plumage coloration in European rollers (*Coracias garrulus*), nestlings from experimental nests were supplemented with methionine, which induces the production of lymphocytes at the expense of growth (Parejo et al., 2010). It was found that the late-hatched nestlings in experimental nests had diminished plumage coloration when compared to late-hatched nestlings from control nests. Meanwhile, late-hatched nestlings from control broods maintained wing development, which implies that when they experience adverse conditions, late-hatched nestlings allocate resources toward wings as they provide direct fitness benefits as opposed to plumage coloration, which provides delayed fitness benefits (Parejo et al., 2010).

In a study of asynchronously hatching zebra finch broods in the wild and in a captive environment, late-hatched wild nestlings reached the fastest point of growth sooner than their larger siblings; there was no difference in domesticated broods. This implies that in the wild, late-hatched nestlings grow rapidly in order to attempt to catch up with their early hatched siblings, while late-hatched nestlings in domestic broods always lagged behind their larger siblings due to differences in age. Wing growth, at least up until the prefledging stage, may be a priority for wild zebra finch nestlings, and particularly those late-hatched nestlings, as it facilitates simultaneous fledging times with their older siblings and minimizes nest predation rates. Additionally, increased flight maneuverability and escape speed against aerial predators are likely to be advantageous when they have fledged. Moreover, the fledglings need to travel to areas of abundant food which is spatially variable in the Australian outback (Zann, 1996). However, it is plausible that late-hatched nestlings in the wild prioritized wing growth at the expense of other fitness-related physiological functions such as immunity (Eraud et al., 2008). Alternatively, wing growth may be a low priority for domestic birds as simultaneous fledging is not a priority in such a benign environment and they are hardly required to fly and consequently, wing growth is likely to be under only weak selection (Mainwaring et al., 2010b).

To summarize, early hatched, dominant nestlings within asynchronously hatched broods allocate resources toward body mass increases and structural growth, which is adaptive as they are often able to access a greater share of the food, primarily by reaching toward provisioning parents more effectively than their younger siblings (e.g., Leonard and Horn, 1996). As parents usually respond to sibling competition, early hatched nestlings beg less intensively than their late-hatched siblings, yet still receive more food for their efforts (Budden and Wright, 2001). These advantages also mean that they reach key developmental landmarks sooner, such as the acquisition of sight, which helps exaggerate their competitive advantage over their younger and smaller siblings (Glassey and Forbes, 2002). Late hatched and subordinate nestlings, however, have been shown to trade-off development of body mass and other morphological characters, such as gape area or wing growth as they facilitate sibling competition or simultaneous fledging, respectively (Lago et al., 2000; Nilsson and Gårdmark, 2001; Nilsson and Svensson, 1996; Zach, 1982a). There is a growing consensus that smaller, marginal nestlings beg more, but receive less for their efforts, than their older and larger siblings (Budden and Wright, 2001), as predicted by theory (Parker et al., 1989, 2002a,b). Glassey and Forbes (2002) suggest four, nonmutually exclusive explanations for this pattern: first, theoretical models of scramble competition predict that competitively handicapped progeny should beg more for smaller rewards; second, variation in maternally derived hormones, such as testosterone, is generally negatively correlated with competitive ability and should allow later hatched offspring to beg more frequently than their relative size within the nest would suggest; third, later hatched nestlings may simply be hungrier; and fourth, later hatched offspring may beg harder to induce parents to increase overall levels of provisioning. Given the unequal struggle for parental investment with broods, it may be expected that older, more superior nestlings use aggression to limit the effectiveness of begging by subordinate siblings by confining the timing, location, or form of begging (Drummond, 2002). Although aggression is a powerful means of confining an inferior sibling's begging, and therefore access to food, it is only likely to evolve if it is both effective and profitable. Parent–offspring conflict over the amount and distribution of parental care results in parents attempting to exert control, for example, by determining the number of offspring and the extent of competitive hierarchies. Offspring can also influence the outcome of family conflict, through mechanisms such as siblicide, solicitation behaviors, or changes to growth rates (Budden and Wright, 2001; Forbes, 1993; Kilner, 2002;

Lessells, 2002a,b; Mock and Parker, 1997; Rodríguez-Gironés, 1996; Royle et al., 1999, 2002b; Trivers, 1974). This is possibly due to the need to fledge at the same time as the majority of the brood, because of the dramatic decrease in food provisioning to those young remaining in the nest once the fledging process has begun (Nilsson, 1990). Late-hatched nestlings that fledge at relatively low body masses are, however, likely to experience high risk of early postfledging mortality, suggesting a trade-off between growth allocation and survival (Nilsson, 1990).

## V. The Consequences of Differential Growth

There is widespread evidence of an effect of early developmental conditions on the subsequent phenotypes, behavioral decision making, and fitness of individuals during adulthood (Birkhead et al., 1999, 2006; Blount et al., 2003, 2006; Criscuolo et al., 2008; Grace et al., 2011; Krause and Naguib, 2011; Lindström, 1999; Naguib et al., 2006, 2008; Royle et al., 2002b; Tschirren et al., 2009), so we would predict that differential growth and development should have similar effects beyond the nestling phase. In contrast to the causes of differential growth, however, the consequences of differential growth are relatively poorly understood. This imbalance in our knowledge is largely due to the logistic problems associated with catching individually marked birds after they have fledged. Consequently, although nestlings employ a variety of resource allocation strategies that are adaptive while in the nest, we know relatively little about the long-term consequences of such growth patterns. This is unfortunate as the growth of different body components is likely to be under antagonistic selection due to shifting pressures at different times within the life history of the individual. Growing nestlings may allocate resources toward a particular morphological character as it is beneficial over at least one of two timescales. First, nestlings have to compete effectively against their siblings for parentally provided food and second, they need to find food and avoid predation after fledging. For example, differentially developing gape area enables nestlings to procure more food from their parents, as they are part of the multicomponent behavioral display that is used to solicit food from parents. However, a large gape area is likely to be of little benefit once the nestlings have fledged the nest, especially if it has been developed at the expense of traits which are more important to postfledging survival.

### A. Differential Growth

Despite the fact that we know relatively little about the long-term consequences of differential growth, we can draw some initial conclusions. Larger and dominant nestlings generally allocate resources toward body mass and structural size increases, which is adaptive in the short term as nestlings are able to compete effectively with their siblings within the nest, and thereby gain access to provisioning parents by jostling or reaching (e.g., Dickens et al., 2008; Lotem, 1998; Mainwaring et al., 2011b; McRae et al., 1993). Moreover, body mass and structural size increases are also beneficial in the long term. In blue tits, for example, there are fitness consequences for male body size. Body size is an important determinant of reproductive success, as larger males have been reported to have higher chances of acquiring larger total fertilization and tend to have greater reproductive success (Kempenaers et al., 1992). This is an important point to consider as there is no postfledging skeletal growth in blue tits and so the skeletal size attained at fledging corresponds to adult size (Merilä and Fry, 1998).

When rearing conditions become unfavorable, smaller male nestlings may reduce investments in some physiological functions, such as the immune system, in favor of growth in order to attain maximal structural body size. It is important to remember that smaller nestlings are still likely to be smaller than the early hatched nestlings, but they are likely to be larger than they would have been if they had followed a similar pattern of resource allocation to early hatched nestlings. Such a pattern of resource allocation may appear optimal if underdevelopment of the immune system at the nestling stage can be compensated later in life (Birkhead et al., 1999). Therefore, when resources are limited it may be more important for male nestlings to sustain skeletal growth at the costs of the investment in immune function, while female nestlings may reduce the allocation of resources to growth in favor of immune defenses. Such a scenario was supported by Dubiec et al. (2006), when tarsus growth in male nestlings was affected by brood size enlargement to a much lesser extent than in female nestlings, and simultaneously, the experimental treatment exerted stronger negative effects on cellular immune response in male nestlings. Additionally, the long-term consequences of differential growth in domesticated zebra finches reared on diets of varying quality showed that nestlings raised in broods reared on a low-quality diet showed slower body mass and tarsus growth rates than those nestlings raised on a standard-quality diet (Arnold et al., 2007). This pattern was reversed, however, when birds on a low-quality diet were switched to an improved diet: they were able to fully catch up their body mass growth and beak color development, but not tarsus length. More specifically, females on low-quality diets had significantly shorter wings than other birds, apparently as a direct result of allocating resources into sex-specific structures instead of feathers and skeletal growth (Arnold et al., 2007). Therefore, there were sex-specific differences in the phenotypic compensation for poor early developmental conditions, with males investing limited resources into structures associated with mate acquisition.

However, a long-term study (Apanius et al., 2008; Müller et al., 2008; Townsend and Anderson, 2007) of obligately siblicidal nazca boobies (*Sula granti*) and facultatively siblicidal blue-footed boobies (*Sula nebouxii*) suggests that negative long-term consequences may exist for aggressive early hatched nestlings. Nazca boobies lay two eggs and the eldest chick unconditionally kills the later hatched chick within a few days of hatching. Both nestlings have higher levels of circulating androgens, which facilitate aggressive behavior, than either of the two blue-footed booby nestlings, which lack the lethal aggression of the nazca boobies. There is now evidence that adult nazca boobies pay long-term costs for their early androgen exposure in terms of aberrant adult behavior. For example, they have been shown to visit unattended nonfamilial nestlings in the colony and direct mixtures of aggression, affiliative, and sexual behavior toward them (Müller et al., 2008). Consequently, early hatched nazca boobies seem to experience long-lasting consequences of androgenic preparation for bouts of aggressiveness early in life (Apanius et al., 2008).

The smaller nestlings, meanwhile, have to employ some sort of trade-off to overcome the adverse conditions they experience in the nest. This is supported by the fact that in species with a clutch size of one, such as the Atlantic puffin, it has been shown that nestlings prioritize the development of body mass during periods of food shortage (Øyan and Anker-Nilssen, 1996). Meanwhile, species with clutch sizes larger than one divert resources into morphological characters such as tarsi, gape areas, and wings that facilitate sibling competition and or simultaneous fledging times (Gil et al., 2008; Nilsson and Gårdmark, 2001; Nilsson and Svensson, 1996). Smaller, subordinate nestlings usually allocate resources toward gape areas and wings and away from body mass increases (Johnson et al., 2003; Lago et al., 2000; Mainwaring et al., 2009, 2010a,b, 2011a; Rosivall et al., 2005; Skagen, 1987; Zach, 1982a,b). Several studies have shown that small and subordinate nestlings forego feather growth. This is likely to negatively impact upon nestlings once they have fledged as stunted feather growth is likely to reduce flight maneuverability and escape speed, thus increasing the risk of predation by aerial predators (Perrins and Geer, 1980).

Late-hatched barn swallow nestlings were found to allocate resources to body mass and skeletal characters similarly to earlier hatched nestlings, but away from wing development (Mainwaring et al., 2009). While maintaining body mass may facilitate competition in the nest, compromising wing development is likely to have costs as underdeveloped wings and reduced flight maneuverability and escape speeds are likely to negatively impact upon nestlings in three ways. First, barn swallows are aerial insectivores which catch their prey on the wing and so underdeveloped wings may lead to reduced foraging success. Second, recently fledged barn swallow nestlings must also avoid fast-moving aerial predators such as Eurasian sparrowhawks. Third, barn swallows are migratory and migrate over the Sahara desert to spend the nonbreeding season in Africa, and underdeveloped wings are likely to have a detrimental impact on an individual's ability to complete the trans-Saharan migration to the wintering areas. However, fledglings may not need to have fully developed wings immediately as their parents continue to provision them for a further 10 days or so after fledging (Medvin and Beecher, 1986). Moreover, fledglings are often accompanied by a parent in flight (Medvin and Beecher, 1986), which could reduce the risk from aerial predators. Consequently, late-hatched nestlings may be able to use the postfledging period to further develop their wings. Nevertheless, such compensatory growth is likely to have a detrimental impact on an individual's ability to complete the trans-Saharan migration to the wintering areas. Such opportunities for compensatory growth might provide an opportunity to redress any deficiencies in their plumage state (Metcalfe and Monaghan, 2001), whereas skeletal growth would be more restricted because it cannot continue once the bones ossify in the first few weeks after hatching (Schew and Ricklefs, 1998).

To summarize, smaller and subordinate nestlings do not have the resources available to grow every morphological character optimally and so have to trade-off the growth of one character against another. The character that is sacrificed is probably a result of the interplay between family conflicts and developmental plasticity, although the underdevelopment of each morphological character probably has separate negative consequences for individual nestlings.

## B. Growth Trade-Offs

Observational studies are useful for observing trends, but not for identifying underlying causes of a behavioral trait. Consequently, only direct manipulations of one trait and the measurement of other traits allow any insight into growth trade-offs, whereby individuals invest more resources into one aspect of growth which directly results in the exhaustion of resources for another aspect of growth (Roff, 1992; Stearns, 1992). Experimental studies of patterns of developmental resource allocation are lacking, and the vast majority of those are experimental manipulations performed on broods, within a species. For example, studies where brood sizes were experimentally enlarged in order to mimic increased levels of sibling competition or subjected to increased levels of detrimental ectoparasites or perceived predation risk report that when compared to control broods, nestlings in experimental broods allocated resources toward their gapes or wings and away from increasing structural body size (Coslovsky and

Richner, 2011; Gil et al., 2008; Miller, 2010; Nilsson and Gårdmark, 2001; Saino et al., 1998). This pattern of developmental resource allocation is likely to have negative long-term consequences for nestlings as when not under stressful conditions, nestlings generally allocate resources toward body mass and structural size increases (Dickens et al., 2008; Lotem, 1998; Mainwaring et al., 2011b).

To summarize, the evidence for trade-offs is sparse when compared to the evidence for patterns of differential growth. Consequently, this is a fruitful area for future research and the development of theoretical models that examine the trade-off between traits that enhance effective sibling competition within the nest and the costs paid later in life may prove illuminating.

## VI. Conclusions and Suggestions for Future Work

Growth is an important stage of an individual's life history. One aspect of developmental plasticity which has received a lot of attention over the past decade or so has been that of compensatory growth, whereby if developing offspring experience a period of nutritional deficit, they can subsequently show accelerated growth should conditions improve, apparently compensating for the initial setback (Auer et al., 2010; Dmitriew, 2011; Richner et al., 1989; Royle et al., 2005; reviews in Lindström, 1999; Metcalfe and Monaghan, 2001, 2003). Meanwhile, another aspect of developmental plasticity which has received less attention is that of differential growth, which is described as the preferential allocation of resources to the growth of body systems and structures according to their importance (Coslovsky and Richner, 2011; Gil et al., 2008; Mainwaring et al., 2010a).

Interspecific studies have clearly demonstrated that natural selection imposes strong selection pressures on growth rates, developmental periods, and fledging ages in birds. Comparative analyses have shown that open-nesting species, which suffer comparatively high levels of nest predation, have evolved to fledge comparatively earlier than hole-nesting species (Remeš and Martin, 2002; Ricklefs, 1969; Ricklefs et al., 1998). However, later studies have shown that levels of sibling competition experienced in the nest, either by conspecifics or brood parasites (Boncoraglio et al., 2008; Ortega and Cruz, 1992; Remeš, 2006; Royle et al., 1999; Werschkul and Jackson, 1979), and levels of parasitism (Møller, 2005) also exert strong pressures on growth rates and patterns of developmental resource allocation. The vast majority of comparative studies only use mass gain as a predictor of growth and developmental rates, and we strongly urge further studies to incorporate other morphological characters, such as a skeletal and wing measurement, into such studies. This is not easy, however, as quantifying the growth of several morphological characters requires more time spent by investigators at the nest. Nevertheless, incorporating a mass, skeletal, and wing growth into comparative analyses should provide great insight into interspecific patterns of developmental resource allocation.

Intraspecific studies, meanwhile, show that large and dominant nestlings preferably allocate resources to body mass increases, which facilitate both short-term sibling competition through effective scramble competition (Budden and Wright, 2001; Leonard and Horn, 1996) and long-term survival and reproductive success. Meanwhile, small and subordinate nestlings employ a variety of trade-offs which are aimed at mediating adverse conditions experienced in the nest. Between-brood studies show that in response to increased levels of sibling competition and ectoparasites, nestlings allocate resources toward gape and wing development, thereby facilitating effective sibling competition and rapid fledging, respectively (Gil et al., 2008; Nilsson and Gårdmark, 2001).

Further studies which examine the causes of growth-trade-offs are required in order to increase the generality of the conclusions which we are able to make. For example, the two studies which have experimentally increased brood sizes, and hence levels of sibling competition, in hole-nesting passerines, report contrasting findings. Spotless starling nestlings were found to prioritize gape area development at the expense of body size (Gil et al., 2008), while marsh tit nestlings developed wing length at the expense of body mass (Nilsson and Gårdmark, 2001). These contrasting findings are interesting as they imply that nestlings of the two species, which should be under similar evolutionary pressures, employ different resource allocation strategies when under periods of stress. The differences may lie in how responsive the parents are to visual stimuli presented by the nestlings, and such behavioral interactions could be usefully explored in the future. Moreover, while one observational study found that late-hatched tree swallow nestlings from asynchronously hatching broods found that later hatched nestlings allocated resources toward wings development and away from mass development (Zach, 1982a), another observational study of tree swallows showed that late-hatched nestlings allocated resources toward the development of body mass and tarsus, but grew shorter wings (Johnson et al., 2003). This suggests that local environmental conditions may influence patterns of developmental resource allocation, and we strongly encourage further studies to increase the generality of the findings.

An exciting avenue for further research involves the roles of sexual size dimorphism and parentage in growth trade-offs, which remain drastically under studied. This is somewhat surprising as the advent of molecular tools for assigning gender and genetic fingerprinting for assigning parentage has made it apparent that males and females in socially monogamous species often mate outside of their pair bonds (Griffith et al., 2002). Given the prevalence of mixed paternity within broods and the knowledge that relatedness strongly influences growth over evolutionary timescales (Royle et al., 1999), then studies examining patterns of developmental resource allocation in within- and extra-pair nestlings would be a fruitful area for future research.

Future research should also concentrate on comprehensively examining the trade-offs between the development of morphological characters, immunity, and their influence on fledging age. While experimentally studies have subjected broods to increased levels of sibling competition and examined growth trade-offs (Gil et al., 2008; Miller, 2010; Nilsson and Gårdmark, 2001), none of those studies examined immunity nor fledging age, which is surprising as nestlings have been shown to trade-off growth and immunity (Parejo et al., 2010; Soler et al., 2003). We hypothesize that nestlings in broods with increased levels of sibling competition will allocate resources toward wing growth at the expense of immunity and will subsequently fledge sooner than control broods. Moreover, one laboratory study elegantly showed that following an experimental period of food shortage during early development, European shag (*Phalacrocorax aristotelis*) nestlings went through a period of compensatory growth which saw them reach comparable structural body sizes to control nestlings. However, experimental nestlings were found to have a lower resting metabolic rate, a lower body temperature, and a reduction in the size of several visceral organs, muscles, and lipid stores, when compared to control nestlings (Moe et al., 2004). Consequently, that study clearly showed a trade-off between morphological characters and physiology and further studies could usefully examine such trade-offs in wild birds.

Patterns of differential growth are generally adaptive in the immediate, and therefore predictable, conditions experienced during growth, rather than in any potential future conditions. Therefore, differential growth is adaptive in the short term, but the limited evidence suggests that they are maladaptive in the long term, as allocating resources away from body mass results in lower postfledging survival. Given that previous studies have shown that compensatory growth has a range of negative long-term consequences for individuals (Auer et al., 2010; Richner et al., 1989; Royle et al., 2005), then further research could usefully examine the long-term consequences of patterns of differential growth. It may well be the case that diverting resources away from body mass increases is maladaptive in the long term. For example, nestlings that allocate resources toward wing development and away from mass gain have lower masses during the nonbreeding season and have lower levels of postfledging survival.

Further research could also usefully examine the consequences of growth trade-offs for fledging age, which while relatively well understood at an interspecific level, remains poorly understood at an intraspecific level. Intergenerational conflicts of interest occur over the proportion of the finite parental effort which is invested in raising the current progeny, with brood demand always exceeding parental effort (Mock and Parker, 1997). Parental food visits are known to dramatically decrease close to fledging, when it is assumed that the parents try to induce the nestlings to fledge and forage for themselves (Roff et al., 2005). However, the influence of family conflicts on fledging age has been almost completely overlooked (but see Davies, 1978; Miller, 2010; Nilsson, 1990), which is surprising as fledging age has previously been shown to be variable. This is surprising as adaptive plasticity in fledging age has recently been reported in two studies. First, barn swallow broods in nests where dipteran ectoparasites were added, developed their wings significantly quicker than control broods in order to fledge sooner and avoid the detrimental ectoparasites (Saino et al., 1998). Second, in colonially nesting pied babblers (*T. bicolor*), those nestlings from solitary nests, which had less adult birds to protect them from predators, fledged more rapidly than those nestlings raised in colonies, where the predation risk was lower (Raihani and Ridley, 2007). Therefore, studies which examine the growth trade-offs which facilitate such plasticity in fledging age may provide important insights into this important life history transition.

**Acknowledgments**

We thank Stuart Bearhop, Megan Dickens, Amanda Gilby, Jonathan Grey, Simon Griffith, David Kelly, and Louise Rowe for collaborating on our research into the causes and consequences of differential growth in birds; the Natural Environment Research Council for funding the majority of our research; Elsevier, John Wiley and Sons, and the Royal Society for granting us permission to reproduce figures from journals to which they hold copyright; Marc Naguib and Jane Brockmann for editorial help; and Scott Forbes, Jan-Åke Nilsson, Vladimír Remeš, Tore Slagsvold, and two anonymous referees for detailed comments which substantially improved the quality of this chapter.

**References**

Amundsen, T., Slagsvold, T., 1991. Hatching asynchrony: facilitating adaptive or maladaptive brood reduction? In: Acta International Ornithological Congress, Christchurch, New Zealand, 19901707–1719.

Andersson, M., 1994. Sexual Selection. Princeton University Press, Princeton, USA.

Apanius, V., Westbrock, M.A., Anderson, D.J., 2008. Parental effort, plastic offspring growth, and immunoglobin G homeostasis in a long-lived seabird, the Nazca Booby *Sula granti*. Ornithol. Monogr. 65, 1–46.

Appleby, B.M., Petty, S.J., Blakey, J.K., Rainey, P., MacDonald, D.W., 1997. Does variation of sex ratio enhance reproductive success of offspring in tawny owls (*Strix aluco*)? Proc. R. Soc. Lond. B 264, 1111–1116.

Ardia, D.R., 2007. Site and sex-level differences in adult feeding behaviour and its consequences to offspring quality in tree swallows (*Tachycineta bicolor*). Can. J. Zool. 85, 847–854.

Arendt, J.D., 1997. Adaptive intrinsic growth rates: an integration across taxa. Q. Rev. Biol. 72, 150–177.

Arnold, K.E., Blount, J.D., Metcalfe, N.B., Orr, K.J., Adam, A., Houston, D., et al., 2007. Sex-specific differences in compensation for poor neonatal nutrition in the zebra finch *Taeniopygia guttata*. J. Avian Biol. 38, 356–366.

Arnqvist, G., Rowe, L., 2005. Sexual Conflict. Princeton University Press, Princeton, USA.

Auer, S.K., Arendt, J.D., Chandramouli, R., Reznick, D.N., 2010. Juvenile compensatory growth has negative consequences for reproduction in Trinidadian guppies (*Poecilia reticulata*). Ecol. Lett. 13, 998–1007.

Badyaev, A.V., Hamstra, T., Oh, K.P., Acevedo Seaman, D., 2006. Sex-biased maternal effects reduce ectoparasite-induced mortality in a passerine bird. Proc. Natl. Acad. Sci. USA 103, 14406–14411.

Beldade, P., Mateus, A.R.A., Keller, R.A., 2011. Evolution and molecular mechanisms of adaptive developmental plasticity. Mol. Ecol. 20, 1347–1363.

Bengtsson, H., Rydén, O., 1983. Parental feeding rate in relation to begging behaviour in asynchronously hatched broods of the great tit *Parus major*. Behav. Ecol. Sociobiol. 12, 243–251.

Bennett, P.M., Owens, I.P.F., 2002. Evolutionary Ecology of Birds: Life Histories, Mating Systems and Extinction. Oxford University Press, Oxford, UK.

Birkhead, T.R., Møller, A.P., 1992. Sperm Competition in Birds: Evolutionary Causes and Consequences. Academic Press, London, UK.

Birkhead, T.R., Fletcher, F., Pellatt, E.J., 1999. Nestling diet, secondary sexual traits and fitness in the zebra finch. Proc. R. Soc. Lond. B 266, 385–390.

Birkhead, T.R., Pellatt, E.J., Matthews, I.M., Roddis, N.J., Hunter, F.M., McPhie, F., et al., 2006. Genic capture and the genetic basis of sexually selected traits in the zebra finch. Evolution 60, 2389–2398.

Bize, P., Roulin, A., Bersier, L.F., Pfluger, D., Richner, H., 2003. Parasitism and developmental plasticity in alpine swift nestlings. J. Anim. Ecol. 72, 633–639.

Bize, P., Roulin, A., Richner, H., 2004. Parasitism, developmental plasticity and bilateral asymmetry of wing feathers in Alpine swift (*Apus melba*) nestlings. Oikos 106, 317–323.

Blount, J.D., Metcalfe, N.B., Arnold, K.E., Surai, P.F., Devevey, G.L., Monaghan, P., 2003. Neonatal nutrition, adult antioxidant defences and sexual attractiveness in the zebra finch. Proc. R. Soc. Lond. B 270, 1691–1696.

Blount, J.D., Metcalfe, N.B., Arnold, K.E., Surai, P.F., Monaghan, P., 2006. Effects of neonatal nutrition on adult reproduction in a passerine bird. Ibis 148, 509–514.

Bonabeau, E., Deneubourg, J.-L., Theraulaz, G., 1998. Within-brood competition and the optimal partitioning of parental investment. Am. Nat. 152, 419–427.

Boncoraglio, G., Saino, N., 2008. Barn swallow chicks beg more loudly when broodmates are unrelated. J. Evol. Biol. 21, 256–262.

Boncoraglio, G., Saino, N., Garamszegi, L.Z., 2008. Begging and cowbirds: brood parasites make hosts scream louder. Behav. Ecol. 20, 215–221.

Boncoraglio, G., Caprioli, M., Saino, N., 2009. Fine-tuned modulation of competitive ability according to kinship in barn swallow nestlings. Proc. R. Soc. Lond. B 276, 2117–2123.

Bortolotti, G.R., 1986. Influence of sibling competition on nestling sex ratios in sexually dimorphic birds. Am. Nat. 127, 495–507.

Bosque, C., Bosque, M.T., 1995. Nest predation as a selective factor in the evolution of developmental rates in altricial birds. Am. Nat. 145, 234–260.
Both, C., Visser, M.E., 2001. Adjustment to climate change is constrained by arrival dates in a long-distance migrant bird. Nature 411, 296–298.
Both, C., Visser, M.E., Verboven, N., 1999. Density dependent recruitment rates in great tits: the importance of being heavier. Proc. R. Soc. Lond. B 266, 465–469.
Briskie, J.V., Naugler, C.T., Leech, S.M., 1994. Begging intensity of nestling birds varies with sibling relatedness. Proc. R. Soc. Lond. B 258, 73–78.
Brown, L.H., Gargett, V., Steyn, P., 1977. Breeding success in some African eagles relative to theories about sibling aggression and its effects. Ostrich 48, 65–71.
Bryant, D.M., 1978. Establishment of weight hierarchies in the broods of House Martins, *Delichon urbica*. Ibis 120, 16–26.
Budden, A.E., Wright, J., 2001. Begging in nestling birds. Curr. Ornithol. 16, 83–118.
Budden, A.E., Wright, J., 2005. Learning during competitive positioning in the nest: do nestlings use ideal free 'foraging' strategies? Behav. Ecol. Sociobiol. 58, 227–236.
Burke, T., 1989. DNA fingerprinting and other methods for the study of mating success. Trends Ecol. Evol. 4, 139–144.
Burke, T., Bruford, M.W., 1987. DNA fingerprinting in birds. Nature 327, 149–152.
Chapman, T., Arnqvist, G., Bangham, J., Rowe, L., 2003. Sexual conflict. Trends Ecol. Evol. 18, 41–47.
Christe, P., Møller, A.P., de Lope, F., 1998. Immunocompetence and nestling survival in the house martin: the tasty chick hypothesis. Oikos 83, 175–179.
Clark, A.B., 1995. Gapes of sexually dimorphic blackbird nestlings do not show sexually dimorphic growth. Auk 112, 364–374.
Clark, A.B., Wilson, D.S., 1981. Avian breeding adaptations: hatching asynchrony, brood reduction, and nest failure. Q. Rev. Biol. 56, 253–277.
Clutton-Brock, T.H. (Ed.), 1988. Reproductive Success: Studies of Individual Variation in Contrasting Breeding Systems. Chicago University Press, Chicago, USA.
Clutton-Brock, T.H., 1991. The Evolution of Parental Care. Princeton University Press, Princeton, USA.
Clutton-Brock, T.H., Vincent, A.C., 1991. Sexual selection and the potential reproductive rates of males and females. Nature 351, 58–60.
Cockburn, A., 2006. Prevalence of different modes of parental care in birds. Proc. R. Soc. Lond. B 273, 1375–1383.
Coslovsky, M., Richner, H., 2011. Predation risk affects offspring growth via maternal effects. Funct. Ecol. 25, 878–888.
Cramp, S., Perrins, C.M. (Eds.), 1990. In: Handbook of the Birds of Europe, the Middle East, and North Africa: The Birds of the Western Palearctic, vol. VII. Oxford University Press, Oxford, UK.
Criscuolo, F., Monaghan, P., Nasir, L., Metcalfe, N.B., 2008. Early nutrition and phenotypic development: 'catch-up' growth leads to elevated metabolic rate in adulthood. Proc. R. Soc. Lond. B 275, 1565–1570.
Darwin, C.R., 1859. On the Origin of Species. John Murray, London, UK.
Davies, N.B., 1978. Parental meanness and offspring independence: an experiment with hand-reared great tits, *Parus major*. Ibis 120, 509–514.
Davies, N.B., 2000. Cuckoos, Cowbirds and Other Cheats. T and AD Poyser, London, UK.
Davies, N.B., Kilner, R.M., Noble, D.G., 1998. Nestling cuckoos *Cuculus canorus* exploit hosts with begging calls that mimic a brood. Proc. R. Soc. Lond. B 265, 673–678.
Dawkins, R., 1999. The Extended Phenotype: The Long Reach of the Gene, second ed. Oxford University Press, Oxford, UK.

Dearborn, D.C., 1998. Begging behaviour and food acquisition by brown-headed cowbird nestlings. Behav. Ecol. Sociobiol. 43, 259–270.

Dearborn, D.C., Lichtenstein, G., 2002. Begging behaviour and host exploitation in parasitic cowbirds. In: Wright, J., Leonard, M.L. (Eds.), The Evolution of Begging: Competition, Cooperation and Communication. Kluwer Academic Publishers, Dordrecht, The Netherlands, pp. 361–387.

Dickens, M., Hartley, I.R., 2007a. Differences in parental food allocation rules: evidence for sexual conflict in the blue tit? Behav. Ecol. 18, 674–679.

Dickens, M., Hartley, I.R., 2007b. Stimuli for nestling begging in blue tits *Cyanistes caeruleus*: hungry nestlings are less discriminating. J. Avian Biol. 38, 421–426.

Dickens, M., Berridge, D., Hartley, I.R., 2008. Biparental care and offspring begging strategies: hungry nestling blue tits move towards the father. Anim. Behav. 75, 167–174.

Dickinson, J.L., Hatchwell, B.J., 2004. Fitness consequences of helping. In: Koenig, W.D., Dickinson, J.L. (Eds.), Ecology and Evolution of Cooperative Breeding in Birds. Cambridge University Press, Cambridge, UK, pp. 48–66.

Dmitriew, C.M., 2011. The evolution of growth trajectories: what limits growth rate? Biol. Rev. 86, 97–116.

Draganoiu, T., Nagle, L., Musseau, R., Kreutzerm, M., 2005. Parental care and brood division in a songbird, the black redstart. Behaviour 142, 1495–1514.

Drummond, H., 2002. Begging versus aggression in avian broodmate competition. In: Wright, J., Leonard, M.L. (Eds.), The Evolution of Begging: Competition, Cooperation and Communication. Kluwer Academic Publishers, Dordrecht, The Netherlands, pp. 337–360.

Dubiec, A., Cichoń, M., Deptuch, K., 2006. Sex-specific development of cell-mediated immunity under experimentally altered rearing conditions in blue tit nestlings. Proc. R. Soc. Lond. B 273, 1759–1764.

Edwards, T.C., 1984. Similarity in the development of foraging mechanics among sibling ospreys. Condor 91, 30–36.

Eraud, C., Trouvé, C., Dano, S., Chastel, O., Faivre, B., 2008. Competition for resources modulates cell-mediated immunity and stress hormone level in nestling collared doves (*Streptopelia decaocto*). Gen. Comp. Endocrinol. 155, 542–551.

Ferrari, R.P., Martinelli, R., Saino, N., 2006. Differential effects of egg albumen content on barn swallow nestlings in relation to hatch order. J. Evol. Biol. 19, 981–993.

Ferree, E.D., Dickinson, J., Rendell, W., Stern, C., Portere, S., 2010. Hatching order explains an extrapair chick advantage in western bluebirds. Behav. Ecol. 21, 802–807.

Fisher, M.O., Nager, R.G., Monaghan, P., 2006. Compensatory growth impairs adult cognitive performance. PLoS Biol. 4, 1462–1466.

Fitze, P.S., Clobert, J., Richner, H., 2004. Long-term life-history consequences of ectoparasite-modulated growth and development. Ecology 85, 2018–2026.

Forbes, S., 1993. Avian brood reduction and parent-offspring 'conflict'. Am. Nat. 142, 82–117.

Forbes, S., 2007. Sibling symbiosis in nestling birds. Auk 124, 1–10.

Forbes, S., 2011. Social rank governs the effective environment of siblings. Biol. Lett. 23, 346–348.

Forbes, S., Wiebe, M., 2010. Egg size and asymmetric sibling rivalry in red-winged blackbirds. Oecologia 163, 361–372.

Gaston, A.J., 2004. Seabirds: A Natural History. T and AD Poyser, London, UK.

Gebhardt-Henrich, S., Richner, H., 1998. Causes of growth variation and its consequences for fitness. In: Starck, J.M., Ricklefs, R.E. (Eds.), Avian Growth and Development: Evolution Within the Altricial-Precocial Spectrum. Oxford University Press, Oxford, UK, pp. 324–339.

Gil, D., Graves, J., Hazon, N., Wells, A., 1999. Male attractiveness and differential testosterone investment in zebra finch eggs. Science 286, 126–128.
Gil, D., Bulmer, E., Celis, P., López, I., 2008. Adaptive developmental plasticity in growing nestlings: sibling competition induces differential gape growth. Proc. R. Soc. Lond. B 275, 549–554.
Glassey, B., Forbes, S., 2002. Begging and asymmetric nestling competition. In: Wright, J., Leonard, M.L. (Eds.), The Evolution of Begging: Competition, Cooperation and Communication. Kluwer Academic Publishers, Dordrecht, The Netherlands, pp. 269–281.
Godfray, H.C.J., 1991. Signalling of need by offspring to their parents. Nature 352, 328–330.
Godfray, H.C.J., 1995. Evolutionary theory of parent-offspring conflict. Nature 376, 133–138.
Grace, J.K., Dean, K.I., Ottinger, M.A., Anderson, D.J., 2011. Hormonal effects of maltreatment in Nazca booby nestlings: implications for the "cycle of violence" Horm. Behav. 60, 78–85.
Griffith, S.C., Owens, I.P.F., Thuman, K.A., 2002. Extra pair paternity in birds: a review of interspecific variation and adaptive function. Mol. Ecol. 11, 2195–2212.
Gustafsson, L., Sutherland, W.J., 1988. The cost of reproduction in the collared flycatcher *Ficedula albicollis*. Nature 335, 813–817.
Hardy, I.C.W. (Ed.), 2002. Sex Ratios: Concepts and Research Methods. Cambridge University Press, Cambridge, UK.
Harrison, C., 1975. A Field Guide to the Nests, Eggs and Nestlings of British and European Birds. Collins, London.
Hartley, I.R., 2007. Sexual dimorphism. In: Jamieson, B.G.M. (Ed.), Reproductive Biology and Phylogeny of Birds. Science Publishers, Enfield, USA, pp. 121–141.
Hartley, I.R., Shepherd, M., 1994. Female reproductive success, provisioning of nestlings and polygyny in corn buntings. Anim. Behav. 48, 717–725.
Hartley, I.R., Royle, N.J., Shepherd, M., 2000. Growth rates of nestling Corn Buntings *Miliaria calandra* in relation to their sex. Ibis 142, 668–671.
Hoi-Leitner, M., Romero-Pujante, M., Hoi, H., Pavlova, A., 2001. Food availability and immune capacity in serin (*Serinus serinus*) nestlings. Behav. Ecol. Sociobiol. 49, 333–339.
Hunt, S., Kilner, R.M., Langmore, N.E., Bennett, A.T.D., 2003. Conspicuous, ultraviolet rich mouth colours in begging chicks. Biol. Lett. 270, S25–S28.
Hussel, D.J.T., 1972. Factors affecting clutch size in Arctic passerines. Ecol. Monogr. 42, 317–364.
Hussel, D.J.T., 1985. On the adaptive basis for hatching asynchrony: brood reduction, nest failure and asynchronous hatching in Snow Buntings. Ornis Scand. 16, 205–212.
Johnson, L.S., Wimmers, L.E., Campbell, S., Hamilton, L., 2003. Growth rate, size, and sex ratio of last-laid, last-hatched offspring in the tree swallow *Tachycineta bicolor*. J. Avian Biol. 34, 35–43.
Johnstone, R.A., Godfray, C.J., 2002. Models of begging as a signal of need. In: Wright, J., Leonard, M.L. (Eds.), The Evolution of Begging: Competition, Cooperation and Communication. Kluwer Academic Publishers, Dordrecht, The Netherlands, pp. 1–20.
Kalmbach, E., Benito, M.M., 2007. Sexual size dimorphism and offspring vulnerability in birds. In: Fairburn, D.J., Blanckenborn, W.U., Székely, T. (Eds.), Sex, Size and Gender Roles: Evolutionary Studies of Sexual Size Dimorphism. Oxford University Press, Oxford, UK, pp. 133–142.
Kempenaers, B., Verheyen, G.R., Van den Broeck, M., Burke, T., Van Broeckhoven, C., Dhont, A.A., 1992. Extra-pair paternity results from female preference for high-quality males in the blue tit. Nature 357, 494–496.
Kilner, R.M., 1999. Family conflicts and the evolution of nestling mouth colour. Behaviour 136, 779–804.

Kilner, R.M., 2002. The evolution of complex begging displays. In: Wright, J., Leonard, M.L. (Eds.), The Evolution of Begging: Competition, Cooperation and Communication. Kluwer Academic Publishers, Dordrecht, The Netherlands, pp. 87–106.

Kilner, R.M., Hinde, C.A., 2008. Information warfare and parent-offspring conflict. Adv. Stud. Behav. 38, 283–336.

Kilner, R.M., Johnstone, R.A., 1997. Begging the question: are offspring solicitation behaviours signals of need? Trends Ecol. Evol. 12, 1–15.

Kilner, R.M., Noble, D.G., Davies, N.B., 1999. Signals of need in parent-offspring communication and their exploitation by the common cuckoo. Nature 6721, 667–672.

Krause, E.T., Naguib, M., 2011. Compensatory growth affects exploratory behaviour in zebra finches (*Taeniopygia guttata*). Anim. Behav. 81, 1295–1300.

Krebs, E.A., 2002. Sibling competition and parental control: patterns of begging in parrots. In: Wright, J., Leonard, M.L. (Eds.), The Evolution of Begging: Competition, Cooperation and Communication. Kluwer Academic Publishers, Dordrecht, The Netherlands, pp. 319–336.

Krist, M., 2011. Egg size and offspring quality: a meta-analysis in birds. Biol. Rev. 86, 692–716.

Krist, M., Remeš, V., Uvírová, L., Nádorník, P., Bureš, S., 2004. Egg size and offspring performance in the collared flycatcher (*Ficedula albicollis*): a within-clutch approach. Oecologia 140, 52–60.

Lack, D., 1947. The significance of clutch size. Ibis 89, 302–352.

Lack, D., 1954. The Natural Regulation of Animal Numbers. Clarendon, Oxford, UK.

Lack, D., 1968. Ecological Adaptations for Breeding in Birds. Methuen and Co, London, UK.

Lago, K., Johnson, L.S., Albrecht, D.J., 2000. Growth of late-hatched, competitively disadvantaged nestling house wrens relative to their older, larger nestmates. J. Field Ornithol. 71, 676–685.

Leonard, M., Horn, A., 1996. Provisioning rules in tree swallows. Behav. Ecol. Sociobiol. 38, 341–347.

Lepczyk, C.A., Caviedes, E., Karasov, W.H., 1998. Digestive responses during food restriction and realimentation in nestling house sparrows (*Passer domesticus*). Physiol. Zool. 71, 561–573.

Lessells, C.M., 2002a. Parental investment in relation to offspring sex. In: Wright, J., Leonard, M.L. (Eds.), The Evolution of Begging: Competition, Cooperation and Communication. Kluwer Academic Publishers, Dordrecht, The Netherlands, pp. 65–85.

Lessells, C.M., 2002b. Parentally biased favouritism: why should parents specialize in caring for different offspring? Philos. Trans. R. Soc. Lond. B Biol. Sci. 357, 381–403.

Lessells, C.M., Riebel, K., Draganoiu, T.D., 2011. Individual benefits of nestling begging: experimental evidence for an immediate effect, but no evidence for a delayed effect. Biol. Lett. 7, 336–338.

Lindström, J., 1999. Early development and fitness in birds and mammals. Trends Ecol. Evol. 14, 343–348.

Lipar, J.L., Ketterson, E.D., Nolan Jr., V., 1999. Intraclutch variation in testosterone content of Red-winged Blackbird eggs. Auk 116, 231–235.

Lotem, A., 1998. Differences in begging behaviour between Barn Swallow, *Hirundo rustica*, nestlings. Anim. Behav. 55, 809–818.

Low, M., Makan, T., Castro, I., 2012. Food availability and offspring demand influence sex-specific patterns and repeatability of parental provisioning. Behav. Ecol. 23, 25–34.

Magrath, R.D., 1990. Hatching asynchrony in altricial birds. Biol. Rev. 65, 587–622.

Magrath, M.J.L., Vedder, O., van der Velde, M., Komdeur, J., 2009. Maternal effects contribute to the superior performance of extra-pair offspring. Curr. Biol. 19, 792–797.

Mainwaring, M.C., Rowe, L.V., Kelly, D.J., Grey, J., Bearhop, S., Hartley, I.R., 2009. Hatching asynchrony and growth trade-offs within barn swallow broods. Condor 111, 668–674.

Mainwaring, M.C., Dickens, M., Hartley, I.R., 2010a. Environmental and not maternal effects determine variation in offspring phenotypes in a passerine bird. J. Evol. Biol. 23, 1302–1311.
Mainwaring, M.C., Hartley, I.R., Gilby, A.J., Griffith, S.C., 2010b. Hatching asynchrony and growth trade-offs within domesticated and wild zebra finch, *Taeniopygia guttata*, broods. Biol. J. Linn. Soc. 100, 763–773.
Mainwaring, M.C., Dickens, M., Hartley, I.R., 2011a. Sexual dimorphism and growth trade-offs in Blue Tit *Cyanistes caeruleus* nestlings. Ibis 153, 175–179.
Mainwaring, M.C., Lucy, D., Hartley, I.R., 2011b. Parentally biased favouritism in relation to offspring sex in zebra finches. Behav. Ecol. Sociobiol. 65, 2261–2268.
Marchant, S., Higgins, P.J. (Eds.), 1993. Handbook of Australian, New Zealand and Antarctic Birds. In: Raptors to Lapwings, vol. 2. Oxford University Press, Melbourne.
Martin, T.E., 1995. Avian life history evolution in relation to nest sites, nest predation and food. Ecol. Monogr. 65, 101–127.
Martin, T.E., 2004. Avian life-history evolution has an eminent past: does it have a bright future? Auk 121, 289–301.
Martin, T.E., Briskie, J.V., 2009. Predation on dependent offspring: a review of the consequences for mean expression and phenotypic plasticity in avian life history traits. Ann. N. Y. Acad. Sci. 1168, 201–217.
Martin, T.E., Scwabl, H., 2008. Variation in maternal effects and embryonic development rates among passerine species. Philos. Trans. R. Soc. Lond. B Biol. Sci. 363, 1663–1674.
Martin, T.E., Arriero, E., Majewska, A., 2011a. A trade-off between embryonic development rate and immune function of avian offspring is revealed by considering embryonic temperature. Biol. Lett. 7, 425–428.
Martin, T.E., Lloyd, P., Bosque, C., Barton, D.C., Biancucci, A.L., Cheng, Y.-R., et al., 2011b. Growth rate variation among passerine species in tropical and temperate sites: an antagonistic interaction between parental food provisioning and nest predation risk. Evolution 65, 1607–1622.
McRae, S.B., Weatherhead, P.J., Montgomerie, R., 1993. American robin nestlings compete by jockeying for position. Behav. Ecol. Sociobiol. 33, 101–106.
Medvin, M.B., Beecher, M.D., 1986. Parent-offspring recognition in the barn swallow (*Hirundo rustica*). Anim. Behav. 34, 1627–1639.
Merilä, J., Fry, J.D., 1998. Genetic variation and causes of genotype-environment interaction in the body size of blue tits (*Parus caeruleus*). Genetics 148, 1233–1244.
Metcalfe, N.B., Monaghan, P., 2001. Compensation for a bad start: grow now, pay later? Trends Ecol. Evol. 16, 254–260.
Metcalfe, N.B., Monaghan, P., 2003. Growth versus lifespan: perspectives from evolutionary ecology. Exp. Gerontol. 38, 935–940.
Miller, D.A., 2010. Morphological plasticity reduces the effect of poor developmental conditions on fledging age in mourning doves. Proc. R. Soc. Lond. B 277, 1659–1665.
Mock, D.W., 1985. Siblicidal brood reduction: the prey-size hypothesis. Am. Nat. 125, 327–343.
Mock, D.W., 2004. More than Kin and Less than Kind: The Evolution of Family Conflict. Harvard University Press, London, UK.
Mock, D.W., Parker, G.A., 1997. The Evolution of Sibling Rivalry. Oxford University Press, Oxford, UK.
Mock, D.W., Ploger, B.J., 1987. Parental manipulation of optimal hatch asynchrony in cattle egrets: an experimental study. Anim. Behav. 35, 150–160.
Moe, B., Brunvoll, S., Mork, D., Brobakk, T.E., Bech, C., 2004. Developmental plasticity of physiology and morphology in diet-restricted European shag nestlings (*Phalacrocorax aristotelis*). J. Exp. Biol. 207, 4067–4076.

Møller, A.P., 2000. Male parental care, female reproductive success, and extrapair paternity. Behav. Ecol. 11, 161–168.
Møller, A.P., 2005. Parasites, predators and the duration of developmental periods. Oikos 111, 291–301.
Møller, A.P., Birkhead, T.R., 1993. Certainty of paternity covaries with paternal care in birds. Behav. Ecol. Sociobiol. 33, 361–368.
Moss, D., 1979. Growth of nestling sparrowhawks (*Accipiter nisus*). J. Zool. 187, 297–314.
Mousseau, T.A., Uller, T., Wapstra, E., Badyaev, A.V., 2009. Evolution of maternal effects: past and present. Philos. Trans. R. Soc. Lond. B Biol. Sci. 364, 1035–1038.
Müller, M.S., Brennecke, J.F., Porter, E.T., Ottinger, M.A., Anderson, D.J., 2008. Perinatal androgens and adult behavior vary with nestling social system in siblicidal boobies. PLoS One 3, e2460.
Naguib, M., Nemitz, A., Gil, D., 2006. Maternal developmental stress reduces reproductive success of female offspring in zebra finches. Proc. R. Soc. Lond. B 273, 1901–1905.
Naguib, M., Heim, C., Gil, D., 2008. Effects of early developmental conditions on male attractiveness in zebra finches. Ethology 114, 255–261.
Newton, I., 1989. Introduction. In: Newton, I. (Ed.), Lifetime Reproduction in Birds. Academic Press, London, UK, pp. 1–11.
Nilsson, J.-Å., 1990. What determines the timing and order of nest-leaving in the marsh tit (*Parus palustris*)? In: Blondel, J., Gosler, A., Lebreton, J.-D., McCleery, R. (Eds.), Population Biology of Passerine Birds. Springer-Verlag, Berlin, Germany, pp. 369–380.
Nilsson, J.-Å., 1993. Energetic constraints on hatching asynchrony. Am. Nat. 141, 158–166.
Nilsson, J.-Å., Gårdmark, A., 2001. Sibling competition affects individual growth strategies in marsh tit, *Parus palustris*, nestlings. Anim. Behav. 61, 357–365.
Nilsson, J.-Å., Svensson, M., 1996. Sibling competition affects nestling growth strategies in marsh tits. J. Anim. Ecol. 65, 825–836.
O'Brien, E.L., Dawson, R.D., 2007. Context-dependent genetic benefits of extra-pair mate choice in a socially monogamous passerine. Behav. Ecol. Sociobiol. 61, 775–782.
O'Connor, R.J., 1984. The Growth and Development of Birds. John Wiley and Sons, New York, USA.
Oppliger, A., Richner, H., Christe, P., 1994. Effect of an ectoparasite on lay date, nest site choice and desertion, and hatching success in the great tit (*Parus major*). Behav. Ecol. 5, 130–134.
Ortega, C.P., Cruz, A., 1992. Differential growth patterns of nestling brown-headed cowbirds and yellow-headed blackbirds. Auk 109, 368–376.
Owens, I.P.F., Bennett, P.M., 1995. Ancient ecological diversification explains life-history variation among living birds. Proc. R. Soc. Lond. B 261, 227–232.
Owens, I.P.F., Burke, T., Thompson, D.B.A., 1994. Extraordinary sex-roles in the Eurasian dotterel—female mating arenas, female-female competition, and female mate choice. Am. Nat. 144, 76–100.
Øyan, H.S., Anker-Nilssen, T., 1996. Allocation of growth in food-stressed Atlantic puffin chicks. Auk 113, 830–841.
Palacios, M.G., Martin, T.E., 2006. Incubation period and immune function: a comparative field study among coexisting birds. Oecologia 146, 505–512.
Parejo, D., Silva, N., Avilés, J.M., Danchin, E., 2010. Developmental plasticity varied with sex and position in hatching hierarchy in nestlings of the asynchronous European roller, *Coracias garrulus*. Biol. J. Linn. Soc. 99, 500–511.
Parker, G.A., 1985. Models of parent-offspring conflict. 5. Effects of the behaviour of the two parents. Anim. Behav. 33, 519–533.

Parker, G.A., Mock, D.W., Lamey, T.C., 1989. How selfish should stronger sibs be? Am. Nat. 133, 846–868.
Parker, G.A., Royle, N.J., Hartley, I.R., 2002a. Intra-familial conflict and parental investment: a synthesis. Philos. Trans. R. Soc. Lond. B Biol. Sci. 357, 295–307.
Parker, G.A., Royle, N.J., Hartley, I.R., 2002b. Begging scrambles with unequal chicks: interactions between need and competitive ability. Ecol. Lett. 5, 206–215.
Perrins, C.M., 1991. Tits and their caterpillar food supply. Ibis 133, 49–54.
Perrins, C.M., Geer, T.A., 1980. The effect of sparrowhawks on tit populations. Ardea 68, 133–142.
Raihani, N.J., Ridley, A.R., 2007. Variable fledging age according to group size: trade-offs in a cooperatively breeding bird. Biol. Lett. 3, 624–627.
Redondo, T., Zuñiga, J.M., 2002. Dishonest begging and host manipulation by Clamator cuckoos. In: Wright, J., Leonard, M.L. (Eds.), The Evolution of Begging: Competition, Cooperation and Communication. Kluwer Academic Publishers, Dordrecht, The Netherlands, pp. 389–412.
Remeš, V., 2006. Growth strategies of passerine birds are related to brood parasitism by the Brown-headed Cowbird (*Molothrus ater*). Evolution 60, 1692–1700.
Remeš, V., 2007. Avian growth and development rates and age-specific mortality: the roles of nest predation and adult mortality. J. Evol. Biol. 20, 320–325.
Remeš, V., Martin, T.E., 2002. Environmental influences on the evolution of growth and developmental rates in passerines. Evolution 56, 2505–2518.
Richner, H., Schneiter, P., Stirnimann, H., 1989. Life-history consequences of growth rate depression: an experimental study on carrion crows (*Corvus corone corone* L.). Funct. Ecol. 3, 617–624.
Richner, H., Oppliger, A., Christe, P., 1993. Effect of an ectoparasite on reproduction in great tits. J. Anim. Ecol. 62, 703–710.
Ricklefs, R.E., 1968. Patterns of growth in birds. Ibis 110, 419–451.
Ricklefs, R.E., 1969. Preliminary models for growth rates in altricial birds. Ecology 50, 1031–1039.
Ricklefs, R.E., 1992. Embryonic development period and the prevalence of avian blood parasites. Proc. Natl. Acad. Sci. USA 89, 4722–4725.
Ricklefs, R.E., 2002. Sibling competition and the evolution of brood size and development rate in birds. In: Wright, J., Leonard, M.L. (Eds.), The Evolution of Begging: Competition, Cooperation and Communication. Kluwer Academic Publishers, Dordrecht, The Netherlands, pp. 283–301.
Ricklefs, R.E., Starck, J.M., Konarzewski, M., 1998. Internal constraints on growth rates in birds. In: Starck, J.M., Ricklefs, R.E. (Eds.), Avian Growth and Development: Evolution Within the Altricial-Precocial Spectrum. Oxford University Press, Oxford, UK, pp. 266–287.
Ridley, A.R., Thompson, A.M., 2012. The effect of Jacobin Cuckoo *Clamator jacobinus* parasitism on the body mass and survival of young in a new host species. Ibis 154, 195–199.
Rivers, J.W., Briskie, J.V., Rothstein, S.I., 2010. Have brood parasitic cowbird nestlings caused the evolution of more intense begging by host nestlings? Anim. Behav. 80, e1–e5.
Rodríguez-Gironés, M.A., 1996. Siblicide: the evolutionary blackmail. Am. Nat. 148, 101–122.
Roff, D.A., 1992. The Evolution of Life Histories: Theory and Adaptation. Chapman and Hall, London, UK.
Roff, D.A., Remeš, V., Martin, T.E., 2005. The evolution of fledging age in songbirds. J. Evol. Biol. 18, 1425–1433.
Rosivall, B., Szöllösi, E., Török, J., 2005. Maternal compensation for hatching asynchrony in the collared flycatcher *Ficedula albicollis*. J. Avian Biol. 36, 531–537.

Royama, T., 1966. Factors governing feeding rate, food requirement and broods size of nestling Great Tits *Parus major*. Ibis 108, 313–347.

Royle, N.J., Hartley, I.R., Owens, I.P.F., Parker, G.A., 1999. Sibling competition and the evolution of growth rates in birds. Proc. R. Soc. Lond. B 266, 923–932.

Royle, N.J., Hartley, I.R., Parker, G.A., 2002a. Begging for control: when are offspring solicitation behaviours honest? Trends Ecol. Evol. 17, 434–440.

Royle, N.J., Hartley, I.R., Parker, G.A., 2002b. Sexual conflict reduces offspring fitness in zebra finches. Nature 416, 733–736.

Royle, N.J., Surai, P.F., Hartley, I.R., 2001. Maternally derived androgens and antioxidants in bird eggs: complimentary but opposing effects? Behav. Ecol. 12, 381–385.

Royle, N.J., Surai, P.F., Hartley, I.R., 2003. The effect of variation in dietary intake on maternal deposition of antioxidants in zebra finch eggs. Funct. Ecol. 17, 472–481.

Royle, N.J., Lindstrom, J., Metcalfe, N.B., 2005. A poor start in life negatively affects dominance status in adulthood independently of body size in green swordtails *Xiphophorus helleri*. Proc. R. Soc. Lond. B 272, 1917–1922.

Royle, N.J., Hartley, I.R., Parker, G.A., 2006. Consequences of biparental care for begging and growth in zebra finches *Taeniopygia guttata*. Anim. Behav. 72, 123–130.

Saino, N., Calza, S., Møller, A.P., 1998. Effects of a dipteran ectoparasite on immune response and growth trade-offs in barn swallow (*Hirundo rustica*) nestlings. Oikos 81, 217–228.

Saino, N., Ninni, P., Calza, S., Martinelli, R., de Bernard, F., Møller, A.P., 2000. Better red than dead: carotenoid based mouth colouration reveals infection in barn swallow nestlings. Proc. R. Soc. Lond. B 267, 57–61.

Saino, N., Romano, M., Ambrosini, R., Ferrari, R.P., Møller, A.P., 2004. Timing of reproduction and egg quality covary with temperature in the insectivorous barn swallow (*Hirundo rustica*). Funct. Ecol. 18, 50–57.

Scheuerlein, A., Gwinner, E., 2006. Reduced nestling growth of East African Stonechats *Saxicola torquata axillaris* in the presence of a predator. Ibis 148, 468–476.

Schew, W.A., Ricklefs, R.E., 1998. Developmental plasticity. In: Starck, J.M., Ricklefs, R.E. (Eds.), Avian Growth and Development: Evolution Within the Altricial-Precocial Spectrum. Oxford University Press, Oxford, UK, pp. 288–304.

Schwabl, H., 1996. Maternal testosterone in the avian egg enhances postnatal growth. Comp. Biochem. Physiol. 114, 271–276.

Schwabl, H., Mock, D.W., Gieg, J.A., 1997. A hormonal mechanism of parental favouritism. Nature 386, 231.

Schwabl, H., Palacios, M.G., Martin, T.E., 2007. Selection for rapid embryo development correlates with embryo exposure to maternal androgens among passerine birds. Am. Nat. 170, 196–206.

Simon, A., Thomas, D.W., Blondel, J., Lambrechts, M.M., Perret, P., 2003. Within-brood distribution of ectoparasite attacks on nestling blue tits: a test of the tasty chick hypothesis using inulin as a tracer. Oikos 102, 551–558.

Skagen, S.K., 1987. Hatching asynchrony in American Goldfinches: an experimental study. Ecology 68, 1747–1759.

Skutch, A.K., 1949. Do tropical birds rear as many young as they can nourish? Ibis 91, 430–455.

Slagsvold, T., 1986. Asynchronous versus synchronous hatching in birds: experiments with the pied flycatcher. J. Anim. Ecol. 55, 1115–1134.

Slagsvold, T., Wiebe, K.L., 2007. Hatching asynchrony and early nestling mortality: the feeding constraint hypothesis. Anim. Behav. 73, 691–700.

Slagsvold, T., Sandvik, J., Rofstad, G., Lorentsen, Ö., Husby, M., 1984. On the adaptive value of intraclutch egg-size variation in birds. Auk 101, 685–697.

Smith, H.G., Montgomerie, R., 1991. Nestling American robins compete with siblings by begging. Behav. Ecol. Sociobiol. 29, 307–312.

Soler, J.J., de Neve, L., Pérez-Contreras, T., Soler, M., 2003. Trade-off between immunocompetence and growth in magpies: an experimental study. Proc. R. Soc. Lond. B 270, 241–248.

Sossinka, R., 1982. Domestication in birds. In: Farner, D.S., King, J.R., Parkes, K.C. (Eds.), Avian Biology. Academic Press, New York, USA, pp. 373–403.

Stearns, S.C., 1992. The Evolution of Life Histories. Oxford University Press, Oxford, UK.

Stoleson, S.H., Beissinger, S.R., 1995. Hatching asynchrony and the onset of incubation in birds, revisted: when is the critical period? Curr. Ornithol. 12, 191–270.

Székely, T., Cuthill, I.C., Kis, J., 1999. Brood desertion in Kentish Plover: sex differences in remating opportunities. Behav. Ecol. 10, 191–197.

Teather, K.L., 1992. An experimental study of competition between male and female nestlings of the red-winged blackbird. Behav. Ecol. Sociobiol. 31, 81–87.

Tella, J.L., Scheurlein, A., Ricklefs, R.E., 2002. Is cell-mediated immunity related to the evolution of life-history strategies in birds? Proc. R. Soc. Lond. B 269, 1059–1066.

Townsend, H.M., Anderson, D.J., 2007. Long-term assessment of costs of reproduction in Nazca boobies (*Sula granti*) using multi-state mark-recapture models. Evolution 61, 1956–1968.

Trivers, R.L., 1974. Parent-offspring conflict. Am. Zool. 14, 249–264.

Tschirren, B., Fitze, P.S., Richner, H., 2003. Sexual dimorphism in susceptibility to parasites and cell-mediated immunity in Great tit nestlings (*Parus major*). J. Anim. Ecol. 72, 839–845.

Tschirren, B., Rutstein, A.N., Postma, E., Mariette, M., Griffith, S.C., 2009. Short- and long-term consequences of early developmental conditions: a case study on wild and domesticated zebra finches. J. Evol. Biol. 22, 387–395.

Uller, T., 2008. Developmental plasticity and the evolution of parental effects. Trends Ecol. Evol. 23, 432–438.

van Noordwijk, A.J., Marks, H.L., 1998. Genetic aspects of growth. In: Starck, J.M., Ricklefs, R.E. (Eds.), Avian Growth and Development: Evolution Within the Altricial-Precocial Spectrum. Oxford University Press, Oxford, UK, pp. 305–323.

Verhulst, C.M., Perrins, C.M., Riddington, R., 1997. Natal dispersal of great tits in a patchy environment. Ecology 78, 864–872.

Visser, G.H., 1998. Development of temperature regulation. In: Starck, J.M., Ricklefs, R.E. (Eds.), Avian Growth and Development: Evolution Within the Altricial-Precocial Spectrum. Oxford University Press, Oxford, UK, pp. 117–156.

Webster, M.S., 1991. Male parental care and polygyny in birds. Am. Nat. 137, 274–280.

Werschkul, D.F., Jackson, J.A., 1979. Sibling competition and avian growth rates. Ibis 121, 97–102.

West-Eberhard, M.J., 2003. Developmental Plasticity and Evolution. Oxford University Press, Oxford, UK.

Westneat, D.F., Clark, A.B., Rambo, K.C., 1995. Within-brood patterns of paternity and paternal behaviour in red-winged blackbirds. Behav. Ecol. Sociobiol. 37, 349–356.

Williams, T.D., 1994. Intraspecific variation in egg composition in birds: effects on offspring fitness. Biol. Rev. 68, 35–59.

Wilson, D.S., Clark, A.B., 2002. Begging and cooperation: an exploratory flight. In: Wright, J., Leonard, M.L. (Eds.), The Evolution of Begging: Competition, Cooperation and Communication. Kluwer Academic Publishers, Dordrecht, The Netherlands, pp. 43–64.

Zach, R., 1982a. Hatching asynchrony, egg size, growth, and fledging in tree swallows. Auk 99, 695–700.

Zach, R., 1982b. Nestling House Wrens: weight and feather growth. Can. J. Zool. 60, 1417–1425.

Zann, R.A., 1996. The Zebra Finch: A Synthesis of Field and Laboratory Studies. Oxford University Press, Oxford, UK.

# Increasing Awareness of Ecosystem Services Provided by Bats

SIMON J. GHANEM[*,†] and CHRISTIAN C. VOIGT[*,†]

[*]LEIBNIZ INSTITUTE FOR ZOO AND WILDLIFE RESEARCH, BERLIN, GERMANY
[†]DEPARTMENT OF BEHAVIORAL BIOLOGY, FREIE UNIVERSITÄT BERLIN, BERLIN, GERMANY

## I. INTRODUCTION

The 2005 synthesis report of the Millennium Ecosystem Assessment made a first attempt to summarize the relevance of animals for ecosystem services (ESs) and the functioning of habitats worldwide (Millennium Ecosystem Assessment, 2005). ESs are defined as benefits to humankind derived from resources and processes supplied by natural ecosystems. The Millennium Ecosystem Assessment recognizes four categories of ESs (1) provisioning services such as the production of fiber, clean water, or food; (2) regulating services such as pollination or pest control; (3) supporting services such as seed dispersal; and (4) cultural services such as intellectual, academic, or spiritual inspiration (Millennium Ecosystem Assessment, 2003).

In many cases, ESs of wild animals such as the provisioning of food for local communities in the form of harvested fish, birds, or mammals are apparent. Other ESs are more subtle but probably equally important for humans. For example, birds and insects have already been widely acknowledged for supporting ecosystems, the agricultural economy, and the well-being of humans. McGregor (1976) estimated that 15–30% of the human diet in North America depends on pollination mediated by insects. Honey bees (*Apis mellifera*) are crucial pollinators of economically important crops such as rape (*Brassica napus*) and many fruits and vegetables. Recently, the significance of bees as pollinators became particularly apparent when populations declined due to Colony Collapse Disorder, which affected negatively the food-producing industry (Klein et al., 2007). Due to the loss of ESs by dwindling populations of *A. mellifera*, wild bee populations have gained recognition for being able to partially substitute the



0065-3454/12 $35.00
DOI: 10.1016/B978-0-12-394288-3.00007-1

pollination services of honey bees (Greenleaf and Kremen, 2006). More than 920 species of birds also provide pollination services worldwide (Whelan et al., 2008). Birds also play an important role in the control of insect and vertebrate pests. For example, many raptors, such as owls, falcons, and hawks, regulate rodent populations that may be detrimental for crop production at high densities (Whelan et al., 2008).

In general, the contribution of bats to the maintenance of ecosystems is rarely acknowledged (but see, e.g., Cleveland et al., 2006; Boyles et al., 2011; Kalka et al., 2008; Kunz et al., 2011; Williams-Guillen et al., 2008). This is most likely due to their nocturnal and cryptic lifestyle and their negative reputation in many human cultures. However, bats are the second largest order of mammals (Nowak, 1994), with more than 1100 currently described species of bats (Simmons, 2005); only Rodentia have more species. Moreover, bats are the most diverse group of mammals in many tropical ecosystems (Bass et al., 2010; Simmons and Voss, 1998). Recently, first attempts have been made to quantify the ESs of Chiroptera in selected geographical regions such as tropical Asia (e.g., Fujita and Tuttle, 1991; Leelapaibul et al., 2005) and North America (Boyles et al., 2011; Cleveland et al., 2006; Kunz et al., 2011; Williams-Guillen et al., 2008). Yet, the overall value of bats for human society in general and human well-being in particular has been largely underestimated (but see, e.g., Cleveland et al., 2006).

Bats provide all four categories of ESs as listed by the Millennium Ecosystem Assessment. Cultures all over the world are intrigued by bats. In western cultures, bats historically have had a negative connotation, as in the story of Dracula or as an ingredient in witches' potions. However, in eastern cultures, the reputation of bats was substantially better; in these cultures, bats are often represented as a symbol of luck and fortune (Allen, 1962; Kunz et al, 2011). Today, bats provide cultural ESs through their appearance in Blockbuster movies, TV series, and books. While bats are not always seen in a positive light, they leave an imprint on the hearts and minds of people all over the world. However, in this review, we focus primarily on ESs that are provisioning, regulating, or supporting services because they are closely related to the daily life and behavior of bats and because they are directly linked to the health and benefit of ecosystems and humankind. The feeding behavior of insectivorous and frugivorous bats, for example, leads to the consumption of insects or fruits and thereby to pest control or seed dispersal. The guano production of many cave-dwelling bats is a direct result of their roosting behavior, as the guano of the roosting bats provides and supports cave ecosystems for many other vertebrate and nonvertebrate species. Moreover, the pollination of plants and the dispersal of their seeds just become effective precisely because of the bats' ranging behavior together with their usually large home ranges.

Here we review information about ESs of bats in relation to three main ecological functions: the regulation of insect populations, dispersal of seeds, and pollination of plants.

## II. Insectivory

Insectivory is considered to be the ancestral diet for bats (Simmons et al., 2008). Insectivorous bats are abundant and distributed throughout all continents, except Antarctica and small and remote oceanic islands (Fig. 1A and B). In some locations, such as caves, they form the largest aggregations of any wild mammal. For example, about 2.6 million wrinkle-lipped free-tailed bats (*Chaerephon plicata*) roost in the Khao Chong Pran Cave in Thailand (Hillman, 1999) and about 100 million Brazilian free-tailed bats (*Tadarida brasiliensis*) roost in a few caves in Mexico and southern USA (McCracken, 2003; Wahl, 1993). The most important ES of insectivorous bats is probably the control of herbivorous arthropods including pest insects, the control of arthropod disease vectors, and the provisioning of bat guano (Fig. 2 and Table I).

### A. Pest Control

Insect herbivory is a major problem affecting plant reproduction and reducing plant biomass production, diversity and distribution (e.g., Fine et al., 2004). The top-down control of populations of herbivorous insects is a valuable ES of insectivorous animals by directly increasing the monetary gain of crop farmers and supporting the food production industry (e.g., Federico et al., 2008; Kalka et al., 2008; Leelapaibul et al., 2005; McCracken et al., 2008; Williams-Guillen et al., 2008). Recently, bats have been identified as important top-down regulators of herbivorous insect populations (Böhm et al., 2011; Kalka et al., 2008; Williams-Guillen et al., 2008). Bats have also been identified as important agents for biological pest control on farmland (Leelapaibul et al., 2005; Williams-Guillen et al., 2008). Interestingly, previous studies on the top-down limitation of insect populations have solely looked at bird predation and have even overestimated the impact of insectivorous birds on insect pest populations by ignoring the additional nocturnal pest control services of bats. For example, researchers have used cage-like exclosures around plants to quantify changes in the rate of herbivory when birds are prevented from feeding on phytophagous insects. However, by excluding vertebrate predators from a given plant or plant community both during the day and night, researchers have quantified accidently the combined effect of birds and bats on

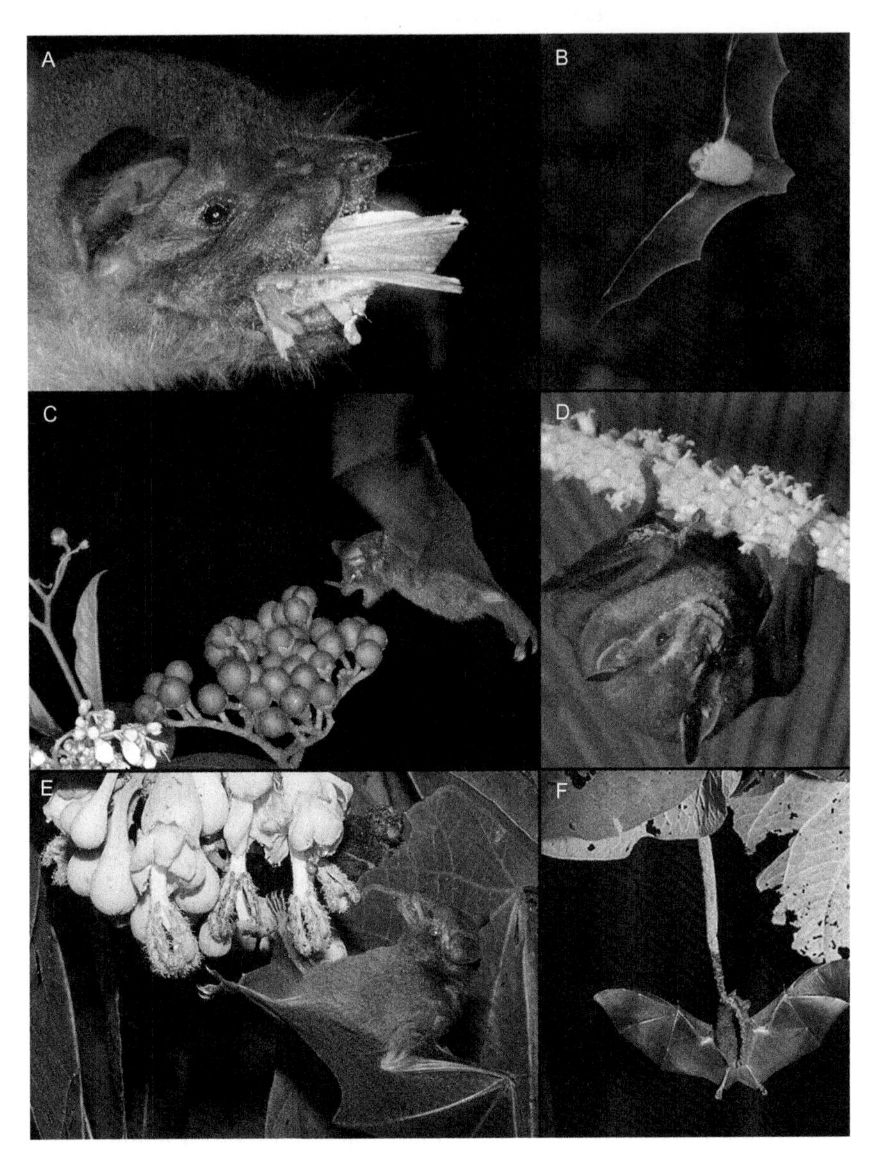


FIG. 1. Five bat species representing the three major feeding guilds and their main ecological function. Insectivorous bats (A and B) primarily provide regulating and supporting ESs by controlling arthropod populations. Frugivorous (C and D) and nectarivorous bats (E and F) mainly provide provisioning and supporting ESs by pollination and dispersal of seeds. A lesser bull dog bat (*Noctilio albiventris*) feeding on a moth (A). An aerial hawking bat (*Centronycteris centralis*) (B). Seba's short-tailed bat (*Carollia perspicillata*) approaching an infructescence of *Solanum rugosum* (C). A Pygmy Fruit-eating bat (*Dermanura phaeotis*) pollinating a *Calyptrogyne ghiesbreghtiana* palm while feeding on its flowers (D). A Pallas's long-tongued bat (*Glossophaga soricina*) pollinating an inflorescence of a Chupa-chupa tree (*Matisia cordata*) (E) and feeding on a neotropical succession plant (*Piper auratum*) (F). (For color version of this figure, the reader is referred to the online version of this chapter.)

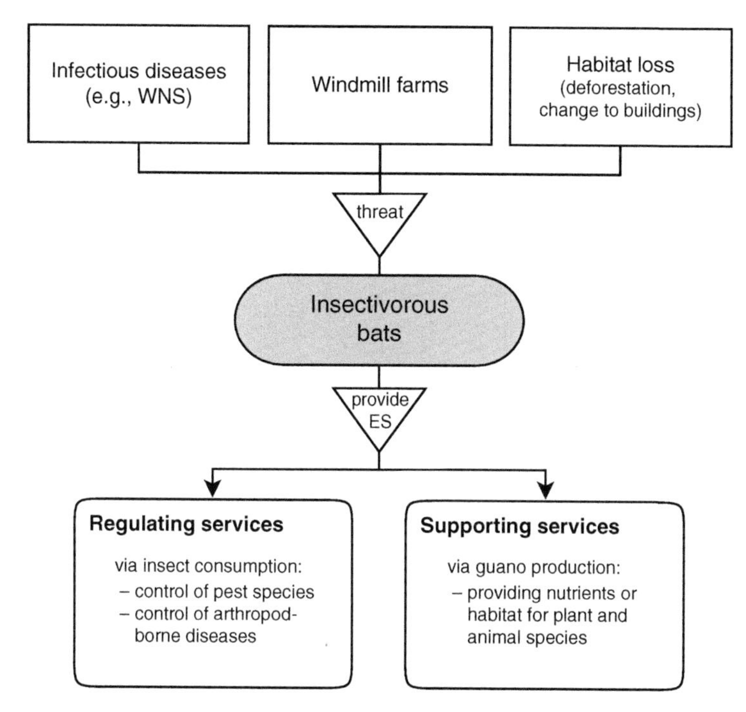


FIG. 2. Illustration of the key ESs provided by insectivorous bats including the major threats that put these services at risk.

herbivorous insect populations and have falsely assigned this effect exclusively to birds (e.g., Greenberg et al., 2000; Holmes et al., 1979; Van Bael et al., 2003). To determine the taxon-specific effect of insectivorous vertebrates on insect populations, it seems more appropriate to conduct night and daytime exclosures separately. Following this experimental protocol, recent studies have highlighted that bats are as important as birds for controlling populations of herbivorous insects (Böhm et al., 2011; Kalka et al., 2008; Williams-Guillen et al., 2008). Moreover, Kalka et al. (2008) provided substantial evidence that bats even have a larger impact on insect populations than birds. They could show that arthropod herbivory increased significantly in plants protected by nocturnal exclosures from foraging bats compared to those protected by diurnal exclosures that only preclude birds from feeding on arthropods.

Worldwide, humans are in competition with a multitude of pest animals for food, fiber, and timber. These pest animals cause large monetary losses when not controlled (Daily, 1997). Losing insect-feeding bats due to habitat destruction or global climate change may put the natural pest control

TABLE I

EXAMPLES OF ECOSYSTEM SERVICES PROVIDED BY INSECTIVOROUS BATS, THEIR USES, AND OCCURRENCES WITH THE YEARLY ESTIMATED ECONOMIC VALUE OF THE RESPECTIVE SERVICES

| Service | Pest or service function | Service provider | Land or reference | Common usage | Approx. yearly value | Source |
|---|---|---|---|---|---|---|
| PC | Prolongation of insect toxin resistance | All insectivorous bats (e.g., *Tadarida* sp.) | Worldwide | Reduction of insecticide use | N.A. | Federico et al. (2008) |
| PC | Cotton bollworm larvae (*Helicoverpa zea*) | Brazilian free-tailed bat (*Tadarida brasiliensis*) | South-central Texas | Insect pest control | 500,000 US$ | Betke et al. (2008) |
| PC | Many insect pest species | All insectivorous bats (e.g., *Tadarida brasiliensis*) | North America | Insect pest control | 22.9 billion US$ (range 3.7–53 billion US$) | Boyles et al. (2011) |
| PC | Plant hopper (*Sogatella* sp.) | Wrinkle-lipped Free-tailed Bat (*Chaerephon plicata*) | Thailand | 20,000 MTs of insect prey yearly | N.A. | Leelapaibul et al. (2005) |
| BG | Bat guano as energy resource for cave ecosystems | For example, gray bats (*Myotis grisescens*) | North America | Energy source | N.A. | Fenolio et al. (2006) |

PC, pest control; BG, bat guano.

services of bats at risk and, thus, may cause large economic, environmental, or human health repercussions. Chemical pest control could function as an alternative technological approach to natural pest control (Cleveland et al., 2006); yet more than 500 insect and mite pest species have already developed resistance against insecticides. In addition, chemicals that act in a nonspecific way toward a variety of animal species may not only affect insects but also vertebrates, including humans. Thus, insectivorous bats can help to reduce the use of chemical pesticides in forestry or agricultural ecosystems (e.g., Cleveland et al., 2006). Moreover, bats can help to keep insect populations below the so-called economic injury level (e.g., Federico et al., 2008; Leelapaibul et al., 2005). The term "economic injury level" depicts the threshold value of lowest pest population density that causes significant economic damage to a crop yield (Pedigo et al., 1986; Stern et al., 1959). Yet, the large-scale application of toxins as well as the use of transgenic crops may lead to a decline in insect populations that may become too small to attract bats to feed on them (Federico et al., 2008). Federico et al. (2008) suggested that bats, together with pesticide application, could have important long-term economic benefits for crop producers. Insects that survive transgenic crops or pesticide toxins may either be resistant or survive by luck. Partially intoxicated insects are possibly more prone to bat predation when their natural behavior is unfavorably influenced. If insectivorous bats feed on these and potentially resistant insects, pesticide resistance may develop and spread at a slower rate in insect populations than without additional bat insectivory (Federico et al., 2008). Therefore, bats may have an impact on crop profitability even when insect populations are low.

Insectivorous bats eat a large variety of insects such as Homoptera, Lepidoptera, Hemiptera, Coleoptera, Diptera, Hymenoptera, Orthoptera, and Psocoptera (e.g., Lee and McCracken, 2005; Leelapaibul et al., 2005). Some of these comprise the most destructive and costly forest and agricultural pests, such as the cabbage looper (*Trichoplusia ni*, Lepidoptera), the army worm (*Spodoptera frugiperda*, Lepidoptera), the tobacco budworm (*Heliothis virescens*, Lepidoptera), the corn earworm or cotton-bollworm (*Helicoverpa zea*, Lepidoptera), and the white-backed or brown plant hopper (*Sogatella* sp., Hemiptera), as well as many Dipterans. However, not all insect species consumed by bats can be considered as pests. Thus, it is probably safe to say that bats have a potential negative effect on insect populations.

On a daily basis, bats are able to consume insect masses equal to about two-thirds of their body mass, especially during pregnancy and lactation when energy requirements increase drastically (Barclay, 1994; Williams-Guillen et al., 2008). Since juvenile bats cannot be weaned before they are almost fully grown, female bats have to build up almost all of the juvenile body via milk (e.g., Korine et al., 2004; Studier and Kunz, 1995). Moreover, most female bats continue to nurse their young after fledging (Jones, 1967).

So far, most of the studies on the effect of bats as pest control have focused on the North American continent, where, for example, Brazilian free-tailed bats, *T. brasiliensis*, offer a large-scale service as natural pest controllers for the agricultural industry. For cotton farmers and related industry, Cleveland et al. (2006) estimated the economic value of pest control services provided by Brazilian free-tailed bats in south-central Texas as 741,000 US$ (range of estimates: 121,000–1,725,000 US$), which is substantial considering that the annual harvest of agricultural goods amounts 4.4–6.4 million US$ each year for the same area. More recently, Betke et al. (2008) recalculated the annual monetary savings of farmers having to use fewer pesticides due to the consumption of pest insects by bats. Based on more recent estimates of populations of Brazilian free-tailed bats (4–6 million as compared for six major caves in the USA), the agricultural industry of southwestern USA saves about 500,000 US$ each year due to the ESs provided by *Tadarida* bats. However, the populations of *T. brasiliensis* have dropped since 1957 when about 54 million bats were estimated to inhabit the same caves. Thus, ESs provided by *Tadarida* bats are already on the decline in North America, and conservation efforts have to be installed to preserve the regulative function of these insectivorous bats.

In a second example from the North American continent, Whitaker (1995) estimated the total insect consumption of an average maternity colony of 150 big brown bats (*Eptesicus fuscus*) based on their foraging days per year, dietary composition, and the number of insects they found in guano pellets. According to this assumptions, colonies of big brown bats may consume up to 600,000 cucumber beetles (*Diabrotica* sp., *Acalymma* sp., Coleoptera), 194,000 june bugs (*Phyllophaga* sp., Coleoptera), 335,000 stink bugs (Heteroptera), and 158,000 leafhoppers (Cicadellidae, Homoptera) during each growing season (Whitaker, 1995). Based on this chapter, Kunz et al. (2011) extrapolated that the consumption of large numbers of adult cucumber beetles protects American farmers from about 33 million larvae, which could have a devastating effect on cultivated cucumber and cause million US$ losses to North American farmers. For 2010, Boyles et al. (2011) estimated that ESs provided by insectivorous bats to the US agricultural industry amount about 22.9 billion US$ (range 3.7–53 billion US$, annually). Consequently, more and more homeowners and farmers are requesting detailed information about how to attract bats to their land (California Agriculture, 1998). Optimal management of artificial roosts for insectivorous bats may promote bats as key actors in integrated pest control.

Insectivorous bats have not only a local effect but also a large-scale effect on pest insects. This is because many bat species, such as *T. brasiliensis*, fly up to an altitude of 3000 m above ground and cover distances of more than 100 km each night (Davis et al., 1962; McCracken et al., 2008; Williams et al., 1973). Many bats such as *Tadarida* and *Chaerephon* feed on high-altitude windborne insects (Fenton and Griffin, 1997; Leelapaibul et al., 2005; McCracken, 1996). The large feeding ranges of these and other bats, in conjunction with a long distance aerial drift of some insect pests, are the main reasons why bats may influence insect populations beyond their actual home range (McCracken et al., 2008). For example, the seasonal high-altitude migration of billions of moths emerging from agricultural crops of northern Mexico and southern Texas toward the north has been well documented (e.g., Wolf et al., 1990). Several studies observed a striking coincidence between the seasonal availability of migratory moths and the relative contribution of moths to bat diets (e.g., Lee and McCracken, 2002, 2005; Whitaker et al., 1996). Cleveland et al. (2006) reported that the activity of *T. brasiliensis* is closely linked in time and space to the major emergence of bollworm moths (*H. zea*), and recent radar monitoring of bats revealed that bats fly actively toward swarming insects over relatively large distances, which implies that they search for and specifically hunt for air-borne insects on a large geographical scale (McCracken et al., 2008).

The ESs of insectivorous bats may apply to all regions where bats live. In Thailand, the wrinkle-lipped free-tailed bat *C. plicata* is a known control agent for plant hoppers, *Sogatella* sp. (Hemiptera), a biological pest insect of rice (Leelapaibul et al., 2005). The 2.6 million *Chaerephon* bats inhabiting the Khao Chong Pran Cave consume at least 17.5 metric tons (MTs) of insects every day (Leelapaibul et al., 2005). For the whole of Thailand, the total mass of insects consumed by *C. plicata* amounts to 20,000 MTs per year. Populations of plant hoppers boom every 8–10 years, often with drastic consequences for local settlements and national economies. In 1990, for example, brown plant hoppers (*Nilaparvata lugens*, Hemiptera) caused damages of up to 200–240 million US$ to the agricultural industry of Thailand (Vungsilabutr, 2001).

Even though current information on ESs of insectivorous bats is missing for ecosystems in Europe and Africa, bats can be expected to have a positive impact on crop production or silviculture in these regions as well. For example, Böhm et al. (2011) recently reported on the impact of vertebrate exclosures on leaf damage in temperate oak (*Quercus robur*) forests in Germany. They showed that the top-down control of herbivory by bats and birds significantly reduced leaf damage and biomass loss in the oak forest canopy. Other studies and this highlight that bats are not only essential members of intact ecosystems but also central agents for functional diversity, which is central to the maintenance of ESs of ecosystems (Tilman et al., 1997). This need for functional diversity is compromised by the decline of bat species in many areas with often unforeseen consequences for pest insect populations in particular and the agricultural economy in general. Down to the present day, the overall impact of many insectivorous bat species on arthropod pests remains unclear. Recently, DNA-barcoding analyses of feces collected from little brown bats (*Myotis lucifugus*, Clare et al., 2011) and eastern red bats (*Lasiurus borealis*, Clare et al., 2009) did not reveal a pronounced consumption of potential arthropod pest species. Thus, until more evidence exists for pest control by species other than *Tadarida* sp., *Eptesicus* sp., or *Charephon* sp., we have to remain careful not to overstate the global impact of bats on pest insects. There is no doubt that bats need protection whether or not they consume arthropod pest species, but one has to be careful to extrapolate ESs across all insectivorous bats.

## B. Control of Arthropod Disease Vectors

Many airborne and nocturnal insects are vectors for pathogens that are relevant for human health, such as the *Anopheles* mosquito which is the vector for the malaria pathogen *Plasmodium*. By feeding on these insects, insectivorous bats may have an indirect impact on disease outbreaks and pathogen prevalence. Up to now, we lack empirical data on how important bats are for controlling insect borne diseases. Around 1900, Charles Campbell established the first artificial daytime roosts for bats with the ultimate goal of controlling mosquito populations in the San Antonio area in Texas. Unfortunately, it is unclear to what extent the disappearance of malaria from this area was supported by his efforts (Murphy, 1989). Until now, only a few studies have demonstrated that free-ranging bats are foraging on mosquitoes. Reiskind and Wund (2009) were among the first to show quantitatively that bats affect populations of mosquitoes (*Culex* spp.). Based on enclosure experiments, the authors demonstrated that aerial predation by northern long-eared bats (*Myotis septentrionalis*) reduced the nightly oviposition rate of mosquitoes by 32%. In a recent study, Bohmann et al. (2011) documented mosquito consumption of the African free-tailed bat *Chaerephon pumilus* using DNA barcoding of fecal samples. The authors revealed that 16 of the 59 sampled fecal pellets (27.1%) contained members of the family Culicidae (mosquitoes). Given these numbers, an impact of bats on disease-carrying mosquito populations seems plausible under certain conditions. Nonetheless, further dietary investigation, such as DNA barcoding of bat feces, is needed to document the large-scale consumption of mosquitoes by bats under natural conditions.

### C. Bat Guano

Cave-roosting bats produce large amounts of guano, and this has been extensively used in many countries as natural fertilizer. In historical times, bat guano was also used for gunpowder production, a highly profitable business during those days. Between 1903 and 1923, companies removed at least 100,000 tons of bat guano from Carlsbad Caverns in New Mexico and used this guano mostly for fruit production in California. Some bat guano is still sold commercially as a natural fertilizer. In the Bracken Cave in Texas, which is now protected by Bat Conservation International, around 85 tons were exploited annually until the late 1980s. The provisioning of guano is not restricted to insectivorous bats, for example, in Thailand or Egypt; guano of fruit bats has been and is still harvested and used as fertilizer. The millions of bacteria and microorganisms in bat guano have not yet been thoroughly monitored for potential use in biotechnologies. However, Tanskul et al. (2009) recently isolated an alkaline proteinase from *Bacillus* sp. that originated from bat feces collected at the Wat Suwankuha cave in Thailand. The authors presumed that this alkaline serine proteinase has a potential use as a detergent additive, due to its high stability against detergents. Thus, some of the unknown guano microorganisms could produce enzymes useful in detoxifying industrial wastes, producing natural insecticides, improving detergents, or developing new antibiotics. Numerous of these organisms may be endemic, and for most of them, we still lack detailed information about their biotechnological potential. Bat guano is also the main energy source in many temperate cave ecosystems, influencing trophic dynamics and community structure of cave specialists (Fenolio et al., 2006; Gnaspini and Trajano, 2000; Harris, 1970; Poulson, 1972). The endemic cave-living salamander *Eurycea spelaea* (Caudata), like many other species, depends directly on the food web based on bat guano (Fenolio et al., 2006). Overexploitation of bat guano or a decline in bat populations may put these cave specialists at risk of local extinction.

In summary, insectivorous bats provide pest control services that can be seen as a direct contribution of bats to agricultural production since they substitute agricultural machines, labor, pesticides, or fertilizers (Cleveland et al., 2006). The monetary benefits for human society may be large but have yet to be quantified.

## III. Plant-Visiting Bats

Mammals are efficient pollinators and seed dispersers and thus have a positive effect on plant reproduction in many habitats worldwide (Fleming and Sosa, 1994). In total, about 250 species of New World leaf-nosed bats

(Phyllostomidae) and Old World fruit bats (Pteropodidae) utilize more than 500 species of trees and shrubs in tropical and subtropical ecosystems from which they consume fruits or nectar (Thomas, 1991).

## A. Seed Dispersal

By dispersing seeds from pioneer plants, neotropical bats promote the quick reforestation of areas such as large cleared areas within forests (Fig. 1C and D; e.g., Gorchov et al., 1993). For example, trees and shrubs of the genera *Cecropia*, *Ficus*, *Vismia*, and others constitute the largest plant biomass in the early years of forest succession (e.g., Mesquita et al., 2001; Saldarriaga et al., 1988). Seeds of these species are predominantly dispersed by phyllostomid or smaller pteropodid bats. Many bats swallow entire fruits, which facilitates the dispersal of several seeds per meal, sometimes even hundreds as in the case of fig fruits (Hodgkison et al., 2003). A fast transit time of about 20–30 min through the bats' digestive tract in combination with the relatively large feeding ranges of bats ensures that seeds are carried far away from their parent plant, which prevents intense seed predation by insects or mammals on the forest floor underneath the parent plant (Heer et al., 2010). The dispersal of seeds by bats is distributed more evenly and over larger areas than that mediated by frugivorous birds. This difference is related to the specific foraging and feeding behavior of bats and birds. While birds often deposit the majority of seeds from a perched position under canopy trees, bats often defecate seeds while flying and thereby rather scatter the seeds along their flight paths (Charles-Dominique, 1986; Galindo-González et al., 2000; Gorchov et al., 1993; Thomas et al., 1988).

The influence of fruit-eating bats on the early succession of forests in the New and Old World was recently reviewed by Muscarella and Fleming (2007). In the Old World, the impact of bats on forest regeneration is thought to be less pronounced than in the New World (Muscarella and Fleming, 2007). This discrepancy stems from the fact that Old World pteropodids consume fruits mostly from trees or shrubs that are considered climax plant species, whereas New World phyllostomids consume fruits from both climax and pioneer plant species.

The important ESs provided by Old and New World bats become particularly apparent when fruit, vegetable, or timber products are considered (Fig. 3 and Table II). Fujita and Tuttle (1991) reported that 23% of the total plant-related products in the Old World tropics depend on services provided by flying foxes. In Africa, about 34% of economically relevant timber species belong to families whose fruits are dispersed by pteropodid bats (Muscarella and Fleming, 2007). Even if tropical timber exploitation is declining, bats will have a crucial impact on the regeneration of populations

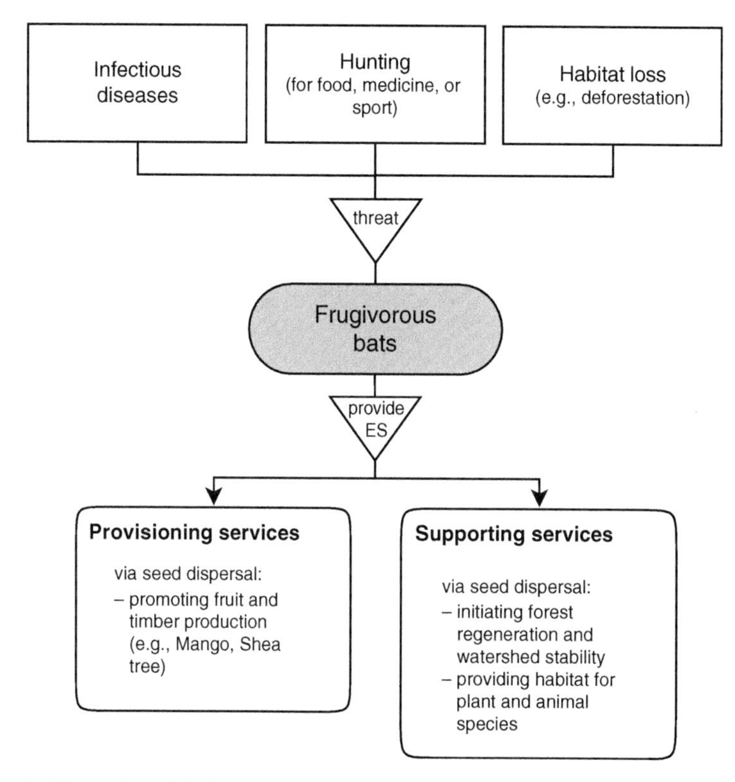


FIG. 3. Illustration of the key ESs provided by frugivorous bats including the major threats that put these services at risk.

of timber tree species in natural forests and the preservation of ecosystems. Indeed, entire ecosystems depend largely on bat-mediated pollination or seed dispersal. In particular, bats play a crucial role for plants on island ecosystems, where the paucity of alternative efficient pollinators and seed dispersers results in a strong dependency of plants on bats for seed dispersal. Flying foxes are the only vertebrate pollinators and seed dispersers for plants on many islands of the Indio-Pacific Ocean (Cox et al., 1991), which contrasts with tropical mainland areas, where monkeys, opossums, rodents, or to a large extent birds usually occupy this niche as well (Thomas, 1991). Bat-dependent plants support the animal diversity of islands by providing food and habitat.

The relevance of bats as seed dispersers is well documented for Mango trees, *Mangifera indica*, and Shea trees, *Butyrospermum parkii*. The seeds of both species are dispersed by various pteropodid bat species (Fujita and Tuttle, 1991). The Shea Belt region in sub-Saharan Africa is known for the

TABLE II

Examples of Ecosystem Services Provided by Fruit-Eating and Nectar-Feeding Bats, Their Uses, and Occurrences with the Yearly Estimated Economic Value of Their Respective Services

| Service | Product | Service provider | Land or reference | Common usage | Approx. yearly value | Source |
|---|---|---|---|---|---|---|
| P | Durian fruit (*Durio* spp.) | Dawn Bat (*Eonycteris spelaea*) | World Market | Fruit, vegetable | 1 billion US$ | Biz Dimension Co Ltd. (2011) |
| P | Agave (*Agave* sp.) | Curaçaoan long-nosed bat (*Leptonycteris curasoae*), Mexican long-tongued bat (*Choeronycteris mexicana*) | Mexico | Tequila, alcoholic beverage | Multimillion US$ market | López-López et al. (2010) |
| SD | Shea tree (*Vitellaria paradoxa*) | Various pteropodid bat species | Africa, Shea Belt region, total production | Charcoal, timber, fuel, oil | 360 million US$ | Lovett (2005) |
| SD | 34% of economically relevant timber species | Various pteropodid bat species | Africa | e.g., timber, food, cosmetics | N.A. | Muscarella and Fleming (2007) |
| SD | 23% of the total plant-related products | Various pteropodid bat species | Old World Tropics | e.g., timber, food, cosmetics | N.A. | Fujita and Tuttle (1991) |
| SD | Pioneer plants (e.g., *Cecropia*, *Ficus*, *Vismia*) | Various Phyllostomid species (e.g., *Carollia*) | Neotropical Forest | Promote quick reforestation | N.A. | Gorchov et al. (1993) |

P, pollination; SD, seed dispersal.

production of shea butter that originates from the fruit kernel. In Western Africa, the dominant variety of Shea tree, *Vitellaria paradoxa*, has a yearly production volume of 800,000 MTs. In 2007, this volume accounts for an estimated monetary value of 115 million US$ for the top seven producing countries (FAO export statistics, 2007). Lovett (2005) estimated the total production of this tree species, based on 500 million productive trees, to be 2.5 million MTs with a value of 360 million US$. This economic value highlights the significance of bats as key seed dispersers in many ecosystems.

An overall calculation of ESs by New and Old World seed-dispersing bats is still lacking. The economic value of goods originating from bat-dispersed plants either grows or declines as local and global ecological conditions change. These global trends are influenced by habitat fragmentation, urbanization, and changes in land use or simply by fluctuating market structures that are driven by economic demands of the human society. These changing conditions make it necessary to recalculate the monetary benefits provided by bats at regular intervals.

### B. Pollination

Many plant species require animal-mediated pollination for a successful seed set. The floral syndrome of pollination by bats is well documented and has been reviewed extensively by several authors (e.g., Baker, 1973; Dobat and Peikert-Holle, 1985). The flowers of bat-pollinated plants are usually large, open at dusk, provide large nectar volumes, and can be recognized by a white- or pale-yellow color and a musky scent (Dobat and Peikert-Holle, 1985). The different genera of nectarivorous bats are generally characterized by long and narrow muzzles, delicate mandibles, reduced dentition, and long tongues (Fig. 1E and F; Freeman, 1995).

Human society benefits from the products of many bat-pollinated plants both in a direct way and in many indirect ways. The successful reproduction of plants pollinated by bats may provide other ESs such as soil stabilization to maintain water catchment areas, or simply habitat for other plant and animal species. Several nectar-feeding and fruit-eating bats in the Old and New World tropics and subtropics are known to provide these valuable ESs (Fig. 4 and Table II). Bat pollination (Chiropterophily) can be found in numerous plant families, but it has mostly been studied in Bombacaceae, Passifloraceae, Mimosaceae, Caesalpiniaceae (reviewed in Bawa, 1990). Chiropterophily is restricted to the tropical and subtropical zone within 30° latitudinal from the equator (Dobat and Peikert-Holle, 1985). Many obligate nectarivorous bat species depend on the continuous phenology of flowering plants throughout the year or are known to migrate seasonally when flowering plants are not available (Fleming et al., 1993).

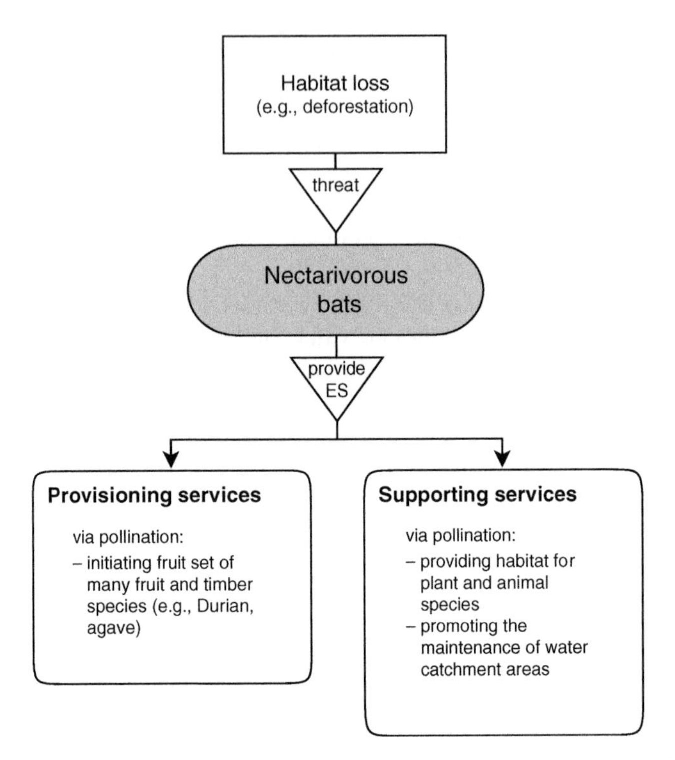


FIG. 4. Illustration of the key ESs provided by nectarivorous bats including the major threats that put these services at risk.

Flying foxes and phyllostomid bats are capable of carrying large amounts of pollen over long distances, which is one of the criteria for efficient pollination (e.g., Law and Lean, 1999). Consequently, bats are able to connect remote plants and forest fragments by covering longer distances than terrestrial animals. In the Mesoamericas, about 60 species of agave plants are pollinated by the lesser long-nosed bat *Leptonycteris curasoae* and the Mexican long-tongued bat *Choeronycteris mexicana* (Ducummon, 2000). The juices of agave plants are fermented and distilled to the alcoholic beverage tequila. In 2008, a total of 227 million liters of tequila were produced in Mexico (López-López et al., 2010), thus linking the ESs of bats to a multimillion dollar market.

In contrast to many islands where flying foxes are the only vertebrate pollinators and seed dispersers, most of the pollination provided by bats in mainland areas should be considered as supporting ESs because bat-pollinated plants are often pollinated by other taxa such as insects, birds, and other small mammals as well. In a recent review, the economic value of the pollinating service of all animal taxa was estimated to vary between 112 and 200 billion US$ globally (Kremen et al., 2007). The specific economic value of pollination services by bats has not yet been quantified. Currently, we know of only a few examples where plant species depend entirely on a single pollinating bat species. In addition, the exact number of plant species pollinated by bats is yet unknown. ESs of bats become most apparent when bat-pollinated plants have a high economic value. The Durian Fruit *Durio* spp., also known as King of Fruits, is a popular fruit from Southeast Asia that can be either consumed fresh or frozen, or processed to chips, powder, paste, cake, or other products. The dawn bat *Eonycteris spelaea* is the only effective pollinator of Durian trees, apart from the honey bee *Apis dorsata* (Bumrungsri et al., 2009). Consequently, this bat species is responsible for the successful cultivation and maintenance of the semi-wild *Durio zibethinus* trees. Durian fruits play an increasing role in the Asian agricultural economy. In 2000, Thailand was the largest producer of Durian fruits with a production volume of more than 826,000 tons and a market value of about 575 million US$ (Biz Dimension Co Ltd., 2011). The world production of Durian amounted 1,400,000 MTs in the same year. We estimate a hypothetical world market value of nearly 1 billion US$ for the ESs of *Eo. spelaea*, when extrapolating the monetary value of Durian fruits in Thailand to other countries. In other markets such as Indonesia, the production and export volumes of Durian are still rising due to the fruit's increasing popularity. Thus, our estimates are probably conservative. Overall, pollination of Durian trees by bats is essential for a significant sector of the Southeast Asian food economy. Therefore, conservation programs should support Asian bats so that they can continue to provide ESs by pollination.

Threats such as habitat fragmentation are problematic for many bat species and consequently pollinating services of bats are at risk in many countries. A lack of pollinators in disturbed forest patches impedes the exchange of pollen between forest fragments and may consequently reduce the reproductive output of plant species (Quesada et al., 2004).

## IV. What Can Be Done to Preserve and Promote Ecosystem Services by Bats?

Despite their crucial role as ES providers, bats are under severe threats. For example, bat populations are rapidly declining in the Old World due to hunting for sport, adventure or meat, to control them as crop pests, or to use them for medical purposes (e.g., Epstein et al., 2009; Fujita and Tuttle, 1991). In addition to hunting and poaching, global bat populations are also suffering from indirect poisoning through the overuse of agricultural pesticides, from habitat fragmentation, from changes in house construction and the related impact on bat roosts, and from recently emerging infectious diseases such as white-nose syndrome (WNS). In the northeastern USA, WNS is responsible for high rates of bat mortality (>75%) in hibernacula (Blehert et al., 2008; Frick et al., 2010b). Many of these diminishing populations may be key contributors to ESs (e.g., *M. lucifugus* Frick et al., 2010a). In a recent article, Boyles et al. (2011) estimated that between 660 and 1320 MTs of insects are no longer being consumed each year, due to the 1 million bats that have already died of WNS. Thus, a continuation of the epidemic spread of WNS throughout the USA will probably produce significant detrimental effects on the ecology of the whole continent. An improved understanding of the conservation status of bat species is therefore pivotal (Williams-Guillen et al., 2008). The current threat of massive population declines in many species around the globe has raised awareness of their importance.

A significant drop in population sizes may put the ESs of bats at risk even when bats may still persist at low densities. For example, ESs of island flying foxes have been observed to collapse when critical thresholds in population numbers have been reached (McConkey and Drake, 2006). Sudden and drastic population declines are particularly problematic for bats since most species produce only one or two offspring each year (Barclay and Harder, 2003; Wilkinson and South, 2002). Such a low reproductive rate delays a recovery from population collapses substantially, and regeneration of disturbed ecosystems may require more time than anticipated (McConkey and Drake, 2006). Thus, it is vital to respond quickly before bat populations become functionally extinct in ecosystems.

There are many practical ways to respond effectively to decreasing bat population sizes and declining ESs of bats. Ecologically important seed dispersers can be attracted to fragmented habitats by offering natural or man-made forest fragments that are planted with the goal of facilitating reforestation in the surrounding area (Guevara et al., 2004; Zahawi and Augspurger, 2006). The survival of plants in such forest fragments could promote the input of plant species into species-depleted areas, leading ultimately to an increase in the biological diversity of larger areas (Muscarella and Fleming, 2007). Kelm et al. (2008) suggested the use of artificial bat cavities to attract and keep important seed dispersers at disturbed areas. Such artificial roosts encourage bats to fly over areas which have been cleared of trees and thus promote seed dispersal into fragmented habitats. Efficient protection of daytime roosts of Old World flying foxes is also of great importance. Although many species are protected in most countries, bats are still vulnerable to illegal hunting, habitat loss, and also invasive species (Allen-Wardell et al., 1998; Epstein et al., 2009; Nyhagen et al., 2004). In the Eastern Amazon lowlands, mineral licks support populations of frugivorous animals, including over 25 bat species (e.g., Voigt et al., 2007). Protection of such hot spots of bat activity may lead to higher abundances of tropical fruit-eating bats, which are essential for forest regeneration (Montenegro, 2004). In an attempt to prevent the decline of insectivorous bat species, protection of cave ecosystems and reduction in the use of agricultural pesticides have also been suggested. At the same time, emerging diseases such as WNS should be studied, and strategies developed to mitigate the negative effects of diseases on bat populations. Most importantly, local people have to be educated to value the importance of bats as providers of ESs. Education and the provisioning of scientifically sound information are crucial in order to maintain and increase awareness of the importance of bats for the global ecosystem.

## Acknowledgments

The authors are grateful to the German Research Foundation for financial assistance (VO 890/15). The authors are also indebted to Marion L. East for valuable comments on an earlier version of this chapter.

## References

Allen, G.M., 1962. Bats: Biology, Behavior, and Folklore. Harvard University Press, Cambridge, USA.

Allen-Wardell, G., Bernhardt, P., Bitner, R., Burquez, A., Buchmann, S., Cane, J., et al., 1998. The potential consequences of pollinator declines on the conservation of biodiversity and stability of food crop yields. Conserv. Biol. 12, 8–17.

Baker, H.G., 1973. Evolutionary relationships between flowering plants and animals in American and African tropical forests. In: Meggers, B.J., Ayensu, E.S., Duckworth, D. (Eds.), Tropical Forest Ecosystems in Africa and South America: A Comparative Review. Smithsonian Institution Press, Washington, DC, pp. 145–159.

Barclay, R.M.R., 1994. Constraints on reproduction by flying vertebrates: energy and calcium. Am. Nat. 144, 1021. doi:10.1086/285723.

Barclay, R.M.R., Harder, L.M., 2003. Life histories of bats: life in the slow lane. In: Kunz, T.H., Fenton, M.B. (Eds.), Bat Ecology. University of Chicago Press, Chicago, IL, pp. 209–253.

Bass, M.S., Finer, M., Jenkins, C.N., Kreft, H., Cisneros-Heredia, D.F., McCracken, S.F., et al., 2010. Global conservation significance of Ecuador's Yasuní National Park. PLoS ONE 5, e8767. doi:10.1371/journal.pone.0008767.

Bawa, K.S., 1990. Plant-pollinator interactions in tropical rain forests. Annu. Rev. Ecol. Syst. 21, 399–422.

Betke, M., Hirsh, D.E., Makris, N.C., McCracken, G.F., Procopio, M., Hristov, N.I., et al., 2008. Thermal imaging reveals significantly smaller Brazilian free-tailed bat colonies than previously estimated. J. Mammal. 89, 18–24.

Biz Dimension Co. Ltd., 2011. Thailand. http://www.foodmarketexchange.com/datacenter/product/fruit/durian/dc_pi_ft_durain_snap.htm.

Blehert, D.S., Hicks, A.C., Behr, M., Meteyer, C.U., Berlowski-Zier, B.M., Buckles, E.L., et al., 2008. Bat white-nose syndrome: an emerging fungal pathogen? Science 323, 227.

Böhm, S.M., Wells, K., Kalko, E.K.V., 2011. Top-down control of herbivory by birds and bats in the canopy of temperate Broad-Leaved Oaks (*Quercus robur*). PLoS ONE 6 (4), e17857. doi:10.1371/journal.pone.0017857.

Bohmann, K., Monadjem, A., Lehmkuhl Noer, C., Rasmussen, M., Zeale, M.R.K., Clare, E., et al., 2011. Molecular diet analysis of two African Free-Tailed bats (Molossidae) using high throughput sequencing. PLoS ONE 6 (6), e21441. doi:10.1371/journal.pone.0021441.

Boyles, J.G., Cryan, P.M., McCracken, G.F., Kunz, T.H., 2011. Economic importance of bats in agriculture. Science 332, 41–42.

Bumrungsri, S., Sripaoraya, E., Chongsiri, T., Sridith, K., Racey, P.A., 2009. The pollination ecology of durian (*Durio zibethinus*, Bombacaceae) in southern Thailand. J. Trop. Ecol. 25, 85–92. doi:10.1017/S0266467408005531.

California Agriculture, 1998. Bats can pack a punch in pest control. California Agric. 52, 6–7. doi:10.3733/ca.v052n01p6b.

Charles-Dominique, P., 1986. Inter-relations between frugivorous vertebrates and pioneer plants: Cecropia, birds and bats in French Guyana. In: Estrada, A., Fleming, T.H. (Eds.), Frugivores and Seed Dispersal. Dr. W. Junk, Dordrecht, The Netherlands, pp. 119–135.

Clare, E., Fraser, E., Braid, H., Fenton, M.B., Hebert, P., 2009. Species on the menu of a generalist predator, the eastern red bat (Lasiurus borealis): using a molecular approach to detect arthropod prey. Mol. Ecol. 18, 2532–2542.

Clare, E., Barber, B., Sweeney, B., Hebert, P., Fenton, B., 2011. Eating local: influences of habitat on the diet of little brown bats (Myotis lucifugus). Mol. Ecol. 20, 1772–1780.

Cleveland, C.J., Betke, M., Federico, P., Frank, J.D., Hallam, T.G., Horn, J., et al., 2006. Economic value of the pest control service provided by Brazilian free-tailed bats in south-central Texas. Front. Ecol. Environ. 4, 238–243. doi:10.1890/1540-9295(2006)004[0238:EVOTPC]2.0.CO;2.

Cox, P.A., Elmqvist, T., Pierson, E.D., Rainey, W.E., 1991. Flying foxes as strong interactors in South Pacific island ecosystems: a conservation hypothesis. Conserv. Biol. 5, 448–454. doi:10.1111/j.1523-1739.1991.tb00351.x.

Daily, G.C., 1997. Introduction: what are ecosystem services? In: Daily, G.C. (Ed.), Nature's Services: Societal Dependence on Natural Ecosystems. Island Press, Washington, DC, pp. 1–10.

Davis, R.B., Herreid, C.F., Short, H.L., 1962. Mexican free-tailed bats in Texas. Ecol. Monogr. 32, 311–346.

Dobat, K., Peikert-Holle, T., 1985. Blüten und Fledermäuse. Bestäubung durch Fledermäuse und Flughunde (Chiropterophilie). Waldemar Kramer, Frankfurt am Main, Germany.

Ducummon, S.L., 2000. Ecological and Economic Importance of Bats. Bat Conservation International, Austin, TX.

Epstein, J.H., Olival, K.J., Pulliam, J.R.C., Smith, C., Westrum, J., Hughes, T., et al., 2009. *Pteropus vampyrus*, a hunted migratory species with a multinational home-range and a need for regional management. J. Appl. Ecol. 46, 991–1002.

FAO export statistics, 2007. Food and Agriculture Organization of the United Nations. http://faostat.fao.org.

Federico, P., Hallam, T.G., McCracken, G.F., Purucker, S.T., Grant, W.E., Correa-Sandoval, A.N., et al., 2008. Brazilian free-tailed bats as insect pest regulators in transgenic and conventional cotton crops. Ecol. Appl. 18, 826–837. doi:10.1890/07-0556.1.

Fenolio, D.B., Graening, G.O., Collier, B.A., Stout, J.F., 2006. Coprophagy in a cave-adapted salamander; the importance of bat guano examined through nutritional and stable isotope analyses. Proc. Biol. Sci. 273, 439–443.

Fenton, M.F., Griffin, D.R., 1997. High altitude pursuit of insects by echolocating bats. J. Mammal. 78, 247–250.

Fine, P.V.A., Mesones, I., Coley, P.D., 2004. Herbivores promote habitat specialization by trees in Amazonian forests. Science 305, 663.

Fleming, T.H., Sosa, V.J., 1994. Effects of nectarivorous and frugivorous mammals on reproductive success of plants. J. Mammal. 75, 845–848.

Fleming, T.H., Nunez, R.A., Sternberg, L.S.L., 1993. Seasonal changes in the diets of migrant and non-migrant nectarivorous bats as revealed by carbon stable isotope analysis. Oecologia 94, 72–75.

Freeman, P.W., 1995. Nectarivorous feeding mechanisms in bats. Biol. J. Linn. Soc. 56, 439–463.

Frick, W.F., Reynolds, D.S., Kunz, T.H., 2010a. Influence of climate and reproductive timing on demography of little brown myotis *Myotis lucifugus*. J. Anim. Ecol. 79, 128–136.

Frick, W.F., Pollock, J.F., Hicks, A., Langwig, K., Reynolds, D.S., Turner, G., et al., 2010b. An emerging disease causes regional population collapse of a common North American bat species. Science 328, 679–682.

Fujita, M.S., Tuttle, M.D., 1991. Flying Foxes (Chiroptera: Pteropodidae): threatened animals of key ecological and economic importance. Conserv. Biol. 5, 455–463. doi:10.1111/j.1523-1739.1991.tb00352.x.

Galindo-González, J., Guevara, S., Sosa, V.J., 2000. Bat- and bird-generated seed rains at isolated trees in pastures in a tropical rainforest. Conserv. Biol. 14, 1693–1703. doi:10.1111/j.1523-1739.2000.99072.x.

Gnaspini, P., Trajano, E., 2000. Guano communities in tropical caves. In: Wilkens, H., Culver, D.C., Humphreys, W.F. (Eds.), Ecosystems of the World—Subterranean Ecosystems. Elsevier, Amsterdam, Netherlands, pp. 251–268.

Gorchov, D.L., Cornejo, F., Ascorra, C., Jaramillo, M., 1993. The role of seed dispersal in the natural regeneration of rain forest after strip-cutting in the Peruvian Amazon. Vegetatio 107 (108), 339–349.

Greenberg, R., Bichier, P., Cruz Angon, A., Macvean, C., Pererz, R., Cano, E., 2000. The impact of avian insectivory on arthropods and leaf damage in some Guatemalan coffee plantations. Ecology 81 (6), 1750–1755.

Greenleaf, S.S., Kremen, C., 2006. Wild bee species increase tomato production but respond differently to surrounding land use in Northern California. Biol. Conserv. 133, 81–87. doi:10.1016/j.biocon.2006.05.025.

Guevara, S., Laborde, J., Sanchez-Rios, G., 2004. Rain forest regeneration beneath the canopy of fig trees isolated in pastures of Los Tuxtlas, Mexico. Biotropica 36, 99–108.

Harris, J.A., 1970. Bat-guano cave environment. Science 169, 1342–1343.

Heer, K., Albrecht, L., Kalko, E.K.V., 2010. Effects of ingestion by neotropical bats on germination parameters of native free-standing and strangler Figs (*Ficus* sp., Moraceae). Oecologia 163, 425–435.

Hillman, A., 1999. The study on wrinkled-lipped free-tailed bats (*Tadarida plicata*) at Khao Chong Pran Non-hunting Area, Ratchaburi Province. Roy. For. Dep. J. 1, 72–83.

Hodgkison, R., Balding, S.T., Zubaid, A., Kunz, T.H., 2003. Fruit Bats (Chiroptera: Pteropodidae) as seed dispersers and pollinators in a lowland Malaysian Rainforest. Biotropica 35, 491–502.

Holmes, R.T., Schultz, J.C., Nothnagle, P., 1979. Bird predation on forest insects: an exclosure experiment. Science 206, 462.

Jones, C., 1967. Growth, development, and wing loading in the evening bat, *Nycticeius humeralis* (Rafinesque). J. Mammal. 48, 1–19.

Kalka, M.B., Smith, A.R., Kalko, E.K.V., 2008. Bats limit arthropods and herbivory in a tropical forest. Science 320, 71. doi:10.1126/science.1153352.

Kelm, D.H., Wiesner, K.R., von Helversen, O., 2008. Effects of artificial roosts for frugivorous bats on seed dispersal in a neotropical forest pasture mosaic. Conserv. Biol. 22, 733–741. doi:10.1111/j.1523-1739.2008.00925.x.

Klein, A.M., Vaissière, B.E., Cane, J.H., Steffan-Dewenter, I., Cunningham, S.A., Kremen, C., et al., 2007. Importance of pollinators in changing landscapes for world crops. Proc. R. Soc. 274, 303–313. doi:10.1098/rspb.2006.3721.

Korine, C., Speakman, J., Arad, Z., 2004. Reproductive energetics of captive and free-ranging Egyptian fruit bats (*Rousettus aegyptiacus*). Ecology 85, 220–230.

Kremen, C., Williams, N.M., Aizen, M.A., Gemmill-Herren, B., LeBuhn, G., Minckley, R., et al., 2007. Pollination and other ecosystem services produced by mobile organisms: a conceptual framework for the effects of land-use change. Ecol. Lett. 10, 299–314. doi:10.1111/j.1461-0248.2007.01018.x.

Kunz, T.H., Braun de Torrez, E., Bauer, D.M., Lobova, T.A., Fleming, T.H., 2011. Ecosystem services provided by bats. In: Ostfeld, R.A., Schlesinger, W.H. (Eds.), The Year in Ecology and Conservation, 2011. Annals of the New York Academy of Sciences, Wiley-Blackwell, New York, USA, pp. 1–38.

Law, B.S., Lean, M., 1999. Common blossom bats (*Syconycteris australis*) as pollinators in fragmented Australian tropical rainforest. Biol. Conserv. 91, 201–212.

Lee, Y.F., McCracken, G.F., 2002. Foraging activity and resource use of Brazilian free-tailed bats *Tadarida brasisliensis* (Molossidae). Ecoscience 9, 306–313.

Lee, Y.F., McCracken, G.F., 2005. Dietary variation of Brazilian free-tailed bats links to migratory populations of pest insects. J. Mammal. 86, 67–76.

Leelapaibul, W., Bumrungsri, S., Pattanawiboon, A., 2005. Diet of wrinkle-lipped free-tailed bat (*Tadarida plicata* Buchannan, 1800) in central Thailand: insectivorous bats potentially act as biological pest control agents. Acta Chiropt. 7, 111–119.

López-López, A., Davila-Vazquez, G., León-Becerril, E., Villegas-García, E., Gallardo-Valdez, J., 2010. Tequila vinasses: generation and full scale treatment processes. Rev. Environ. Sci. Biotechnol. 9, 109–116. doi:10.1007/s11157-010-9204-9.

Lovett, P.N., 2005. Shea butter industry expanding in West Africa. INFORM 16, 273–275.

McConkey, K.R., Drake, D.R., 2006. Flying foxes cease to function as seed dispersers long before they become rare. Ecology 87, 271–276. doi:10.1890/05-0386.

McCracken, G.F., 1996. Bats aloft: a study of high altitude feeding. BATS 14, 7–10.

McCracken, G.F., 2003. Estimates of population sizes in summer colonies of Brazilian free-tailed bats. In: O'Shea, T.J., Bogan, M.A. (Eds.), Monitoring Trends in Bat Populations of the United States and Territories: Problems and Prospects. Information and Technology Report USGS/BRD/ITR-2003-0003. Washington, DC, pp. 21–30.

McCracken, G.F., Gillam, E.H., Westbrook, J.K., Lee, Y., Jensen, M.L., Balsley, B.B., 2008. Brazilian free-tailed bats (*Tadarida brasiliensis*: Molossidae, Chiroptera) at high altitude: links to migratory insect populations. Integrat. Comp. Biol. 48, 107–118. doi:10.1093/icb/icn033.

McGregor, 1976McGregor, S.E., 1976. Insect pollination of cultivated crop plants. Agriculture Handbook 496. US Department of Agriculture, Washington, DC.

Mesquita, R.C.G., Ickes, K., Ganade, G., Williamson, G.B., 2001. Alternative successional pathways in the Amazon Basin. J. Ecol. 89, 528–537.

Millennium Ecosystem Assessment, 2003. Ecosystems and Human Well-being: A Framework for Assessment. Island Press, Washington, DC.

Millennium Ecosystem Assessment, 2005. Ecosystems and Human Well-being: Synthesis. Island Press, Washington, DC.

Montenegro, O.L., 2004. Natural licks as keystone resources for wildlife and people in Amazonia. PhD thesis. University of Florida, Gainesville, Florida, USA.

Murphy, M., 1989. Dr. Campbell's "Malaria-Eradicating, Guano-Producing Bat Roosts". One doctor's vision to control malaria led to a novel idea. BATS 7, 2.

Muscarella, R., Fleming, T.H., 2007. The role of frugivorous bats in tropical forest succession. Biol. Rev. 82, 573–590.

Nowak, R.M., 1994. Walker's Bats of the World. Johns Hopkins University Press, Baltimore.

Nyhagen, D.F., Turnbull, S.D., Olesen, J.M., Jones, C.G., 2004. An investigation into the role of the Mauritian flying fox, *Pteropus niger*, in forest regeneration. Biol. Conserv. 122, 491–497.

Pedigo, L.P., Hutchins, S.H., Higley, L.G., 1986. Economic injury levels in theory and practice. Ann. Rev. Entomol. 31, 341–368.

Poulson, T.L., 1972. Bat guano ecosystems. Bull. Natl. Speleol. Soc. 34, 55–59.

Quesada, M., Stoner, K.E., Lobo, J.A., Herrerías-Diego, Y., Palacios-Guevara, C., Munguía-Rosas, M.A., et al., 2004. Effects of forest fragmentation on pollinator activity and consequences for plant reproductive success and mating patterns in bat-pollinated Bombacaceous trees. Biotropica 36, 131–138.

Reiskind, M.H., Wund, M.A., 2009. Experimental assessment of the impacts of northern long-eared bats on ovipositing *Culex* (Diptera: Culicidae) mosquitoes. J. Med. Entomol. 46, 1037–1044.

Saldarriaga, J.G., West, D.C., Tharp, M.L., Uhl, C., 1988. Long-term chronosequence of forest succession in the upper Rio Negro of Columbia and Venezuela. J. Ecol. 76, 938–958.

Simmons, N.B., Voss, R.S., 1998. The mammals of Paracaou, French Guiana: a Neotropical lowland rainforest fauna. Part 1. Bats. Bull. Am. Mus. Nat. Hist. 237, 1–219.

Simmons, N.B., 2005. Order Chiroptera. In: Wilson, D.E., Reeder, D.M. (Eds.), Mammal species of the world: a taxonomic and geographic reference, Third Edition, Volume 1, John Hopkins University Press, Baltimore, USA, pp. 312–529.

Simmons, N.B., Seymour, K.L., Habersetzer, J., Gunnell, G.F., 2008. Primitive early Eocene bat from Wyoming and the evolution of flight and echolocation. Nature 451, 818–821. doi:10.1038/nature06549.

Stern, V.M., Smith, R.F., Bosch, R., Hagen, K.S., 1959. The integrated control concept. Hilgardia 29, 81–101.

Studier, E.H., Kunz, T.H., 1995. Accretion of nitrogen and minerals in suckling bats, *Myotis velifer* and *Tadarida brasiliensis*. J. Mammal. 76, 32–42. doi:10.2307/1382312.

Tanskul, S., Higara, K., Takada, K., Rungratchote, S., Suntinanalert, P., Oda, K., 2009. An alkaline serine-proteinase from a bacterium isolated from bat feces: purification and characterization. Biosci. Biotechnol. Biochem. 73, 2393–2398. doi:10.1271/bbb.90289.

Thomas, D.W., 1991. On fruits, seeds, and bats. BATS 9, 4.

Thomas, D.W., Cloutier, D., Provencher, M., Houle, C., 1988. The shape of bird- and bat-generated seed shadows around a tropical fruiting tree. Biotropica 20, 347–348.

Tilman, D., Knops, J., Wedin, D., Reich, P., Ritchie, M., Siemann, E., 1997. The influence of functional diversity and composition on ecosystem processes. Science 277, 1300–1302. doi:10.1126/science.277.5330.1300.

Van Bael, S.A., Brawn, J.D., Robinson, S.K., 2003. Birds defend trees from herbivores in a Neotropical forest canopy. Proc. Natl. Acad. Sci. USA 100, 8304–8307.

Voigt, C.C., Dechmann, D.K.N., Bender, J., Rinehart, B.J., Michener, R.H., Kunz, T.H., 2007. Mineral licks attract neotropical seed-dispersing bats. Res. Lett. Ecol. Article ID 34212, 4 pages. doi:10.1155/2007/34212.

Vungsilabutr, P., 2001. Population management of the rice brown planthopper in Thailand. In: Paper presented at the Inter-Country Forecasting System and Management for Brown Planthopper in East Asia, 13–15 November 2001, Hanoi, Vietnam.

Wahl, R., 1993. Important Mexican free-tailed bat colonies in Texas. In: Jordan, J., Obele, R. (Eds.), 1989 National Cave Management Symposium Proceedings. Texas Parks and Wildlife Department, Austin, TX, pp. 47–50.

Whelan, C.J., Wenny, D.G., Marquis, R.J., 2008. Ecosystem services provided by birds. Ann. N. Y. Acad. Sci. 1134, 25–60. doi:10.1196/annals.1439.003.

Whitaker Jr., J.O., 1995. Food of the big brown bat *Eptesicus fuscus* from maternity colonies in Indiana and Illinois. Am. Midl. Nat. 134, 346–360.

Whitaker Jr., J.O., Neefus, C., Kunz, T.H., 1996. Dietary variation in the Mexican free-tailed bat (*Tadarida brasiliensis*). J. Mammal. 77, 716–724.

Wilkinson, G.S., South, J.M., 2002. Life history, ecology and longevity in bats. Aging Cell 1, 124–131. doi:10.1046/j.1474-9728.2002.00020.x.

Williams, T.C., Ireland, L.C., Williams, J.M., 1973. High altitude flights of the free-tailed bat, *Tadarida brasiliensis*, observed with radar. J. Mammal. 54, 807–821.

Williams-Guillen, K., Perfecto, I., Vandermeer, J., 2008. Bats Limit Insects in a Neotropical Agroforestry System. Science 320, 70. doi:10.1126/science.1152944.

Wolf, W.W., Westbrook, J.K., Raulston, J., Pair, S.D., Hobbs, S.E., Riley, J.R., et al., 1990. Recent airborne radar observations of migrant pests in the United States [and Discussion]. Philos. Trans. R. Soc. Lond. Ser. B 328, 619–630. doi:10.1098/rstb.1990.0132.

Zahawi, R.A., Augspurger, C.K., 2006. Tropical forest restoration: tree islands as recruitment foci in degraded lands of Honduras. Ecol. Appl. 16, 463–478. doi:10.1890/1051-0761(2006)016[0464:TFRTIA]2.0.CO;2.

# Index

Note: The letters '*f*' and '*t*' following the locators refer to figures and tables respectively.

## S



## T



## U

**V**



**W**



**X**



**Z**

# Contents of Previous Volumes

**Volume 21**



**Volume 22**



**Volume 23**

**Volume 29**



**Volume 30**

**Volume 34**



**Volume 35**

**Volume 36**



**Volume 37**

**Volume 41**



**Volume 42**

**Volume 43**

Printed and bound by CPI Group (UK) Ltd, Croydon, CR0 4YY

08/05/2025

01864954-0001